The Nature of the Future

The Nature of the Future

Agriculture, Science, and Capitalism
in the Antebellum North

EMILY PAWLEY

The University of Chicago Press
Chicago and London

PUBLICATION OF THIS BOOK HAS BEEN AIDED BY A GRANT FROM THE
BEVINGTON FUND

The University of Chicago Press, Chicago 60637
The University of Chicago Press, Ltd., London
© 2020 by The University of Chicago
All rights reserved. No part of this book may be used or reproduced in any manner whatsoever without written permission, except in the case of brief quotations in critical articles and reviews. For more information, contact the University of Chicago Press, 1427 E. 60th St., Chicago, IL 60637.
Published 2020
Paperback edition 2022
Printed in the United States of America

31 30 29 28 27 26 25 24 23 22   1 2 3 4 5

ISBN-13: 978-0-226-69383-5 (cloth)
ISBN-13: 978-0-226-82002-6 (paper)
ISBN-13: 978-0-226-69397-2 (e-book)
DOI: https://doi.org/10.7208/chicago/9780226693972.001.0001

Library of Congress Cataloging-in-Publication Data

Names: Pawley, Emily, author.
Title: The nature of the future: agriculture, science, and capitalism in the antebellum North / Emily Pawley.
Description: Chicago: The University of Chicago Press, 2020. | Includes bibliographical references and index.
Identifiers: LCCN 2019041925 | ISBN 9780226693835 (cloth) | ISBN 9780226693972 (ebook)
Subjects: LCSH: Agriculture—New York (State)—History—19th century. | Agriculture—Capital productivity—New York (State)—History—19th century. | Crop science—New York (State)—History—19th century.
Classification: LCC S451.N7 P39 2020 | DDC 338.10974—dc23
LC record available at https://lccn.loc.gov/2019041925

♾ This paper meets the requirements of ANSI/NISO Z39.48-1992 (Permanence of Paper).

*To my parents
who taught me to ask questions
and to Roger
who looks for answers with me*

Contents

Introduction: Bending Reality with Large Strawberries   1

PART 1 Performances

1 Capitalist Aristocracy   23
2 No Ordinary Farmers   39

PART 2 Experiments

3 Experiments All for Worldly Gain   63
4 Trying Machines   81

PART 3 Futures

5 Coining Foliage into Gold   103
6 Divining Adaptation   130

PART 4 Values

7 Truth in Fruit   161
8 The Balance-Sheet of Nature   189

Epilogue   219

Acknowledgments 231
List of Abbreviations 237
Notes 239
Index 289

Introduction

Bending Reality with Large Strawberries

At least if the agricultural journals were to be believed, giants walked (or sometimes just grew) in the fields of antebellum New York. Wedged between columns on agricultural chemistry and soil degradation, new designs for machinery, and recipes for the destruction of insects are a surprising number of little stories about precisely measured big bodies: an ox weighing "2,546 pounds"; the carcass of a pig, two years old, weighing "975 lbs!"[1] The agricultural journal the *Cultivator* reflected with pleasure on a strawberry, eight and a half inches around, "fairly measured in the presence of a number of citizens" and "a splendid beet" received from "Mr. L. Hotchkiss" weighing "14 pounds closely trimmed . . . a very smooth handsome root for one of such great size."[2] Displayed in newspaper offices, on shop counters, and at the new agricultural fairs during the 1830s and 1840s, what we might call agricultural giants were becoming an increasingly familiar source of astonishment. They joined a crowd of other bodies on display in the antebellum United States: in Barnum's American Museum, and in the menageries, circuses, and sideshows trundling over newly built rural roads, bringing animals and humans rendered strange by distance, unusual forms, or artifice.[3]

Historians have thought a lot about the meaning of unusual bodies, sometimes called "monsters," "wonders," "curiosities," or "freaks." Doing so can tell us a lot about what the slippery word "nature" means in any particular culture. Unusual bodies have been read as messages from an angry God, jokes of a playful Nature, or failures of natural laws that, in their malformation, revealed the shape of those laws. Nineteenth-century American bodies sometimes retained these meanings. The hairy-faced families displayed in Renaissance courts appeared again in the bearded ladies of antebellum fairs. The carefully crafted "basilisks" that formed a staple of early modern apothecaries' displays were miracles of virtuosic fakery like the "Feejee Mermaid" exhibited at the American Museum.[4] Agricultural exhibitions sometimes borrowed from these cultures of display: the six-eyed piglet reported in the *American Agriculturist* was perhaps a joke of nature; the flamingo displayed by the American Institute in Manhattan brought the neighboring chickens closer to the menagerie; and when the young John B. Weeks went to the Mechanicsville Fair in 1855, he was delighted by "A Bearded Lady from Geneva, Switzerland and her bearded son, 2 1/2 yrs old, hair an inch long or more on his face and back."[5]

However, gigantic pigs, enormous strawberries, and other wonders produced at home were not like other monsters. For one thing, peculiar numbers rather than peculiar forms infused their stories: a Persian squash was more than six feet around; a single watermelon seed had produced 504 pounds of melons.[6] Sometimes witnesses attested to the truth of these figures in signed affidavits from the newspaper offices where huge vegetables were brought to be measured or from the butchers who slaughtered cattle and weighed out their skin, fat, and muscle. Such numbers had a meaning grounded in work—504 pounds of melons took a lot of lifting. But they got their exactitude from the scales and bushel baskets that Americans were learning to use to measure their farms' produce and from the markets that converted those measurements into prices.

To us, an old sheep whose yearly profit was *twenty-five dollars* or a corn yield of *ninety bushels an acre* may not read as prodigies straining the bounds of credibility. However, among antebellum Northerners, these numbers were expected to produce a thrill of wonder, even disbelief—the *Boston Cultivator* passed on the account of the surprising old sheep with a note observing that it looked "a *leetle*" like "a fish story" a term that, then as now, referred to a class of jokes about scale.[7] Even some of the more credible monsters were funny—impractical practical jokes. There are reasons not to grow six-foot squash, as becomes apparent if you imagine storing, carrying, or cooking them. However, both the aesthetics of quantified scale and the humor in inflated creatures depended, not on rarity, but on implied and familiar profit.

Antebellum American agricultural monsters were often also products in a different way, part of a commercial wave of organisms that promised to upgrade the domestic animals and plants that had come with early colonists. Images of Berkshire pigs, round with tiny trotters, seemed exotic to those antebellum Americans still used to the hairier, tusked, semiferal pigs (known by their detractors as "land-pikes") that still roamed forests and towns, reportedly fighting bears, wolves, and sometimes humans. Berkshires, by contrast, moved not through their own explosive reproductivity, but through

FIGURE 1. The astonishingly tall Cochin China chicken, subject of a speculative craze in the 1850s, depicted as a joke of scale in George Pickering Burnham's *A History of the Hen Fever: A Humorous Record* (1855). Author's collection.

a commercial network.[8] Other giants like the *multicaulis* mulberry and the "Chinese Tree Corn" did the same thing. These commercial monsters could also be fish stories. The *Genesee Farmer* joked about a "Gigantic Clover" being displayed in London under the headline "The Newest and Greatest Humbug yet Announced." Purportedly from Central Asia, the clover grew to "the enormous height of twelve or fifteen feet, and can be cut every month." Its vast stems could be a superior substitute for hemp, and "it is said that each grain will produce 300,000 seeds," perhaps rendering British farmers "independent of foreign supplies of clover seed."[9] But the great range and diversity of biological novelties and their rapid changes meant that the natural boundaries of belief were not generally clear.

Agricultural monsters were also unusual in signifying human rather than natural capacity. The landscape of agriculture is human-built, populated by organisms shaped powerfully by human intervention: seeds or flesh swelled to extraordinary sizes, colors, and flavors, carried thousands of miles from where their ancestors evolved. To antebellum Americans, agricultural monsters were obviously artifacts of what farmers called "cultivation" or "culture."[10] Giant strawberries or turnips evidenced an intentional regime of selection, manuring, and weeding, in short, a feat of skilled production. This meant that they could become foci of competition. At the agricultural fairs of the 1840s, giant pigs and cattle were used to demonstrate superior breed choice, or selection, or excellence in the art of feeding (perhaps careful formulations of corn, turnips, and whey, or pumpkins in the new steamers and "masticators"). The freighted politics of agricultural practice turned these competitions into proxies for other battles—extraordinary bodies could be made to represent labor arrangements or social prescriptions.

The swelled bodies of animals and plants could also be deployed in more fundamental and violent narratives. "Before they were touched by the finger of culture," announced an article circulating through New York's agricultural journals, "corn, potatoes, cabbages, fruits, &c." were "as unlike what they now are, as different species are unlike."[11] To the author, as to many white New Yorkers, this power of transformation belonged peculiarly to white people. "Compare the maize or Indian Corn, as first seen in the feeble stalk and slender roasting ear around the wigwam," with its new form, he demanded, "its hundred varieties in its present maturity, yielding in value its countless thousands to national wealth."[12] Fattened and split into new varieties, he argued, corn had been brought to maturity by white ingenuity. (It seems worthwhile noting here that Haudenosaunee corn grown using Haudenosaunee techniques was likely rather larger than the corn grown by whites.)[13] Building on indigenous techniques and planting indigenous plants on indigenous land, New Yorkers used and sometimes fabricated stories of scale that hid their indebtedness and justified further territorial appropriation.

Past change could also suggest future human-driven transformation: the same article, entitled "No End to Improvement," promised that cultivated animals and plants remained "susceptible of continual improvement, all ever running into new varieties."[14] Agricultural prodigies might also be harbingers of a world to come, ancestors from which a new family of giants, no longer remarkable, might be made to spring. "No End to Improvement" was not just a claim about an abstraction, however. "Improvement" or "agricultural improvement" was by this period a powerful system of knowledge making that manifested itself most clearly in the agricultural journals and fairs where agricultural monsters were displayed.

Agricultural Improvement

This book takes agricultural improvement as its subject in order to understand how markets, speculation, and knowledge came together in a place and period where agriculture was rapidly being remade: New York State during the years between the opening of the Erie Canal and the Civil War. It follows a group of market enthusiasts, the thousands of self-proclaimed "improving agriculturists," to illuminate the calculations, storytelling, and volatility that shaped the landscapes of the antebellum rural North.

In doing so it makes some claims about the nature of rural Northern capitalism. For many historians, Northern farming remains the "traditional" precapitalist foil to both the vicious calculations of the slave plantation and the unpredictability and social instabilities of the Northern city.[15] The powerful modernizing efforts of New York's improvers can help us provide a counternarrative. To create a landscape built for markets, improving New Yorkers elaborated on a form of knowledge in which profit was not only the goal but a crucial form of evidence, and, indeed, was understood to be the underlying purpose of the natural world. Far from producing a simpler, more rational vision of nature, improvers created a contested field in which they fought for their own reputations and for the reputations of regions, machines, and organisms—an area of science in which convincing visions of the future were built and evaporated in quick succession and in which the nature of commercial value itself was the subject of constant theorizing and debate. In tracing the rapid expansion of improvement in the 1830s, 1840s, and 1850s, the book highlights crucial transformations in American societies and landscapes. It also uncovers a system that lies at the root of both the modern system of industrial agriculture and the alternative forms of agriculture that today critique the industrial food system.

By the mid-nineteenth century, improvement was a global phenomenon, shaping social and physical landscapes from Great Britain and Prussia to Chile, Russia, Australia, and the Cape Colony.[16] Agricultural journals like the *Cultivator* were being published around the globe; agricultural societies were proliferating just as quickly, as were agricultural schools, experimental farms, agricultural publishers and warehouses, agricultural surveys, and most of all, agricultural fairs, places where remarkable animals, plants, machines, and techniques were being exhibited to audiences of unheard of sizes. Agricultural improvers' institutions and print networks would come to constitute one of the largest systems of scientific knowledge making in the antebellum United States and around the world.

Though avowedly scientific, agricultural improvement was not beholden to a simple or single science or focused on a simple or single object. Farms, particularly the mixed farms of the antebellum North, were complex assemblages of species, technologies, and systems of labor. A table of contents from one number of the *Cultivator* for 1841 gives a taste of the great variety of knowledge they required. "Making Wine—Growth of Trees—Sagacity of the Dog—The Agricultural Art—Bloody Murrain—Butter" comes near "A letter from a Conservative Bee-Keeper—Application of Clay to Soils—Agricultural Schools—Plum Tree Blight," followed by "Reasons for Engaging the Silk Culture—Ayrshire Cattle—Wheat Statistics—Crushed Bone" and inevitably "Large Vegetables."[17] Reading this now is a little disorienting—an apparent jumble of trivia or maybe found poetry. Improvers making sense of their own variety broke agriculture down into "branches" or "departments" linked to particular farm tasks—breeding, or soil amelioration, or architecture—which could be followed through the journals in long threads.

Improvement drew heavily on emerging forms of natural science like chemistry, geology, botany, and entomology. Like them it generated museums and lecture series, public experiments, new laboratories, specialized journals and, increasingly, paid experts.[18] However, improvement was more than just an attempt to apply the theories of "real" sciences. Improvers developed their own experimental culture as well as specialized forms of knowledge particular to the cultivated landscape—the balancing of "points" necessary to create a well-bred cow, for example, or the practices of naming and description required by "pomology," the science of fruit.[19]

In the decades before the Civil War, tens of thousands of Americans involved themselves in improvement as authors, readers, experimenters, consumers, and members of the agricultural societies.[20] Many historians use "improving farmers" to describe them, a term sometimes used in the antebellum period as well. In this work, I refer to them as "improvers" for two reasons. First, many improvers were *not* farmers; limiting improvement to farmers limits our understanding of its scope. Second, the definition of "farmer" was in flux at this time, a main theme of chapter 2. Regardless of their title, there were a lot of them. By 1860 the number of agricultural societies and farmers' clubs in the United States had reached 912. One-ninth of these were in New York, one of the most significant centers of improvement in the United States, where it is likely that at least fifteen thousand people were active society members.[21] Certainly, they were a minority of farmers: the 1855 New York State Census counted about 320,000 farmers.[22] However, they were a minority that made a great deal of noise. State fairs in the late 1840s and 1850s were huge public spectacles and massive commercial opportunities for hosting cities. During the four-day Rochester Fair of 1851, for example, the New York State Agricultural Society displayed more than two thousand animals to an estimated one hundred thousand visitors.[23] Driving along the canal road to the state fair in Albany, farmer's son Herman Coons complained, "Thousands were on there [sic] way; the Troy road was occupied by one unbroken line of vehicles of every description conveying passengers to the Grounds."[24]

Agricultural journals pushed the public profile of improvement even farther. Approximately four hundred new journals appeared between 1829 and 1859, and dozens joined a national conversation that likely reached hundreds of thousands. While journals likely inflated their circulation figures of tens of thousands, rural communal reading practices meant that they likely underestimated the number of actual readers.[25] The journals also promised and sometimes delivered forums for public debate. Each volume of the *Cultivator*, its editors advertised in 1847, contained "contributions from over 300 correspondents."[26] Print was so key to improvement that its critics sometimes called its adherents "book farmers"—by the 1840s, some improvers had begun to claim the name themselves, proudly inhabiting a wider world of print.

New York as a Center of American Improvement

New York's centrality to antebellum improvement is not surprising; it was both agriculturally powerful and wracked by changes that would fundamentally transform its agricultural landscape. The most visible of these changes was, of course, its extraordinary growth. By the Civil War, New York would be the most populous state in the nation and the wealthiest agriculturally, with farms valued at more than half a billion dollars. Its farmers raised twice as much fruit as any other state, had the largest number of market gardens, raised the most livestock, and purchased the most machinery.[27] This dominance was attributed at the time to the Erie Canal, a projector's "ditch" that after 1825 connected the waters of the Great Lakes to the Hudson and the Atlantic and fulfilled the hopes of its projectors to an unsettling degree.[28] "Surely, the water of this Canal must be the most fertilizing of all fluids," wrote Nathaniel Hawthorne, traveling down the canal five years after it opened, "for it causes towns with their masses of brick and stone, their churches and theaters, their business and hubbub, their luxury and refinement, their gay dames and polished citizens, to spring up, till in time the wondrous stream may flow between two continuous lines of buildings, through one thronged street, from Buffalo to Albany."[29]

The canal's completion prompted equally dizzying shifts in rural landscapes. It sparked a population boom in canal-side counties while casting southern counties, along the slightly older turnpike, into depression. Its success was used to justify a much larger network of feeder canals and rail lines, giving more and more rural New Yorkers swift access to eastern markets.[30] In western New York, speculators gave new towns names like "Wheatland" while "farm makers" cut stands of massive trees and plowed up river flats for waving fields of grain, which would soon outcompete wheat growing in the eastern counties. Simultaneously, the growing wealth of New York City flowed out into new kinds of agricultural landscapes—the "dairy zone" around Herkimer and Oneida Counties, the market gardens of Suffolk and Queens Counties, and the rapidly spreading orchards around Lake Ontario—while hay and oat fields sprang up to feed the burgeoning population of urban horses.[31] Sheep farming rose and fell, and new crops like hops generated new rural ways of life.[32] The canal moved goods the other way as well, providing New York farmers with clocks, silks, oysters, cast-iron stoves, agricultural machinery, and sometimes debt. Intensifying uncertainty and dislocation, commodity prices fluctuated with the Panics of 1819, 1837, 1839, and 1857, the ending of the Corn Laws in Britain in 1846, and the Mexican and Crimean wars. By 1865 most farmers were producing a surplus for the market not only to purchase goods, but also to pay off mortgages and IOUs.[33]

Economic shifts had massive ecological consequences. During this period, New Yorkers would fell much of their portion of the vast Eastern Woodlands, leveling a mosaic of old and new growth trees. Floating softwoods downstream to the cities and burning hardwoods where they stood, they changed the cover and the composition of the soil, the winds, and patterns of moisture.[34] As trees fell, the soil their roots had held began to wash away, filling in rivers already sped up by the slaughter of beaver and the collapse of the slow pools held by their dams. The habitats of wolves, bears, cougars, and deer were also cut to pieces. In destroying environments, white settlers struck at people too, doing their best to wipe away (or occupy) the carefully built patchwork of hunting runs, berry patches, corn and squash fields, fishing weirs, trails, and home places established over generations by Algonquian and Haudenosaunee peoples.[35] Expanded fields attracted new pest and disease invasions. The wheat midge, the potato blight, and peach yellows kept New York farming in a state of flux.³⁶ Declensionist futures of agricultural degradation thus ran alongside the stories of impending agricultural wealth. These visible changes made future changes seem more plausible and undermined efforts to sustain old ways.

White evangelicals saw the inevitable pull of the millennium in many of these shifts. While only a minority of New Yorkers joined William Miller in awaiting the Second Coming in 1844, many more confidently expected a thousand years of prosperity to precede the physical return of Jesus and saw New York's growing wealth as a portent of that change. These ideas joined the more general ferment of prophesy in upstate New York, where projected futures, bubbling up, crowded each other before popping, or, even more disorientingly, reshaping reality. The Fourierists and the Oneida Community imagined and then built seemingly perfectible communities in Skaneateles and Oneida; the phrenologist Orson Fowler's adherents dotted the landscape with life-rationalizing octagonal houses; and the Grahamites worked to build futures free from liquor, meat, and masturbation. In radical Rochester, a Venn diagram of white and Black abolitionists, feminists, and spiritualists agitated for a world without slavery or the oppression of women, but also for a future where the veil between the living and the dead had been pierced.³⁷

It is customary to separate the economic landscape, particularly the rural economic landscape, from this sort of cultural dreaming. This is a mistake.³⁸ Local boosters were being no less dreamy when they built Greek columns onto their new banks and churches in towns named "Troy," "Rome," and "Syracuse." The painter Thomas Cole's cycle *The Course of Empire* made these same columns the warning symbols of imperial degeneracy. Imperial expansion suited another foundational vision—the domination of New York by white settlers—a vision aided by the developers devising semilegal methods to push the Seneca and Oneida from their remaining lands and by authors like James Fenimore Cooper, a land developer's son who mourned *The Last of the Mohicans* even as the real Mohicans, still very much alive, had taken refuge with the threatened Oneida.³⁹

The landscape of agricultural improvement fit in with these visions. Shaped by conflicting ideas of capitalist development, driven by urban taste and rural refinement, and wracked by repeated "fevers" and "manias," it partook of many of the qualities we are beginning to recognize in urban capitalism: it was volatile, dependent on uncertain projections of value, and on conflicting ideas about what markets could or should do. As in other parts of the marketplace, consuming improving goods and imagining rural futures demanded impossible new kinds of judgment: sorting speculators from visionaries, quackery from new science, fads from trends, humbugs from true giants.

Improvement, Revolution, and British Agricultural Capitalism

Like many modernizers, New York's agricultural improvers built their stories of the future on other countries' pasts, turning to a parade of exemplars stretching from the Roman Empire, imperial China, and the kingdoms of the Bible to the Dutch Republic and the states of Germany and northern Italy.[40] Their most important source of ideas, however, was in agricultural improvement's international center: Great Britain. Specifically, they hoped to emulate the dramatic changes they had seen in Great Britain since about the 1760s.

Historians also see this place and period, sometimes termed the "agricultural revolution," as key to the history of capitalism. During the eighteenth century, increasingly powerful British landowners appropriated improving techniques, building them into projects of profit-seeking internal colonialism. British landlords enclosed common and "waste" land; extinguished commoners' customary rights to subsistence; shifted small tenants and subtenants off the land to create a class of landless laborers; and, simultaneously, backed banks, cut timber, dug canals, and searched for coal. These activities peaked during the agricultural boom of the Napoleonic Wars, when they were extravagantly publicized in print, in public exhibitions, and in the new agricultural societies.[41]

To many improving Americans, British-style agricultural capitalism was an attractive model: intensive, regionalized, highly productive, and seemingly profitable, fueled by metropolitan desires and extraordinary wealth. In Britain, cattle raised in Scotland walked hundreds of miles to London, where "graziers," professional fatteners, fed them leisurely on the sweet sown grasses of the home counties, until they were sold to butchers who held them in ever more elaborate stalls and pastures while they waited for the knife. In the counties around London, asparagus sprouted under bottles, and peaches, grapes, and oranges ripened under acres of coal-heated glass. More distant counties became "natural districts" famous for cider or cheese.[42] British improving institutions and principles were also crucial to imperial projects as colonial authorities exported European organisms to bolster settler colonialism and reorganized international trade around new sources of tropical and plantation goods. In the nineteenth century, British agricultural societies helped spread merino sheep and their fodder grasses to South Africa, Australia, and New Zealand. They laid out new tea plantations in Assam to undermine Chinese tea monopolies, while Indian opium plantations pried open Chinese markets from the consumer side. Since the most important articles of global trade—sugar, tea, indigo, coffee, cotton, opium, silk—came from plants, economies depended on this massive botanical reshuffling.[43] Both imperial and home branches of British improvement would prove enticing to Americans.

When Hezekiah Hull, farming in Berlin, New York, in 1837, spread lime on his fields and sowed them with clover, covered one field in ashes, and planted another with turnips and rutabaga, he was drawing from the standard repertoire of the British agricultural revolution, a repertoire that to many New Yorkers was synonymous with improvement. Hull's spelling of rutabaga, "root of bago," suggests that he heard of them before reading about them.[44] But Hull also had a direct textual link to British improvers through his intermittent subscription to a New York agricultural journal: the *Cultivator*. In 1842, the *Cultivator's* editors would boast, "Such is the ease and certainty of intercourse between G. Britain and this country, that the foreign journals are received almost as quickly, and with as much regularity, as our own."[45]

Lax copyright laws and journal exchanges made it easy for Americans to turn the British boom in improving print into an American boom. American journals and publishers padded their pages with copied British texts, sent correspondents to inspect British farms, and invited British agricultural experts to tour and speak in the United States.[46] The earliest agricultural societies in the US modeled themselves directly on British counterparts. Wealthy American improvers built estates that resembled and were sometimes named after British models; poorer ones borrowed rhetoric from British agrarian radicals.[47] Many of the new breeds that filled out the agricultural fairs still bore the names of the British places they had come from—Berkshire, Hereford, Durham, Ayrshire. Even reports of agricultural giants had their British counterparts—particularly enormous American oxen were placed in printed competition with the long-dead but still famous Durham Ox; wealthy American improvers bid high for his living relatives.[48]

In some ways, New York seemed like fertile territory for the proliferation of British-style improvement because of its unusual settlement patterns. During the colonial period, the Hudson River Valley had been divided up into "manors" granted to influential families like the Rensselaers, the Livingstons, and the Cortlandts. From the 1790s to the 1820s, improvement was their province, as they worked to imitate the profit-oriented large-scale estates of their British counterparts. Improvement's imperial elements also had their place. White New Yorkers, who already referred to the "Empire State," were industrious practitioners of settler colonialism. The oldest New Yorkers could still remember 1779 when General Sullivan's expedition had destroyed the base of Haudenosaunee power in what would become central New York, burning fields, girdling fruit trees, killing men, women, and children, and converting two million acres of land into scrip with which to pay the Continental Army. Though active fighting stopped with the end of the War of 1812, eager speculators, many embedded in the state government, continued to extract land from the Haudenosaunee throughout the nineteenth century.[49] Integrating these territories into larger markets, they used practices of landscape evaluation and transformation developed in the imperial networks of global improvement.

This did not mean that New Yorkers simply copied British or British imperial landscapes. For one thing, even as they imported fashionable new animals and plants, American settlers continued to grow and eat indigenous crops like corn, squash, beans, and pumpkins using techniques developed by Haudenosaunee and Algonquian peoples. In Russell Menard's phrase, New York was not a "neo-Europe" but a site of "mestizo agriculture," where a mixture of Eurasian, American, and African organisms and techniques was developing. Indeed, American plants were rapidly spreading globally as well: maize had a foothold in Italy, potatoes in Scotland and Ireland, and British aristocratic parks were dotted with American trees purchased from colonial botanists.[50]

For another thing, when New Yorkers did try to copy distant landscapes, they found that agricultural ideas and structures did not move easily. Techniques developed in fields ploughed for centuries did not work on the rocky, stump-dotted fields of New York. Huge herds of cattle were harder to maintain in New York, where longer winters required more hay and, consequently, much larger hayfields. Social structures differed even more starkly. Where British improvers complained about the masses of agricultural workers who needed to be provided for, in New York, agricultural labor was scarce and precious. More symbolically, practices literally learned from aristocrats often required republican masking before they could be imported to the United States. Sometimes this masking failed.

This last point is particularly evident in a central arc of this book, which traces the role New York's landlords played in establishing strong institutions of agricultural improvement before their abrupt decline and replacement by a much larger coalition of improvers in the 1840s. In the late eighteenth and early nineteenth centuries, New York landlords, emulating their aristocratic counterparts in Great Britain, twice secured state funding for the earliest agricultural societies. When funding faltered and these societies collapsed, landlords carried improving work forward, planning development projects, establishing scientific institutions, and sometimes forcing improving practices on their tenants. In the 1830s, landlords' efforts to promote improvement appeared to bear fruit—the New York State Agricultural Society, founded in 1832, was pushing hard for state funding, aided by new agricultural journals out of Albany and Rochester. This Albany-centered network had a metropolitan mirror in the American Institute, a manufacturer-sponsored institution that ran agricultural fairs at Niblo's Garden in New York City itself.[51] As the financial panics of 1837 and 1839 began to make farming look appealingly solid, the state legislature seemed increasingly interested in improvement.

However, as improving institutions flourished, landlords were thrown into turmoil and then permanently weakened by debt and the Anti-Rent movement of the 1840s.[52] The state legislature would take their place. In 1841, the legislature finally voted for funding that allowed the state society to hold annual fairs and to support dozens of county agricultural societies and county fairs.[53] The state would soon construct a purpose-built museum of agricultural science and support a cadre of agricultural scientists in Albany. Legislators' fondness for corrupt state printing contracts produced a cascade of agricultural print including the massive annual *Transactions* of the state society and geological and agricultural surveys printed in thousands of illustrated multivolume sets that were sent to heads of state around the world.[54]

Though this support was remarkable, it was not enough to explain the great expansion of improvement during the 1840s. As improvement slipped from landlords' grasp, a much more diverse collection of adherents picked it up: angry tenants, politicians, local boosters and land speculators, retiring urbanites, and middling farmers aspiring to gentility. Each of these groups fostered different ideas about the future structure of rural society. Landlords had dreamt of a British-style landscape of deferential, profitable tenants; tenants dreamed of a republic of free soil and free white labor. Bankers retired to the country to nurse political ambitions, but also envisioned sheep farms on a "capitalist scale" and acres of mulberry trees converted into silk by worms and young women in home cocooneries. Middling farmers dreamed of (and worried about) refined landscapes, where white fencing, named fruit trees, and fountains signified the mental elevation to be found indoors.[55]

Binding this diverse group together was a new commercial system. In rural New York, the "market revolution" of the 1830s, 1840s, and 1850s manifested itself in a new world of relatively affordable ready-made goods—democratized though far from democratic—that could be used to transform the landscape: agricultural implements and machines, improved breeding stock, fertilizers, fruit trees, new seeds, books, and journals. Purchasing these goods exposed farmers to debt and (sometimes) increased their profits. It also marked the landscape in ways that neighbors could read—tidy rows, for example, indicated the households that had seed drills and horse-hoes. New plows made recognizable cuts in fields and vibrant green foliage showed where guano had been spread. Just as magazines like *Godey's* told middling women how to create and read indoor spaces, agricultural journals encouraged farm men and women to produce and read outdoor spaces, animals, and plants for evidences of virtue, prosperity, and refinement.

The great expansion and endurance of agricultural improvement rested on this system of goods. Nurserymen, seedsmen, machinery makers, fertilizer manufacturers, clover millers, lime diggers, cattle importers, and guano merchants influenced improvement profoundly, though they have been studied much less than landed interests, state institutions, and presumed gentlemanly reformers. Warehouses and seed stores not only supplied products; they also operated informal museums and testing facilities; acted as libraries and distributors of pamphlets, books, and journals; and occasionally joined the agricultural societies as members.[56] Perched advantageously between established coastal markets and expanding midwestern ones, New York's agricultural businesses were in a strong position to amplify New York's improving voices, by selling goods and by promoting agricultural journals, many of which issued directly from warehouses and seed stores. At least eight New York journals would reach a national audience.[57] This complex of warehouses and journals would help New York's improvers leave indelible marks on American agricultural science and practice.

Expanding commercial networks did more than amplify improvement; they also fundamentally changed the character of that expansion. Fundamental and unexpectedly slippery questions about the value of fertilizers, fruit trees, seeds, bloodstock, or swill boilers would become and remain a central improving concern. Moreover, some of the most significant forms of expertise were dominated by those selling goods—nurserymen who were pomologists or chemists selling patent fertilizers. They join this story not just as beneficiaries of the expansion of improvement but as significant knowledge makers in their own right.

Improvement as a Way of Knowing

This book is a history of agricultural improvement and the rural North, but it is also a history of science and its relationship to capitalism. At its most basic level, it aims to show how people made knowledge about the living world around them. It looks not at the wilderness or the distant landscapes of exploration, but at the landscapes that most Americans knew intimately and tried to make profitable, the farms where they managed relationships between species, nourished bodies and butchered them, collected and sometimes sold their excretions and where they dug in the soil and read rocks and trees to try to imagine the potential of land. While far from encompassing all agricultural knowledge, agricultural improvement was an expansive and influential knowledge-making system, one that resembles better-studied forms of natural history from a distance, but on closer inspection differs from them profoundly. Thinking hard about these differences, about the things that make agricultural improvement not a weakened application of natural history or chemistry or geology but a different beast altogether, matters very much for understanding what knowledge and capitalism meant in the countryside.

First, unlike natural history, improving knowledge was futuristic. Rather than mapping existing landscapes, it prophesied the creation of new ones and tried to bring them into being. Where geologists stared back into deep time using fossils and overlapping strata, improvers used the same information to make claims about the hidden future purposes built into territory by a benevolent and designing God. Where naturalists described the structure of skeletons and shells so as to sort the seeming chaos of nature into manageable and exploitable categories, improving agriculturists attempted to turn individuals into populations. They sought, not typical structures, but extraordinary ones—the wobbling hams on a Berkshire pig, the shining red of a Baldwin apple—that could be spread into broader populations of pigs and fruit by dominating markets in blood or cuttings.

Improvement occupies a peculiar position among the sciences of the future. Some sciences, like meteorology, have no effect on the futures they describe but instead attempt to perceive the unfolding of natural laws. Others, like economics, can be performative, changing the future as they describe it—as when a prediction of higher stock prices produces higher stock prices. Even more than economic predictions, improvers' predictions were clearly consciously performative. They actively attempted to bring the futures they described into being, predicted their successes, claimed credit for them, and when they could, profited from them. At the same time, they expected that landscapes were divinely intended for particular forms of profit, casting their activities as the consequences of inevitable natural law. Their oscillation between determinacy and performativity more resembles the "fictional expectations" or active storytelling that Jens Beckert sees as central to capitalist development, which are both claims about a "real future" and open attempts to bring a particular imagined future into being.[58]

Improving knowledge always pushed hard at the bounds of credibility, since the landscapes it aimed at did not yet exist. Many, indeed, would never exist—or would exist only briefly. The boom in mulberry trees described in chapter 5 never brought a projected silk empire to the North. It never even got past the mulberry trees to silkworms. Others, like the butter districts established in chapter 6, would come to seem permanent features of the landscape. We tend to sort these successes and failures retrospectively: we tell stories in which new crops and techniques that succeed are essentially practical, while stories of failure are also stories of folly. This has created (or perhaps stems from) an internally illogical way of describing farmers. Just as antebellum Americans did, we often characterize farmers as hardheaded, innately conservative, and resistant to change. When they do change (which they do), their success is attributed to this hardheadedness, a sturdy grasp of physical and market realities that results in an almost prophetic rationality. Seeing a rising demand for butter, and seeing good dairy land, they founded dairy farms, goes the reasoning. This sense of farmer rationality makes failure seem like an aberration rather than a standard feature of capitalism—the most dramatic failed futures are sometimes described as "manias" or "fevers" disordering rural minds, making them unable to perceive otherwise clearly apparent market or environmental truths.

But prophecy on the ground is not clear. Improving New Yorkers did not know themselves if their visions were real or illusory, if their plans would work, their science would last, or the forms of consumer desire on which they based their farms would prove durable. Indeed, even knowing the outcome does not help us distinguish between possible and impossible schemes. Many entirely possible agricultural futures did not become concrete reality. Silkworms can live reasonably well in upstate New York; the mulberries planted for them pop up in suburban gardens today. Butter production was surely not the only potential fate of the Catskills. Moreover, while New Yorkers called mulberries a mania at the time, they also saw mania in things that became substantial crops: sugar beets, for example. In sorting through improving claims, we should be cautious in sorting the "rational" from the "fevered" ideas of the future, acknowledging that both successful and failed schemes sometimes emerged from the same institutions and kinds of evidence.[59] We should notice that claims of rationality and accusations of irrationality were, then as now, often gambits to attract credit and discredit competitors, or attempts to attribute widely differing fates in a volatile economy to differences in personal virtue. And we should remember that for all its interpretive flexibility, the present is a more unitary phenomenon than the future, which has more branches than we ever see.

Second, improvement was fundamentally concerned with urgent questions of value determination. As builders of landscapes meant for profit, improvers would attempt to calculate the value of everything from the color of apple skins to the smell of manure. Concern about the nature of value was fundamental to antebellum American culture. Americans worried about the fragile and fluctuating value of paper money tied to their unstable banking system; about the rising and falling prices that could make for a lively marketplace or kill one dead; and about the potential for frauds, adulteration, counterfeits, confidence men, and humbugs. Improving debates merely brought this concern for value to new realms: the hidden flows of valuable nutriment that connected fields to barns to manure piles and the speculative possibilities of mulberry trees, which multiplied from their cuttings at an enticing annual rate. Some of the new products of the New York agricultural landscape derived from ephemeral new concepts of taste, luxury, and refinement: the particular flavors of butter from particular kinds of land, for example, or the great if confusing array of specific flavors available from grafted fruit trees. At the most basic level, demonstrating the construction and consequences of these culturally determined categories of value can help disperse that air of economic inevitability that often hangs around narratives of agricultural (and indeed all) capitalist development.

Thinking about the volatility of value in improvement can also help us see new aspects of the relationship between science and capitalism. One story we sometimes tell about this relationship is that science provides capitalism with stable sets of values. By providing outside expert opinion, disinterested judgment rationalizes markets, enables trust, inventories natural resources, and renders them available to calculation. This presumes, however, that markets are rational, that they require stable forms of value to operate, and indeed that stable forms of value exist. In antebellum markets, at least, this was not the case. Often, monetary value was unbearably difficult to pin down, in part because it depended on future characteristics that were unknowable; in part because it depended on complex and uncertain landscapes, bodies, and practices; and in part because some of the sources of value were stories: stories about flavor, about luxury, or novelty. More than one form of value regularly coexisted. Individual improvers often benefited from this instability, promoting the forms of value that elevated whichever particular goods they were selling. Like everyone dealing with markets, then and now, improving farmers navigated market claims without a reliable map, assembling knowledge out of experience and reports of experiments, but also out of knowledge gleaned from advertisements and commercial exhibitions. They weighed claims to value based on a shifting set of criteria that never settled.

We sometimes also assume that experts themselves aspired to disinterestedness (or at least feigned it) and that laypeople expected them to do so.[60] However, as we have seen, New York's major experts were increasingly directly financially interested.[61] They sold improving goods or boosted particular landscapes by attempting to tell particularly persuasive stories. The maps through which states saw the landscape often drew their data from a competing crowd of improving boosters, making individual bids to claim a reputation for productive value for their own place.[62] The unresolvable proliferation of many of these stories and the decentralized expertise they represented contrasted with the simultaneous rise of state-sponsored, centralized, "disinterested" knowledge.

Far from aiming at knowledge separated from money, improvers developed forms of varietal identification that resembled anticounterfeiting techniques and systems of credit ratings; used sales figures and profits as evidence of the value of particular techniques; and kept "accounts" with animals, plants, and fields in their experiments. Their theories spread and warred through advertisements as well as through journals. Indeed, commercial texts sometimes provided an alternative home for theories that state and academic experts hoped to extirpate. Because profit was the explicit aim of improvement, sometimes improvers' connections to the market became, not damning evidence of unreliability, but the source of their credibility. These links between ideas about money and knowledge of environments, animals, and plants were not transgressions of an expected boundary. Rather, they had at their root the belief that profit was itself an underlying feature of the nonhuman world, that hidden wealth had been divinely laid up for those willing to seek it. The improving vision of nature headed inexorably toward postmillennial prosperity. Looking at the book of nature, improvers expected to see a balance sheet.

By focusing our attention on New York's antebellum improvers, this book shows that the markets and ideas of value reshaping the rural North were as various, as volatile, and as story-driven as those shaping urban capitalism. It argues that improvement, science, and capitalism twined with and fed on each other to produce, not a rational, calculated, centralizable vision, but multiple market futures and multiple definitions of value that depended on systems of credibility drawing both on natural scientific and market practices of knowledge making. Some of these knowledge-making practices were used to build state institutions; others spread into the system of decentralized commercial experts that operated in parallel with improving and state networks of science. Seeing this can illuminate important features of modern agricultural science and reveal fundamental ways in which markets and landscapes interact. Commodification is one such interaction, but so are other less-examined forms of economic practice like accounting, management, economic projection, and the creation of consumer desire. Exploring these, the book creates a richer story of capitalism in the American landscape, helping us to understand not only the voracious western edge of the grain frontier and the industrialized violence of the South, but fundamental changes that continued (and continue) to unsettle the East and Midwest.

★

If you have ever been to a state fair, you've seen the modern descendants of the agricultural giants in the truck-hauled hulks of the giant pumpkin competition or in the pen of the massive prize boar. By now they are a tradition, representing a self-consciously "old-fashioned" country culture along with pie contests and displays of quilts. It is therefore easy to imagine that this has always been their meaning, just as it is easy on a larger scale to make even modern "family" farms seem like artifacts of the past. Despite their best efforts, agricultural historians have not been able to eradicate American myths of the Northern "family farm" as a pastoral, unchanging, and innocent space, a safe haven from industrialism and markets, a naturalized response to landscape, and a slightly boring but worthy repository for tradition and authenticity.[63] Variations on these myths infuse everything from the idealized farms on milk cartons, to political speeches about the "Heartland," to "heirloom" fruits and "heritage" breeds, to the astonishingly large number of children's books, toys, and songs intended to teach animal sounds. The nostalgic mist surrounding farms makes it difficult to write their history, freezing rural Northerners into stasis, rendering their changing features "traditional," and encrusting them with all of the virtues and stereotypes Americans associate with the countryside.

In fact, the agricultural landscape that produced both giants and improvers in the antebellum period was new and unstable in ways that the giants help us to see. It was populated by exotic species imported as a part of conscious efforts to recreate distant European landscapes and to violently replace indigenous peoples and landscapes. Creating these new environments, white Northerners argued for and built multiple social structures—manors with tenants, large commercial operations, radical small farms, and genteel sites of retirement. They imagined silk empires and regions of hops as well as the now familiar dairy farms and apple orchards. To see them clearly, we must fight our comfortable sense of familiarity and restore their place in the weird, calculating, speculative, and futuristic world of antebellum agricultural science.

PART ONE

Performances

1

Capitalist Aristocracy

In its first decades, agricultural improvement was the domain of the landlords who held surprisingly large parts of New York. It can be hard, frankly, to know what to think of these landlords. With their sometimes county-sized "manors," their great houses, and their clear aristocratic pretensions, they do not look like Northern rural society is supposed to look. Were they feudal relics or profit-seeking capitalists? Representatives of a dying order or a path not taken?[1] Certainly they were a colonial product, their manors parceled out to them by Dutch and British authorities. But if they were a vestige, they were far from powerless. While the loyalist landlords of the southern Hudson had their lands confiscated after the Revolution, on the northern Hudson, powerful families like the Rensselaers, the Schuylers, and the Livingstons would form the backbone of the New York Federalist elite.[2] Nor were they declining. In 1824, Stephen Van Rensselaer III was the largest landholder in the United States, with five thousand tenants on holdings of more than two hundred square miles—all of two counties and half of a third.[3] Post-Revolutionary developments, moreover, expanded this class of large landholders. As the state legislature clawed territory piecemeal away from the Haudenosaunee, they shunted it into the hands of companies of investors or individual developers. Many of these aspired to the status of more established landlords, competing with them for tenants and buyers among the wave of New Englanders that would quadruple the population of New York between 1790 and 1820.[4]

Tenants themselves often had strong feelings on the question of landlord modernity, coming down hard on the side of feudal anachronism. "Anti-Renters" described the manors as "relics," analogous to the feudal holdings of "barbarian chiefs." One tenant newspaper promised readers, as a sort of how-to text, an account of "the first effort made in the middle ages, the time of the feudal tyranny, and popular servility, to throw off the feudal yoke."[5] Another commentator suggested that the manors had been planned as "a royal stud to propagate and supply the kingdoms of the New World with princes and princesses."[6] Charges of aristocratic pretension were perfectly justifiable. Hudson Valley landlords often spent more than they could afford on liveried servants, coats of arms, carriages, trips to Europe, and named "country seats." Stephen Van Rensselaer III had given tenants particularly good ammunition by maintaining his title, "Patroon," persisting in extending two fingers to tenants rather than his whole hand, and having rents paid in wheat rather than cash.[7] Like others in his social circle, moreover, he managed his tenants through leases modeled on European leases and sometimes enforced more strictly.[8]

Then and now, however, accusations of anachronism obscured New York landlords' focus on the profitable future. In imitating British aristocrats and members of the gentry, New York landlords hoped to re-create the success of classes that had recently risen enormously in wealth, stability, and political clout, classes moreover on which modern definitions of agricultural capitalism have long depended. Agricultural improvement was key to this plan. Copying British institutions, New York's landlords would lay crucial groundwork for American agricultural improvement and American scientific culture more generally, establishing institutions and agendas that would long outlast their heyday. However, the bases of New York landlords' power would prove less certain than those of their European counterparts. During the 1840s, New York landlords would be hamstrung and sometimes ruined by the combination of economic volatility, debt, tenant resistance, and electoral politics. As other groups began to take the reins of improving institutions, tenants themselves would appropriate the arguments of improvement for their own purposes, drawing on their own connections to British agrarian networks.

Aristocratic Modernity and the Agricultural Revolution

Far from representing the traces of a dying feudalism, in eighteenth- and early nineteenth-century Britain, landed power was rising. During the mid-eighteenth century, new rules of landownership combined with a demographic crisis had consolidated landholdings into the hands of a smaller, richer set of families. In subsequent decades, the pace of enclosure accelerated, converting common grazing lands, "wastes," and forests into fenceable private property, undermining rural subsistence while enriching a relative few. In the 1790s, the value of these expanded estates skyrocketed, driven up by wartime food prices. Landlords also profited from the mineral rights granted by enclosure, in particular from the exploitation of new coalfields on their lands and from the new canals that carried coal and food between the expanding cities.[9] This increased wealth translated into greater control over political life—a smaller and smaller number of families held the hundreds of "rotten boroughs" that sent representatives to Parliament. By 1816, one contemporary commentator estimated that three-quarters of the lower house was in the hands of 140 peers and 120 landed commoners. Unsurprisingly, these politicians created legislation supporting landed interests: new laws in the 1810s and 1820s punished poaching with death or transportation, and, as prices dropped with the end of the Napoleonic Wars, the Corn Laws preserved landholders' profits (at the expense of hungry urbanites). As the Erie Canal was being cut in New York, in Britain new "Ejectment Acts" allowed landlords to evict tenants en masse and to replace them with larger, profit-driven enterprises. These developments stabilized and enriched Britain's landed elite, creating what David Cannadine has called "a distinct group of new super-rich grandees."[10]

Theirs was a capitalist landscape. Farms on the estates of wealthy Britons were profit-driven, regionally specialized, and dependent on prices in distant markets. They were administered by a new managerial class of "agents" who produced increasingly elaborate forms of accounting and profit calculation to satisfy their distant employers. They were often worked by a large class of dependent landless laborers, including an annual influx of Irish migrant labor.[11] The accelerating pace of enclosures had devastated common usufruct rights and expanded private property in land. These changes in British landscapes were financed and supported by a different branch of agricultural capitalism: the plantations of sugar, tea, opium, tobacco, and cattle spreading from Ireland as far as the Caribbean and India and worked by dispossessed, convict, and enslaved people.[12]

The features of agricultural capitalism are so visible in this period in part because this place and period have been used to define them. The "agricultural revolution" has long been characterized as a necessary precursor to the Industrial Revolution. Initially historians credited famous wealthy improvers like Coke of Norfolk, "Turnip" Townshend, and Robert Bakewell with inventions and techniques—intensive breeding, labor-saving machinery, new fodder crops, and systems of crop rotation—that released rural labor, created food surpluses, and amassed capital that would fund industrialization. More recent work has complicated this story, showing that eighteenth-century gains in production in fact capitalized on a more fundamental "yeoman's revolution" in farming practice beginning as far back as the fifteenth century.[13] However, though they may not have invented the techniques with which they were credited, Britain's landed elites capitalized on them, forming a rising, development-minded, profit-seeking class, and publicized them, through agricultural journalists like the famous Arthur Young.

Interested observers, New York landlords among them, mostly perceived British agricultural development filtered through the mass of print that people like Young produced, but many major political figures of the founding generation communicated with British improvers directly.[14] Antebellum improvers still purchased print editions of the correspondence between George Washington, Arthur Young, and the Scottish agriculturist and politician Sir John Sinclair.[15] Projecting an enticing image of abrupt, top-down social transformation, British improving print provided a model for the first iterations of American improvement.[16] Among New York landlords, these efforts at emulation predated the American Revolution. During the 1760s, in imitation of London's Society for the Encouragement of Arts, Manufactures, and Commerce, the Schuylers, Livingstons, and Rensselaers had established the Society for the Promotion of Arts, Agriculture, and Oeconomy, which offered premiums for agricultural products before it collapsed under the pressure of political rivalries.[17]

Following the Revolution, land speculation in western New York would enrich landlords and add to their numbers. Federalist New Yorkers, upper Hudson landlords included, had been heavily invested in western expansion. In the 1810s and 1820s, they increasingly merged with western developers who built roads, performed surveys, planned towns, and in general acted in lieu of a state.[18] Getting western lands to meet the fevered expectations of their new owners became the project of developers. Wealthy land speculators like the Wadsworths of the Genesee Valley and William Cooper of Cooperstown aimed directly for the kinds of status enjoyed by older landlords, retaining large estates for themselves, struggling to attract tenants, and building manor houses and social connections.[19] As it became embedded in the manors of the upper Hudson, agricultural improvement also shaped the developmental push toward western lands. The same moralizing vision of profit that had justified the continuing enclosure of common land in Britain and Ireland, and the mass displacements and appropriations of the British Empire, was also applied to the Haudenosaunee, whose lands came into speculating American hands from the late eighteenth century onward.[20]

This newly enlarged landed class created New York's first post-Revolutionary agricultural society, the Society for the Promotion of Agriculture, Arts, and Manufactures (SPAAM). Established in 1791 and chartered by the state in 1793, this society included many of the same political stars who were promoting the Western Inland Lock Navigation Company, the first serious attempt to drive a canal through western lands.[21] Borrowing a style of

Baconian inquiry from the Royal Society of London, the SPAAM sent out circulars full of inquiries about New York's potential for profit: "Can the silkworm be profitably introduced into your neighborhood?" "Can you suggest any thing capable of raising the reputation of our flour in foreign markets?" "Are there any coal-mines?"[22] In their *Transactions*, they outlined projects to expand the capacities of their holdings. In the first volume, William Cooper sketched a future maple sugar industry to rival the West Indies, surveyor Simeon De Witt planned meteorological maps to reveal western New York's agricultural potential, and Robert Livingston touted lime as a soil-reviving panacea for the worn soils of the Hudson. Like their counterparts in Philadelphia, Massachusetts, and Britain, they also developed correspondence networks, throwing out feelers to British luminaries like Young and Royal Society president Joseph Banks.[23]

The exclusivity of these efforts limited their scale. Even after receiving a corporate charter from the state legislature in 1792, the SPAAM remained essentially a private club, with seventy-two members admitted by internal nomination and election. Despite constant lobbying, they were unable to carry out their larger plans: county societies, a botanical garden, or more than one professorship at Columbia.[24] After Livingston's death in 1813, the SPAAM petered out. Even as it did so, a new wave of societies, tracking a more successful canal effort, was on the way.

In the 1810s, small landlord-sponsored societies in Otsego, Jefferson, and Rensselaer Counties appeared on a new model evangelized by Elkanah Watson, a speculator in western New York land and a zealous canal promoter. In 1819, these were voted state financial support, to be coordinated by a new institution: the New York Board of Agriculture.[25] This body mirrored its namesake, the British Board of Agriculture, established by Sinclair and Young during the food shortages and high agricultural prices of the Napoleonic Wars.[26] It drew its membership from the same elevated circles as its predecessors, landlords and land speculators excited by the impending completion of the Erie Canal and worried by the precipitous collapse of wheat prices during the Panic of 1819 (particularly alarming to those who, like the Rensselaers, took their rent in wheat).[27]

Briefly, at least, this second wave spread more broadly than the first had, focusing on "cattle shows" where farmers would be encouraged to develop more profitable practices through example and healthy competition. By 1822, there were thirty-six societies in New York.[28] However, these rapidly came under fire; critics pointed out that only wealthy farmers received the premiums for excellent farming or for expensive imported animals. Worse, in a time of partisan political strife, they had been sponsored by canal-promoting

Governor DeWitt Clinton and, at the local level, were led by Clintonian politicians. Martin Van Buren's anti-Clintonian party, the "Bucktails," supported by newly enfranchised voters, cut off state support for the societies in 1824.[29] Without a regular influx of state money, all but one withered.[30]

Organizers, who expected their own virtuous example to spark a conflagration of grateful, deferential emulation, were repeatedly surprised and disappointed when it did not. However, while struggling to establish agricultural societies, landlords forged durable ties with each other and laid down an agenda that later institutions would partially take up. More importantly, even as their calls for state aid faltered, New York landlords promoted improvement by other means, importing expensive new breeds and varieties, conducting agricultural experiments on their estates, and building new improving institutions.

Most famously, between 1802 and 1815, Chancellor Robert Livingston and a group of fellow projectors launched a project to import the fine-wooled merino sheep. An estimated twenty thousand were carried to the US, and prices for good rams reached astonishing levels—reportedly over $1,000 for a single ram when the normal price was around three dollars. By the end of the War of 1812, however, prices and the project had largely collapsed. Historians have sometimes characterized this period as one of madness, a "craze" or "mania," language that the merino's detractors used at the time.[31] But Livingston and others were participating in a much broader international developmental push that, in some places, succeeded astonishingly well. Merinos were both a marker of participation in cosmopolitan scientific and aristocratic circles and the subject of massive investment. British textile manufacturing depended on their incomparably fine, long fibers for the fine woolen trade. However, the animals themselves had long been controlled by the Spanish crown, which raised them on an epic scale. Flocks of ten thousand or more migrated annually on a great multiprovince circuit from the lowlands to the highlands on dedicated highways. During the eighteenth century, celebrated moments of espionage and diplomacy had spread merinos into France, Germany, Sweden, and Great Britain. As the Napoleonic Wars reached Spain, they increased both the vulnerability of the merino wool supply and the scale of merino seizures and sales. In Britain, Sir Joseph Banks had alternatingly smuggled and purchased thousands of merinos for George III and members of the "Merino Society."[32] As US minister to France, during the Napoleonic Wars, Livingston was well aware of these machinations, indeed he had used his connections in Paris to acquire two pairs from the celebrated flock established by Louis XVI.[33] While efforts to introduce the merino permanently to Great Britain failed, efforts in Australia succeeded spectacularly; millions of pounds of merino wool would pour into Britain from Australia by the mid-1820s.[34]

Livingston's other developmental efforts would be more successful, helping to convert the Hudson (and the Mississippi) into a thoroughfare for trade. After years of considering new designs for horse treadmill and steam-driven boats, Livingston famously encountered the American inventor Robert Fulton in Paris, where Fulton was attempting to sell submarine designs to Napoleon's navy. After Fulton's submarine ventures failed in both Britain and France, Livingston became Fulton's partner in steamboat development. The first commercial steamboat consequently ran on the Hudson, not the Thames or the Seine, with a monopoly from the New York State Legislature. It would be known as the *Clermont* in honor of Livingston's country estate, its first destination, which it significantly enriched by opening the Hudson to bidirectional trade.[35]

Stephen Van Rensselaer III's private efforts left even more enduring marks on American improvement and American science. In the 1820s, Rensselaer commissioned the geologist Amos Eaton to perform agricultural and geological surveys first of Rensselaer's own counties and then of the land along the track of the Erie Canal.[36] When published, these surveys were explicitly intended to provide an American analogue to British improving literature. A circular sent out by Eaton soliciting the observations of successful farmers along the route made this evident. "What a speculating visionary agriculturalist may have read in European books, will not be enquired after," Eaton wrote, acerbically, "We can read these books at home." Indeed, Eaton pointed out, such books needed adjustment before moving to the United States. The seasons were different, altering "the times of ploughing, seeding, hoeing and manuring." Moreover, demographically and politically the advice in British texts was inappropriate for the American landscape. Eaton remarked that British techniques often required a "state of poverty and abject servility among the laboring class, which does not comport with the nature of our civil institutions."[37]

This last point shows the negotiations that Rensselaer was making as he attempted to translate texts and economic systems from across the Atlantic. Such republican sentiments notwithstanding, in an 1820 speech to the Board of Agriculture, Rensselaer openly hankered after the much greater wealth of his British counterparts, sketching an idealized vision of the prosperous market-oriented tenants who formed the middling classes of the British countryside. "What has been the history of Great-Britain, may at some future day become applicable to the condition of the fine country in which we live,"

FIGURE 2. Chancellor Robert R. Livingston (1746–1813). Courtesy of the Albany Institute of History and Art.

he suggested. "In Great-Britain, the farmer can afford to pay as many dollars annual rent for an acre of ground as would purchase the fee simple here."[38]

During the 1820s, Rensselaer became a crucial patron of science, establishing the science-focused Rensselaer Institute and the Rensselaer Flotilla, a geological summer school on a chain of canal boats. Both were run by Amos Eaton, and both were linked to improving goals. Rensselaer claimed his main aim in establishing the institute was to "qualify teachers for instructing the sons and daughters of farmers and mechanics . . . in the application of experimental chemistry, philosophy, and natural history, to agriculture, domestic economy and the arts."[39] The nearby Troy Female Seminary already had a significant scientific presence (in 1856 former student Eunice Foote would be the first person to demonstrate the warming effect of atmospheric "Carbonic Acid Gas" now termed carbon dioxide). These schools would turn Albany

FIGURE 3. The Patroon: Stephen Van Rensselaer III (1764–1839). Courtesy of the Albany Institute of History and Art.

into an unlikely scientific powerhouse that would educate an astonishing number of well-known men and women of science.[40]

Landlords also employed more direct forms of control. "It is now understood by all intelligent farmers," ran one clause in an 1834 lease to a farm on the Wadsworth lands, "that stocking down ploughed land constantly with clover seed is one of the greatest modern improvements in husbandry." The same lease also specified details of fence construction, fruit tree planting and pruning, stock numbers, weed control, and the spreading of all barn, stable, and chip manure in a manner "as shall be most proper and farmerlike."[41] Other leases specified crop rotations to be applied in specific fields and required the use of plaster as well as stable manure. George Clarke, landlord in Otsego County, went further, specifying that his tenants should shift to the more profitable and capital-intensive hops and dairy businesses (with Clarke providing the initial investment and taking a share of the profits).[42]

In the 1830s and 1840s, improvement finally began to expand on a scale resembling landlords' early hopes. In 1832, a convention in Albany established a new statewide organization—the New York State Agricultural Society—and lobbied for state support. Some members of the new society were familiar names from among landlordly families; others were part of Rensselaer's scientific coterie. However, membership was open to anyone who could write a letter and send two dollars a year, opening the ranks of membership considerably. After years of campaigning, and the Whig victory in 1840, the society finally succeeded in getting a new subsidy of $8,000 a year for five years. With this money, they created a semicentralized system—money reached county societies through the state agricultural society, which soon accumulated dedicated state rooms and a state museum. However, despite such aid, local societies had to supply half their own funding, leaving them beholden to larger local donors.[43]

This new system grew rapidly. Within a year there would be thirty-two societies; in another, forty-two. Their popularity was enabled, in part, by a much more powerful publicity arm: a new generation of agricultural journals. Established in the 1820s and 1830s, the *New York Farmer*, the *Genesee Farmer*, and the *Cultivator* had lobbied hard for the new state society. New commercial networks of agricultural warehouses, nurseries, and fertilizer and machinery manufacturers expanded their reach further. Where in earlier phases, prominent improvers had often relied on personal transatlantic connections for animals, plants, machinery, and texts (as when Livingston acquired his merinos on his diplomatic mission to France) or on a relatively small number of well-connected dealers, this new commercial network domestically produced or bulk imported and distributed the goods, machines, and organisms that improving practices required. Selling agricultural journals and promoting fairs as a form of advertising, manufacturers and warehouse operators became noisy agents of agricultural improvement and provided new, cheaper ways to participate, by buying a book, a silkworm cocoon, a bag of guano, a stack of drainage tiles, or a new variety of seed.

In this period of expansion, however, landlords themselves were in decline. Their insistence on improving practices had been most effective during the mid-1830s land-price bubble, which made it difficult for tenants to refuse stringent controls in leases. Subsequent crashes loosened these controls.[44] More fundamentally, while the British landlords they admired had (and used) the power to ship desperate crofters to foreign shores during the Scottish famines of the 1840s, New York's great landholders struggled just to collect rent.[45] Instead of controlling an agricultural population whose labor could be exploited or dispensed with, landlords competed for tenants and purchasers and, once they had them, struggled to keep them from moving west.[46] While landlords might fondly imagine a future of combined profit and deference, their tenants had different ambitions. During the 1840s, tenants developed new tactics and new unexpected levels of political power.[47]

Tenant Resistance and Freeholder Modernity

The limitations of the power of New York landlords became clear during the Anti-Rent Wars of the 1840s.[48] The proximate cause of the dispute was the landlords' morass of debt—land speculation and expensive acts of patronage had taken their toll.[49] When Stephen Van Rensselaer III died in 1839, he left approximately $400,000 in debt, an amount equaled by a mountainous pile of back rents, forgiven as part of the delicate dance of paternalism during the Panics of 1819 and 1837. Attempting to collect rents and evict defaulters during a general depression, Rensselaer's heirs broke an uneasy truce. Other landlords, caught in similar speculation traps, followed suit, provoking a multipronged resistance, which spread from the Rensselaer estate across the upper Hudson and from there to tenants in central New York. Tenants organized themselves into Anti-Rent Committees, refused to pay rent, and either demanded the right to purchase their lands outright or contested landlords' titles and argued that their labor on the land meant that they owned it already. More radical bands of tenants, "Calico Indians," disguised in Indian costumes, offered more violent resistance, killing three people.[50]

British landlords also encountered agrarian revolt during this period, including the Swing Riots of the 1830s and the Chartist agrarianism of the 1840s. In fact, British and American agrarians were directly linked. Under the banner of the Anti-Rent *Albany Freeholder*, the editors printed lines from Oliver Goldsmith's anti-enclosure lament "The Deserted Village" (1770):

> The man of wealth and pride
> Takes up space that many poor supplied
> Space for his lake, his park's extended bounds
> Space for his horses, equipage, and hounds
> The robe that wraps his limbs in silken sloth
> Has robbed the neighbr'ing fields of half their growth.[51]

People linked these movements even more clearly. George Evans, editor of the *Working Man's Advocate* and the *Radical*, sold the works of famed British rural critic (and former Long Island farmer) William Cobbett. Thomas Devyr, immigrant editor of the Albany *Anti-Renter* and briefly of the *Freeholder* had rallied against Irish landlords during the famine of the early 1830s and marched against Newcastle in a failed Chartist uprising before fleeing to the United States.[52] Very different rural hierarchies meant that this translation was not straightforward. Where in the US, tenants were the focus of radicalism, in Britain, the "rural wars" of the 1830s pitted laborers against the tenant farmers who employed them. However, the common experience of labor on the land gave British laborers and American tenant farmers a common alternative language of property, improvement, and natural law.[53]

These parallels notwithstanding, British rural elites wielded much more power than their American counterparts. After the Swing Riots of southern and eastern England in the 1830s, for example, cavalry, "forest associations," and local constabularies rounded up 2,000 prisoners. Of these, 233 were sentenced to death (only 19 were actually executed), 505 were transported, 644 were imprisoned, 7 were fined, and 1 was whipped.[54] By contrast, in New York close races between the Democrats and Whigs in the 1840s meant that both sides were forced to court Anti-Renter votes, particularly once they had organized a third "Anti-Renter" Party. When Anti-Renters shot and killed a deputy sheriff in the summer of 1845, local posses and militiamen sent by Democratic governor Silas Wright arrested almost four hundred people. But Wright, concerned about the upcoming election, arranged for most charges to be dropped. Thirty "Calico Indians" were fined, fifteen were sentenced to life in prison, and two men were convicted of murder (Wright commuted their death sentences).[55] This attempt at leniency was fruitless—when Wright came up for election, Anti-Renters combined with Whigs to defeat him. As part of this new coalition, Anti-Renters were able to significantly shape the new state constitution of 1846—building in powerful incentives for landlords to sell off their lands and reshaping the conditions of tenancy. Though some landlords survived this crisis (often with diminished revenues), most did not. The financially beleaguered Rensselaers sold their remaining lands to a speculator who, enmeshed in continuing lawsuits, abandoned them.

Landlords' centrality to agricultural societies had not escaped Anti-Renter attention. Even the less radical Anti-Rent paper, the *Albany Freeholder*, referred scornfully to the state agricultural society "with Presidents, Vice Presidents, and Secretaries, all of them retired merchants turned amateur farmers, or great landlords who never ate the bread of their own labors." The *Freeholder* continued, "Have our agricultural societies, State or county ever lisped a syllable on the subject of land tenures? Look over the lists of their officers and you cannot be at a loss to know why."[56] Despite such criticisms, however, Anti-Renters did not abandon the language or the institutions of agricultural improvement to landlords. Instead, they worked to occupy them on their own terms.

As they turned to party politics, some Anti-Renters sought legitimacy, demanding, as the members of an Anti-Renter convention put it, "the same right to form associations . . . as for the diffusion of knowledge, the distribution of the Bible, or the conversion of the heathen."[57] Improvement had this air of respectable association: the *Albany Freeholder* regularly printed advertisements for improving machinery and articles from its Albany neighbor, the *Cultivator*.[58] Even the *Freeholder*'s call was not to abandon improvement but to demand from the agricultural societies "some expression of opinion in favor of the extinguishment of perpetual leases, and the commutation of ground-rents."[59] More broadly, Anti-Rent candidates, like other candidates for office, found that agricultural fairs doubled as political rallies. Thus, for example, William Van Schoonhoven, the Whig/Anti-Rent candidate, placed himself prominently on the committee of reception for the Rensselaer County Fair in 1847. Jeremiah Allen, a farmer of East Grafton, Rensselaer County, took his chickens to that same fair. He recorded it in his diary sandwiched between an Anti-Rent meeting and a "Calico Indian" training without comment or apparent sense of contradiction.[60] Sixty years after the founding of the original agricultural society, improvement no longer seemed to be the vehicle of landlord interests.

Anti-Renters described their own future of landscape transformation, appropriation, and ownership. "The man who enters the wilderness and removes the wild growth of nature and cultivates it fit to live on, restricting himself to a limited number of acres sufficient to draw a subsistence from, ought of right to be the sole and only owner," declared one representative to an 1846 meeting.[61] Building on the eighteenth-century legacy of Thomas Paine and Thomas Spence, and stretching accepted versions of the labor theory of value,

Anti-Renters argued that initial cultivation should mean ownership. However, new developments had begun to complicate this definition of property. Landlords, forced by tenants' resistance, had begun to compensate departing tenants for orchards planted or fields cleared, separating improvements from ownership of the land itself.[62] Moreover, by the 1840s, tenants too were increasingly socially stratified. Better-off tenants, who had begun to purchase cheap wood lots for fuel and building material, objected when their neighbors claimed these lots as "unimproved."[63]

The principles of agricultural improvement offered tenants new ammunition for attacking landlords' claims of beneficent development. Landlordism, argued Anti-Rent leader Ira Harris, "prevents agricultural improvement." Even the most "ordinary observer" visiting tenant-held Columbia County would "find the buildings dilapidated—the farms in bad repair—bad husbandry, and everything that indicates bad thrift," where in Dutchess County, held in freehold, "he will find everywhere evidences of a high state of prosperity." Far from enforcing improving practices, Harris argued, leases discouraged them by making it in tenants' interest to gain quick profits and move on: "Who, that holds a farm liable to forfeiture at any moment, at the mere will of the landlord—and who, in case of sale, is required to pay a portion of all he receives to the landlord—would improve such a farm?"[64] Furthermore, Harris claimed, the capital necessary for improvements melted away in rent. Berkshire farms looked better than those in tenants' counties, "because the farmers of Berkshire are not compelled to pay a tax of 200,000 a year to absentee nabobs, to be spent in building palaces, in lavish prodigality and luxurious indulgence."[65] Specific improving practices could also be twisted against landlords. Standard exhortations to intensively cultivate small farms rather than extensively cultivate large ones (exhortations that landlords intended to keep tenants from clearing more land than they could manure) fit neatly into the idealized landscape of small farms described by agrarian radicals. "Large estates must be cultivated by tenants, or slaves," remarked the *Freeholder*, and "inevitably deteriorate in quality and value."[66]

Though they might have destroyed the most storied landlords of the upper Hudson, tenant demands for a freeholder landscape did not produce one. Though tenancy rates decreased, tenancy itself persisted particularly in central and western New York, becoming more stringently shaped by landlordly controls in commercial districts. The historiographical debate about Northern tenancy, focusing mostly on the Midwest, has sometimes suggested that tenancy was simply a low rung in a ladder to farm ownership, rather than a speculators' trap. The continuance of sporadic "Anti-Rent" wars, arson, rent strikes, and property violence in central and western New York until well into the 1870s suggests that tenants in New York at least did not see the ladder stretching ahead of them so clearly.[67]

However, Anti-Renter rhetoric about agricultural futures strongly influenced the world outside New York. The fine balance between parties made attacking landlordism a part of political orthodoxy with surprising speed. Whig governor William Seward assured the crowd at the 1842 State Agricultural Fair in Albany of his commitment to "perpetual adherence to the policy which by laws regulating descents, devises and trusts, prevents the undue accumulation of estates." He cast America as a "community of freeholders, in contrast with the systems of other countries, under which lands are cultivated by tenants, the rewards of whose labor pass to the benefit of landlords."[68] Such sentiments, declaimed within earshot of Rensselaer land, surely helped upend the sense of landlord control. As Reeve Huston has shown, such arguments, along with much Anti-Renter rhetoric, would form a substrate for later Free Soiler arguments against slavery.[69]

★

Agricultural societies, fairs, and journals had begun in New York as instruments serving a particular, British-influenced vision of agricultural capitalism. Landlords, developers, and speculators had expected to spearhead the same kind of massive period of development that had enriched their titled and genteel British counterparts. They imagined coal mines, canals, new forms of transportation, industrially supportive populations of merino sheep, and a prosperous and hopefully grateful tenantry. Their investment in this vision, particularly in the form of western lands but also in agricultural science would bind them in debt and undermine their authority. Tenants in turn made improvement into a platform both rhetorically, strengthening their case that landlordism was literally destructive to the republic, leading to practices that eroded its soil, and practically, by sending candidates to harangue the agricultural fairs.

Landlords and tenants both left durable marks on improvement. The Rensselaer Institute long outlived its founder, maintaining a permanent group of geologists and chemists within crucial reach of the agricultural journals and of the state legislature. Rensselaer's agricultural surveys would be picked up and echoed by a developmental state, which would shortly embark on surveys on a much grander scale. The most seemingly hopeless dream, the smallholder agrarianism of the Anti-Renters would arguably have an even larger effect. Feeding into the Free Soil and then into the Republican parties, it would undergird the Homestead Act, the US Department of Agriculture, and the land-grant colleges.[70]

As the power of landlords declined, the ranks of agricultural improvers suddenly became much larger and more diverse. Where the statewide society of the 1790s had had a total of seventy-two members, by 1860, New York agricultural societies would have perhaps fifteen thousand; tens of thousands would subscribe to the agricultural journals, and combined fair attendance would reach the hundreds of thousands.[71] As it grew, improvement was taken up by large and small farmers, tenants, laborers, urbanites hoping for recreational gentility and investors looking for stable profits. It thus became a repository for more than one vision of agricultural modernity. As improvement expanded, its ideas, practices, and institutions altered, shifting and splitting to suit the varying needs of different segments of a complicated, tense, rapidly changing, and fuzzily defined rural hierarchy.[72] Increasingly, the category "improving farmer" concealed the many identities that composed it.

2

No Ordinary Farmers

Only three years after the New York State Agricultural Society's first fair, organizers of the society's Cattle Show at Poughkeepsie in 1844 were maybe a little astonished by the hundreds of competitors and the crowd of thirty thousand that packed the displays. At least this is suggested by the most frequently published image of the show; any actual cattle are obscured by a sea of indistinct, tiny figures, stacked four layers high in frail stands and blending into invisibility around the Floral Hall.[1] New Yorkers were not the only ones to be amazed: news of the Poughkeepsie Fair also reached London as a specimen of a "Monster Meeting."[2] The scale of participation in this wave of New York improvement outstripped its predecessors in other arenas as well. By 1844 the state supported more than forty county societies, each holding its own fair, and journals were "rising up," as one society organizer wrote to another, "like mushrooms," buoyed by cheap steam printing, improved postal networks, and a much wider group of readers.[3] Agricultural improvement had never been so popular, and this time its popularity would grow and last.

Who were all these people? The Poughkeepsie Fair's main speaker, the Democratic politician and historian George Bancroft, spent much of his speech rendering them as a by-then familiar monolith: "the farmer." Bancroft's farmer was composed of virtues. He was intellectual: "the farmer's mind is exalted," Bancroft announced. He was moral: "his principles stand as firm as your own Highlands; his good deeds flow like self-moving waters." And patriotic: "He loves America—is the depository of her glory and the guardian of her freedom." He was independent: "the farmer is independent." Yet he operated in harmony with the other major social interests: "with the mechanic and manufacturer . . . he makes our country safe against foreign

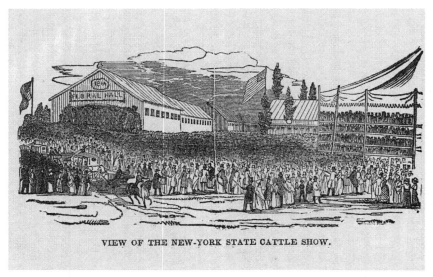

FIGURE 4. The Poughkeepsie Fair. A cattle show where the cattle are rendered invisible by the crowd. Courtesy of the Albany Institute of History and Art.

foes, for it becomes perfect by its own resources." In fact, he was possessed of something approaching divinity: "No occupation is nearer heaven. The social angel, when he descended to converse with men, broke bread with the husbandman beneath the tree."[4]

Except for the angelic visitation, none of this will sound particularly surprising—Bancroft wove together strains of admiration that were hardening into clichés of antebellum political speech. "The farmer" was the fundamental imagined unit of multiple political philosophies, a central symbol of American distinctiveness, an essential constituency, and already the focus both of imagined pasts and imagined futures.[5] Many of these ideas survive in modern political rhetoric about family farms and the "heartland." Sometimes they even creep into modern historiography, in the guise of the "ordinary" Northern farmer who sturdily resisted the incursions of the market, and who defended tradition in the face of change. This is understandable. As historians pay greater attention to the appalling blood-etched hierarchies of antebellum Southern agriculture, the differences between antebellum Northern farmers seem insignificant—making it easier to believe the stories that some Northerners told themselves about their unity. Indeed, it would be wrong to deny that Northern farmers were, as a group, wealthier and more socially mobile than their contemporaries in other countries. To describe them, however, we sometimes erase their differences by reaching for a fiction built around an average: the comfortingly fuzzy but oddly solid-sounding "ordinary farmer," a phrase still vibrating with the ideas of agrarian virtue that Bancroft tried to make real.[6]

In fact, the Poughkeepsie Fair occurred at a moment when New York rural power structures were shifting profoundly. The landlord most involved in improvement, Stephen Van Rensselaer III, had died, and his heirs and fellow landlords were being fundamentally challenged by the Anti-Rent War, which would soon break up their estates and limit their power, making Bancroft's invocations of unified rural identity particularly marked and ironic. As landlords clashed with tenants, other forms of rural inequality were intensifying. New wealth from cities and manufacturing fed a class of landlord-imitators: merchants, manufacturers, and professionals who dotted New York with "rural retreats," engaged in new kinds of rural investment and tried to build political careers on ideals of rural virtue. At the same time, middling farmers experienced new levels of social division. They both aspired to rural refinement and worried about it and alternately elevated and denigrated the physical labor that few of them could escape.

While Bancroft sang a paean to "the farmer," the members of his audience likely had widely different and unsettled ideas about what a farmer was or should be. In print, New Yorkers wrote about "Patroons" and "tenants" but also about "gentleman farmers," "practical farmers," "agriculturists," "hobby farmers," "dirt farmers," "plough joggers," "sheep masters," "working farmers," "book farmers," "anti-book farmers," and "scientific farmers." That many of the terms above were simultaneously derogatory and proudly claimed makes them both more interesting and almost unbearably slippery in actual use. The overlay of ill-fitting categories borrowed from British agrarian capitalism complicated things further—"tenant," "laborer," "manor," and "gentleman" meant very different things in the American context. Urbanites' nostalgia about the countryside added another layer of complexity, casting all classes of rural people as more solid, quieter, and more innocent.[7] Rural New Yorkers sometimes eagerly cultivated this image, though accounts of rural innocence could also be reversed to depict farmers as yokels blundering into the city, easy prey for the confidence man.[8] Perhaps most revealingly, New Yorkers also wrote about "real farmers." Realness mattered in agricultural improvement in the 1840s, partly because improvers were so often accused of unreality. Improving societies during the early republic had withered in the face of accusations of improper politics but also of impracticality, and of hobby or "book" farming. But authenticity was coming to matter for other reasons as well. Increasingly, acting like a "real farmer" was a moving target.

New participants in improvement came from many walks of life, occupied multiple farming identities, and wrestled with their meaning. For some, like merchants, bankers, and manufacturers, "rural retirement" promised economic security, investment opportunities, and political advancement in the country but also threatened political perils and accusations of inauthenticity. Yet so-called "real" middling farmers also consciously performed multiple farming identities. They wrestled with questions of political legitimacy and authenticity raised by new social divisions, by the new demands of and concerns about rural refinement, and by the new dignity and cultural risks associated with physical labor. The class that was closest to the field and ostensibly the most authentic of all—the broadening class of paid and family laborers—surfaced in improving texts only occasionally; most vanished into the reports of employers and fathers, leaving their marks only on the fields.

The idea of the "farmer" became increasingly valuable as a form of political performance, not because it referred to a uniform lump, but precisely because it allowed politicians to shift between registers and implied classes. As the site where farming was most visibly displayed, where farms and ways of farming could be depicted in print and their products shown to thousands, improvement was the stage upon which rural New Yorkers could fashion themselves in new ways, drawing shifting, convenient, and contested boundaries around the class that most Americans agreed was the most solid.[9]

Performance and Profit: Merchants, Bankers, and Manufacturers

The clearest performance of farmer identity came from the class that had most visibly changed their role. During their dominance, Hudson Valley landlords had expressed their status through the use of space, by building "seats" and "manors." As their power waned, however, seats, manors, and "villas" multiplied. Traveling on the upper Hudson in 1848, the horticulturist Andrew Jackson Downing noted that for twenty miles the riverbanks were "nearly a continuous succession of fine seats."[10] Such places were not new. In 1831, the British agricultural traveler John Fowler commented on the Long Island prevalence of "Retired, or half retired merchants . . . farming about as much for amusement as profit."[11] Even in the colonial period, American merchants had built themselves country retreats just as generations of British merchants had. During the 1830s and 1840s, however, money from the cotton trade, real estate, railroad investments, manufacturing, and the newly dominant New York banks expanded the ranks of wealthy urbanites.[12] The same 1846 state constitution that undermined landlords' power would further empower this class, weakening legislative controls on the formation of corporations and opening the state to a whirlwind of speculation.[13]

These wealthy urbanites are sometimes left out of historical accounts of the countryside, perhaps because, as extensions of urbanity, they don't feel like a real presence in rural society. To other rural New Yorkers, though, they felt present enough: they bought farms, sometimes at high prices; hired laborers; backed local banks and railroads; and occasionally demanded (without always receiving) a measure of local deference. As the *Albany Freeholder* bitterly noted, they would also become a highly visible force in agricultural societies. Two retired urbanites probably lingered most freshly in the memories of *Freeholder* readers. The first was Elkanah Watson, a merchant turned land speculator, canal promoter, and failed merino farmer. In the 1800s and 1810s, Watson promoted what he called the "Modern Berkshire System," the pairing of county societies with agricultural fairs that kicked off the 1810s wave of improvement. On these grounds, Watson claimed loudly (if somewhat inaccurately) to be the inventor of the agricultural fair.[14] The second, Jesse Buel, had been a printer and editor in Albany. After retiring at forty-three, Buel turned eighty acres of sandy land into a working farm, a transfiguration that created a standard attraction for visitors to Albany. Buel was a primary force in improvement in the 1830s, becoming corresponding secretary of the new state society and founding the *Cultivator*, soon the most influential agricultural journal in the nation.[15] While Buel and Watson were exceptional, former or current lawyers, printers, bankers, and merchants were common in improvement, particularly in the visible upper reaches of the state society.

Motivations for the move to farming were complex. Most obviously, rural estates offered opportunities for conspicuous consumption. An advertisement for a "Farm and beautiful Country Seat" posted in the *American Agriculturist* in 1846 promised sixty acres "in the best state of cultivation," but also views of the bay, marble mantles, a carriage house, a greenhouse, a hothouse, an orchard with 250 varieties of fruit, a vinery containing more than a hundred varieties of grape, and trellised nectarines, apricots, and greengages, all within easy reach of the city by steamboat. The land was to be worked, conveniently, by the farming couple advertised with the property.[16] Participation in improvement allowed the owners of such spaces to make their consumption even more conspicuous by circulating images and descriptions of their villas and sending expensive animals and foreign grafted fruits to the agricultural fairs. This performance also had more practical functions: marble mantles described in print communicated stability to potential business partners, to creditors, and to the network of credit bureau correspondents who monitored men of business to determine their creditworthiness starting in 1841.[17]

Retreats to such luxurious landscapes were sometimes described using the standard rhetoric of "rural retirement." This was an ancient story. Classically educated New Yorkers would have known Horace's *Beatus Ille*, written in the first century BCE, in which a moneylender muses sentimentally about whether to turn farmer ("Happy the man free of business cares / who, like the men of olden days / ploughs the family fields with his own oxen / and neither lends nor borrows") before, two weeks later, choosing to continue at moneylending.[18] In the early days of the republic, members of the new political elite had recast their rural seats as signs of republican virtue and as evidences of civilization that could compete with their counterparts in Britain.[19] As Tamara Plakins Thornton has shown, merchants and bankers used country living to link themselves to older notions of landed political legitimacy, evidencing a disinterested gentlemanly identity, spiked with references to Cincinnatus and George Washington.[20]

But particular developments of the 1830s and 1840s gave old stories new inflections. Ideas of the countryside as a place of sentiment, health, and virtue where one could periodically recover from urban evils clearly fed on the sentimental literature of domesticity and were sometimes seen as feminizing. An essay in the *Albany Freeholder* mocked those who, persuaded by daughters, "nearly crazed with the delightful idea of living in the country" moved there, enticed by visions of cream, butter, fresh vegetables, "and newly-mown hay the perfume of which is so agreeable."[21] Sentimental and nationalist visions of home and countryside were strengthened by changes in the increasingly crowded cities. Miasmatic theory, combined with the cholera epidemics that raged through New York City in 1832, 1848, and 1854, lent urgency to accounts of distance and agreeable rural scents. Some antebellum authors postulated a "death-line" on the Hudson, where brackish miasmatic water gave way to the safe pine-scented air of the countryside.[22]

Familiarity with improving science could also convert rural retirement into a marker of mental cultivation. "A little familiarity with the easily acquired knowledge of the thermometer, barometer, hygrometer, wind-gauge, and rain gauge," chemical lecturer William MacNeven informed the patrons of the New York Athenaeum in 1825, "may provide [the man of fortune] daily some hours of very entertaining employment."[23] Mental culture, in turn, could link aspiring gentlemen to a cosmopolitan web of cultivated enthusiasts, while at the same time, promising public benefit in the form of agricultural experiments or imported stock to be added to American bloodlines. The language of rural retreat could also provide an honorable mask for commercial failure.[24] John Delafield, for example, worked as president of the Phoenix Bank, and then as president of the New York Banking Company, but retired to his country retreat "Oaklands," near Geneva, after losing most of his fortune for the second time in the Panic of 1837.[25]

While recreational farming was real and growing, not all merchants and bankers were playing or particularly retiring. The terms of rural retirement survive in our culture, but a less familiar group of narratives about the country clearly elaborated the connection between expanding rural production and urban fortunes. Improving journals consistently characterized agricultural pursuits as sources of profit worthy of urban capital, complete with calculated returns on investment. Attempting to negotiate these new realms of social distinction and profit, many improvers cast themselves as directors of capital and labor. The term "agriculturist," which had emerged in mid-eighteenth-century Britain to refer to students of formal agricultural knowledge, became more common in the nineteenth century.[26] Regularly used as a substitute for "farmers," it conveyed an aura of material wealth, and expert knowledge, analogous to the dignity of "industrialist" over the still hazy "manufacturer," or "capitalist" over "speculator." Other title choices marked out positions in a managerial structure. Those attempting to commercialize sheepherding on the large scale sometimes referred to themselves as "flockmasters," a term that separated them from the more plebian "shepherds" whom they employed.[27]

High rates of mercantile failure made heartfelt appeals to the assumed certainties of farming life a standard genre. "Banks may fail—merchants' notes may be protested, and their drafts dishonored," declared the *New-Yorker* in 1839, "but 'Seed-time and Harvest,' that old and stable firm, shall never 'fail'; drafts upon them are answered at sight, and the book of nature, where the farmer makes his deposits, is 'good as gold.'"[28] In 1839, the *Journal of Commerce* advertised rural lands as investment opportunities—real estate on the borders of Seneca Falls, broken into village lots, millworks, and "Wheat land of the first quality," promised to become a mixed agricultural and industrial investment with labor supplied by "six houses for tenants."[29] William Cobbett's old farm on Long Island reappeared in the *Journal of Commerce* not only as a large and elegant house but also as a well-manured potential mulberry farm.[30]

Though commodity prices slumped slowly across the antebellum period, the spike in food prices and the subsequent riots that accompanied the Panic of 1837 may also have left a durable if inaccurate impression of the easy wealth to be made from food.[31] Visions of futures sustained by stable, high, agricultural profits were fed both by the mid-1830s land bubble, which produced remarkable expectations about the possibilities of the New York landscape, and by the subsequent crash. As investors scrambled for new, more secure forms of profit, they looked to cattle, to fine fruit, and, as we shall see in chapter 5, to silk.

As some urbanites dreamed of mulberry farms, others were working out other ways to harvest profits from their rural neighbors. Significant numbers of merchants from the 1830s onward turned from trade relationships with Southern planters toward a hinterland of Northern farmers that would eventually reach deep into northwestern states like Wisconsin and Illinois. The American Institute, established by manufacturers in 1828, provided an independent metropolitan strand of New York improvement, exhibiting the agricultural implements that were becoming central to American manufacturing.[32] In the 1850s, these efforts would reach a pinnacle in the triumphant trials of American agricultural machinery at the London Crystal Palace in 1851 and the New York Crystal Palace in 1853.[33] Crowded with producers who were learning to become consumers, this hinterland was also crisscrossed more and more densely with the railroads and canals that funneled agricultural wealth to stockholders. Spectacular displays of prosperity made at agricultural fairs asserted the economic potential of the Northern farm landscape to investors in transportation.[34] Promising to increase agricultural wealth, improving institutions made their future profits look more realistic.

Both rural retirement and rural investment sometimes offered another kind of future to Northern urbanites. For visible numbers of elite New Yorkers, participation in improvement became a stage on the way to a future in politics. Improving societies offered powerful connections to political, mercantile, and landowning circles. Timed for harvest, agricultural fairs drew together a crucial constituency right before elections. After admiring "a ponderous pumpkin, a mammoth squash, [and] a Giant cabbage-head," at the Norwich Fair in 1856, the school teacher Eliakim Weld dryly remarked in his diary, "I saw other *cabbage heads* listening to a series of Blackguard arguments from *D. S. Dickinson*, in favor of the styled Democratic party, against the holy principles of the Republicans."[35]

While owning an estate signaled gentlemanly status, the flexibility of the word "farmer" also allowed politicians to claim a plebeian identity that was increasingly attractive in the 1840s. When Jesse Buel ran as Whig candidate for governor of New York in 1836, the supportive Whig press praising him split in their definition of farmer along these lines. In the *Long Island Star*, he was a model of expert agricultural supervision. "Under his skilful management and the discreet application of capital," it gushed, "[his farm] has become a PATTERN for the State, and has been visited by thousands of citizens, as well as foreigners, as a bright evidence of the capacities of our mother Earth." While reprinting the *Star*'s article, Thurlow Weed's *Evening Journal* ran a picture of a plowman at their masthead and depicted Buel as a model of physical labor. "Is there any reason why a FARMER, who regularly 'mows his swarth' and '*keeps up his row*' and '*rakes and binds his day's work*' should not be elected Governor?" they inquired.[36]

However, this public symbolic work was also perilous—claims of farmer identity were frequently assailed. Enraged by Buel's move to the Whigs, the Democratic *Argus* attacked him as a "misnamed farmer" and "political agriculturist" and claimed that his real hold on farmers was more literal and oppressive; "he holds the farmers," they argued, "by holding their farms." Tenants who had failed to pay their rent or taxes would find their farms bought up by "the great *Farmer Buel*." William Henry Harrison, then running for president as "the Farmer of South Bend," likewise came in for skepticism: the *Argus* delighted over a story put out by the Whigs that Harrison had needed a glove after shaking hands with too large a crowd. "What will the 'huge paws' say," the *Argus* inquired, "of making one's hand sore with squeezing the soft and delicate hands of a few hundred, or thousand if you please, of the *Whig aristocracy*."[37] Harrison's efforts to convert himself into a farmer candidate would be more successful in 1840, despite the "Rough Hewers' Associations" that Democrats established to combat stories that he had been threshing his own grain on his own barn floor.[38]

If attacks on "political farmers" were the most venomous, attacks on novice improving farmers as a whole were common. "Book farmer," "gentleman farmer," and "aerialist" could all connote an overly theoretical novice, a city man humorously out of place, the obverse of the country oaf stories told in the city. In book farmer stories, neighbors' jokes often took the place of the tricks of the confidence man. Such stories were particularly stinging because they mimicked (and surely provoked) real incidents. After starting to farm, Elkanah Watson scribbled a furious testimony in his diary against his neighbor, the farmer, lawyer, and improver Thomas Gold. Gold had convinced Watson, apparently without difficulty, that he owned a hen with enormous talons perfect for digging potatoes, which she did with alacrity, throwing "out potatoes as fast as four men can pick them up—she springs from hill to hill." Watson was even more annoyed when his other neighbors laughed, refusing to see Gold's story as a lie and a matter of honor.[39]

Questions of authenticity had a more than abstract significance for improvement. When the state society made an early bid for state funding in 1834, the state senate's agricultural committee, led by Jehiel Halsey, scathingly rejected both the society's petitions and the governor's pro-improvement message in part on this ground. The 1819 act, Halsey declared, had been a failure because "practical farmers" had been superseded by "persons who were wholly destitute of practical knowledge, . . . [who] rendering it subservient to individual or sectional purposes, prevented it from exercising that benign influence which had been expected."[40] Already braced for such complaints, agricultural journalists had begun to make them themselves. The *Genesee*

Farmer derided "aerialists," "castle builders," and "Men who would pass themselves off for scientific farmers, before they have even learned to be familiar with the most common terms made use of by practical men."[41]

Certainly, even a little investigation makes it tempting to dismiss this class of improvers as "hobbyists"—it is difficult, for example, not to snigger at society members Jacob Ten Eyck and Erastus Corning, who sometimes described themselves as farmers, but who appeared in the 1844 *Albany Directory* as a senator (Corning), a director of insurance companies (Ten Eyck), a merchant (Corning), and as bank presidents (both).[42] However, in general, the lines dividing rural and urban Americans were hard to draw. Many seeming "urbanites" had relatives or roots in the countryside or moved back and forth. Even "political agriculturist" Jesse Buel had been born into a farm family.[43] The financially unsteady ranks of doctors and lawyers also often still relied on farms for a significant part of their income, perhaps following the advice of Benjamin Rush who had assured medical students in 1807 that "the resources of a farm, will prevent your cherishing, even for a moment, an impious wish for the prevalence of sickness in your neighborhood."[44]

We should not contrast this back-and-forth movement with an imagined group of exclusive working-class farmers. Poorer New Yorkers, pressed by the need for cash income, often kept their farms only by alternating farming with lumbering or local industries, or worked off the farm to raise funds to set up housekeeping.[45] Thus, Salmon Bostwick worked as a butcher and then a distiller in Cooperstown before buying a farm in 1829; Solomon Northup financed his farming with years rafting timber and with his wife Anne's work cooking at the Eagle Tavern in Kingsbury; and Oliver Tillson, a farmer's son, financed his soft fruit-growing business as a mapmaker, tracing the roads of Ulster County with a compass, a wheelbarrow odometer, and future financier and robber baron Jay Gould.[46] Though symbolic lines between country and town hardened in the 1840s, the economic precarity of rural life meant that in practice they remained permeable. Many people found themselves in positions from which it was hard to argue rural (or urban) authenticity or purity. This problem would also complicate life for that group of improving farmers that we might be most likely to see as "real" or "ordinary."

The Many Ways of Middling

Of all rural New Yorkers, the farmers who owned and worked on their own farms are perhaps the hardest to analyze, hidden behind ideologically freighted names: "ordinary farmer," "practical farmer," and sometimes "*real* farmer." Middling farmers were the clearest target of the agricultural journals and of the rhetoric of people like Bancroft. At the same time, opponents of improvement sometimes denied the existence of middling improving farmers, accusing societies of serving as a gentleman's hobby or a mask for bankers heading into politics (which, as we have seen, they sometimes did). However, the diaries and account books of middling farmers show that that some indeed subscribed to improving journals, joined agricultural societies, or experimented with bee feeding, subsoil plowing, and bone dust as a manure. They bought Berkshire pigs, Merino sheep, and part-bred shorthorn cattle. They planted "artificial grasses" like clover and timothy and new varieties of grain; grafted fruit trees; and spread plaster, swamp muck, and sometimes guano on their fields. They tested agricultural machinery and tinkered with new accounting forms.[47]

What proportion of agricultural improvers were middling farmers? This is difficult to know. Journals did not generally preserve lists of subscribers, but Sally McMurry's analysis of a rare *Cultivator* list found that more than two-thirds of these subscribers were working farmers, most farming exclusively. Rural-exchange networks likely widened journals' reach. The *American Agriculturist* would eventually claim to have found that the twenty-two copies of their journal sent to a single rural post office had reached 107 families, passing through the hands of 506 people.[48]

Society lists are just as scarce. The Dutchess County Agricultural Society, very unusually, kept a list of members for its first two years, 1842 and 1843, offering an imperfect window into the first moments of participation. The list of 138 members is impossible to trace definitively (a surprising number of Vails, Adriances, and Hoags lived in Dutchess County). However, the fifty-four town representatives were listed with their hometowns, making it possible to link almost all of them to specific entries in the 1850 census and in county histories, showing us at least what kinds of people took up the task of organization.[49] Some representatives, it is clear, fit the stereotype. The richest, William Augustus Davies, was about to be elected president of the Farmers' and Manufacturers' National Bank of Poughkeepsie and lived on the proceeds of several thousand inherited acres. However, not all banking connections were created equal. Representatives Abraham Dibble and Culver Backus were also bankers, but their "Pine Plains Bank" was a tiny operation run out of Dibble's store.[50] Other representatives only aimed at urbanity: farmer's son Henry Lambert hoped to become a merchant but died a farmer.[51]

Of the forty-nine traceable representatives, forty were listed as "farmers" in the 1850 census, and efforts to trace them in county histories suggest that this was their main identity. While a few were retired, their average age was only forty-one. Overall, the town representatives were certainly wealthier than their neighbors; the value of their land averaged about $16,500 (though only $14,000 if we discount the wealthy Davies), which made them about twice as rich as the average local freeholding farmer. This does not tell us the wealth of society members as a group, since representatives were likely selected for their local prominence. Moreover, it conceals great variability. Nine percent of representatives had lands worth less than the local average: people like John Adriance, with only $2,000 in land, or twenty-two-year-old Elnathan Miller, who still lived with his father on a farm valued at only $8,000.[52]

In some ways, improvement's appeal to middling farmers seems obvious. Journals and fairs were stuffed with praises for farmers, public ceremonies in which "the agricultural interest" was celebrated with pageantry and parades. Accounts of improving practices promised profit, freedom from want, uncertainty, and drudgery; strategies for managing labor scarcity; and defenses against soil exhaustion, insects, and blights. Even more concretely, fair prizes could become premium stamps on barrels or sacks, allowing farmers to bargain for higher prices, translating honor into cash. Participants in agricultural societies established nodding acquaintance with bank presidents, storekeepers, and justices of the peace, particularly important in what Paul Johnson has identified as the "sponsored" character of upward mobility in upstate New York.[53] Moreover, as farmers increasingly acquired tenants of their own, landlords' tools became more widely useful.[54] In 1857, Heman Chapin, a Monroe County farmer, would prescribe the same kinds of soil treatments to his tenant—alternations of wheat, clover seed, and two tons of plaster—that the wealthy Wadsworths required of theirs.[55]

However, even if we accept the estimate of fifteen thousand participants in the agricultural societies and the tens of thousands whom journal editors claimed as readers, it is clear that most farmers did not participate—after all, there were 231,730 farms in New York according to the 1855 state census.[56] Here too, possible reasons are not hard to find. The happiest accounts of future profit were countered by the memory of some of the more dramatic failed visions. Farmers digging bird-borne white mulberry trees out of their fields, for example, had reason to curse the silk enthusiasms that had brought them to North America. Added to this was the insult to existing practices and people that was often embedded, thorn-like, in the most extravagant praise for "the farmer." Improvers often wrote at length about their disbelieving, superstitious neighbors. While chapter 3 will complicate these characterizations, there were plenty of reasons for neighbors to be offended or suspicious. Finally, and perhaps most importantly, improving practices were often expensive and, for farmers without access to capital, potentially disastrous—mortgages taken on to purchase machinery or fertilizers could bring a previously successful farmer to ruin when commodity prices dropped, a classic boom-and-bust cycle that has continued to dog agricultural development.

To understand the choice to participate or not and the meaning of this participation, we have to pay attention to a finer-grained form of social stratification: that occurring within the ranks of farmers. Records from the small town of Amenia in Dutchess County from 1850, the year of the first detailed federal agricultural census, give an intriguing glimpse of the kinds of differences that could divide freeholders. During the antebellum period, Dutchess County was known for wealth.[57] The 161 Amenia farmers listed in the agricultural portion of the census owned, on average, about $7,600 worth of land, more than their contemporaries in the Connecticut Valley who averaged between $2,000 and $4,000. These averages, again, conceal differences that surely felt large. They lump Calvin Chamberlin, who held $25,000 in real estate, with John and Jeanette Thorpe, who supported four young children on land worth $400.[58] The sixteen farmers (10 percent) whose farms were worth $2,000 knew fifty-three people in their town whose farm was worth at least five times as much. Following kinship networks reveals further variation: the Beldings controlled $74,000 worth of land to the Guthries' $2,400. Other differences between neighbors, like the sharp bite of mortgages or the hidden flows of money between kin, are invisible in these records.[59]

As elsewhere in New York, Amenia's local variation may have been intensified by its landscape; cut through by two mountain chains, it ranged between rich bottomland and poorer, more recent hill farms.[60] Slaveholding had produced further hierarchies. During the 1790s, 40 percent of households in Dutch-populated southern Hudson Valley counties like Dutchess County had held enslaved people, a greater percentage than the Carolinas.[61] Slaveholding families extracted more from the land and expanded their holdings and, as the date of emancipation approached, put pulses of cash into their farms by illegally selling people South. African American farmers that stayed in Dutchess County were likely excluded from the credit relationships that buoyed white New Yorkers; many were forced into the small but growing class of permanent laborers.[62]

The level of inequality in Amenia was normal. Martin Bruegel calculates that farm revenues in the mid–Hudson Valley ranged from $200 to $2,000 per year, a spectrum of difference comparable to that among artisans and storekeepers.[63] We should not see such divisions as marking hard-and-fast lines between social groups. Northern antebellum rural social circles were more inclusive than the rigidly divided visiting patterns developing in cities. Small farmers dined with large farmers in a way that small shopkeepers would not do with large merchants, in part because of the links forged at barn-raisings, quilting circles, husking bees, and long-term debt relationships; in part because in small communities rigid social exclusivity spelled social isolation; and in part because provincial people came to pride themselves on a distinctive, more egalitarian, sociability.[64] However, such differences in wealth would have been written across the landscape in clear signs. The fertility of the fields appeared in the size of the barn; the shape of the scars in mowed hay showed who used scythes and who used machinery; drilled wheat fields displayed more wealth than hoed corn hills or scattered buckwheat. Stone or brick houses transmitted different messages than wooden ones.[65]

Differences in wealth also determined access to improving practices. Life sketches of Herman Ten Eyck Foster and the Weeks brothers, John B. and Levi, make this evident. Born into a wealthy mercantile family in New York, Herman Ten Eyck Foster had attended Columbia before deciding on the "profession" of farming and apprenticing with Aaron Owen, a Seneca County farmer. His first farm, bought for him by his father for $10,000 cash down in 1845, was valued at $20,000 five years later.[66] Though he lived there until his death, Foster retained his wealthy urban connections, eventually marrying Pauline Lentilhon, a relative of the Du Ponts. He copied French phrases and poetry in his diary, shot birds sportingly, played the flute, and moved in a similarly refined society of well-off farmers in Seneca County.[67] Though profitable (reportedly clearing over a thousand dollars a year), his farming style required serious capital. When Foster won the state society's farm management prize in 1848, he reported using Hussey's reaper and Emery's seed planter. His cattle, part-Devons, ate cornstalks mashed by a horse-powered cutting machine, and his fields received four hundred wagonloads of manure per year.[68]

In their diaries from the 1850s, the Weeks brothers also recorded improving practices, plastering their corn and composting. Both took agricultural journals, and attended nearby fairs as well as lectures on electricity, magnetism, physiology, and chemistry. However, where Foster had left his family to farm, Levi Weeks confided to his diary, "I ardently wish that I had a place of my own for then we should be in no ones way. . . . There are many things unpleasant and trying & which we would avoid by absenting ourselves from the place where some would wish us to be."[69] Unfortunately for Levi, he and John B. split their farm, their house, and their profits with their father, packing thirteen family members and two hired hands into one house. They were not poor—they received rent and labor from a tenant, Oliver Sisco. But where Foster bought agricultural machinery from well-known manufacturers, the Weeks brothers made their own Geddes harrow and their own children's shoes. When they bought a reaper-mower in 1855, Levi borrowed money from his wife and his aunt to make up his share. Their improving practices did not win any prizes.[70]

Buoyed by his wealthy relatives, Foster could make experiments and investments that would ruin the Weeks brothers, let alone their tenant and poorer neighbors. Such gaps were a continual sore spot in the letters sent to agricultural journals. As one correspondent to the *Cultivator* bitterly remarked, "Gentlemen of large property, or high salaries, owning from 50 to 200 acres of land near a good market, may farm it *according to the book*, and talk learnedly of 'rural architecture.'"[71]

Performing Refinement, Performing Labor

Universalizing invocations of "the farmer" notwithstanding, improving institutions made rural difference more visible not less. Two areas in particular offered sources of permanent tension and activity: the expansion of rural refinement and gentility and the dignity (or not) of physical labor. Middling farmers would use improving institutions to assert their refinement and to rail against it, to perform labor, and to separate themselves from a growing class of laborers.

Where in the eighteenth and early nineteenth centuries, refinement had been the concern of landlords and their social circle, and then of merchants who aspired to join them, in the antebellum period, new ideas of refinement profoundly affected the lives of a broader swathe of rural New Yorkers. While their parents might have contented themselves with a wooden clock or painted chest, by the 1840s, well-off rural New Yorkers were building parlors, planting ornamental gardens, buying gloves to conceal work-reddened hands, and covering their rooms in hard-to-protect new curtains, tablecloths, and rugs.[72] As with merchants' rural retreats, this public aesthetic performance was not simply a matter of aesthetics. By tastefully ornamenting their homes and farms, cultivating sentiment, and adopting new codes of manners, rural New Yorkers could lay claim to an inward life and moral and political standing as "ladies" and "gentlemen." If they failed to do so, if their performance was unconvincing, they could lose their standing as decent people.[73] The same sentimental assumptions about rural lack of artifice that made the countryside an attractive refuge could justify stinging portrayals of country rubes: coarse, unfashionable, and easily fooled.

As rural Northerners improvised distinctive rural forms of refined sociability and consumption, improving institutions offered advice. The *Cultivator*'s "Design for a Genteel Farm House" gives a good sense of the mixture of insecurity, pride, and insult that often characterized such efforts. Its design was, the *Cultivator* suggested, better than the "shingle palaces," in which many farmers became "the just object of ridicule." At $2,000, a price "of moderate pretensions," it mingled producing and consuming spaces: a columned verandah in the Greek style, a parlor and a library but also a dairy room, a root cellar, and, revealingly, separate stairs for the farm servants.[74] Improving architectural guides for houses, and even barns, built in the "correct taste," became common in the 1840s.[75] Agricultural warehouses sold flowering trees and fountains "highly ornamental for the garden and lawns" priced from ten dollars to $1,500 alongside corn shellers and hay rakes.[76] The act of reading improving texts itself could demonstrate refinement. When Moses Eames or the Weeks brothers made bookcases or took their journals to the binder for a more beautiful spine, they gave their improving reading practices a public face.[77]

Such efforts at refinement were not simply imposed from above. As fairs expanded, middling participants' enthusiasm for refined culture sometimes unsettled wealthier organizers. As Catherine Kelly has shown, fair organizers hoping to promote local textile production expected and encouraged rural women to produce domestic manufactures—rough fabrics and yarn, but quickly found displays overrun with paintings; floral arrangements; wax flowers, fruit, and birds; and bouquets made from braided hair. By the early 1850s, agricultural societies had been forced to include special categories for the fancy work that inundated them. By the late 1850s, improving commentators used these ornaments to farm homes as evidence of Northern progressive distinctiveness.[78]

As refined images met rural realities, however, performances were difficult to sustain. In a system of production that increasingly depended on women's labor for cash income, it was difficult for rural New Yorkers to sustain many illusions about separate spheres, for example.[79] Even the physical signs of refinement were difficult to manage. In 1857, "A Mother" sent the *Genesee Farmer* a series of prescriptions for keeping young men from fleeing to the city, by then a regular theme. Women, she argued, should help men to participate in gentility, by easing their way into the fabric-covered spaces of the genteel interior. Rather than complaining of their brothers' and husbands' dirty boots and smell of manure, they should keep "slippers always at hand" and make cotton overalls that could be slipped off, leaving young farmers "in as fit condition to enjoy a book or pleasant conversation, as a lawyer from his office."[80] On the farm, the dictates of refinement made ever-present dirt more shameful as well as more time-consuming.

Even as they offered instruction in refinement, agricultural journals also reflected disquiet with the idea of refinement. The same accusations of inauthenticity that might be applied to merchants "retiring" to farm could just as easily be applied to farm families who built a parlor or planted an ornamental tree. Just like urban New Yorkers, middling farmers deprecated artificiality, pretension, and excess, the unavoidable flip side of refined performance.[81] The *Genesee Farmer* caricatured an imagined local farmer who rejected his homespun neighbors, allowed his daughters to be "instructed in music or painting" and his sons "in dancing," and ultimately succumbed to dyspepsia, gout, and mortgages.[82] As the presumed bearers of refined culture, women came under particular scrutiny in such texts. In his bachelor youth, one reader wrote wistfully to the *New York Farmer* in 1831, women had not been "*eddycated*" to be made ashamed of labor, "Instead of stealing off silently to milking, as if ashamed of having cows to milk, they used to accept our help; and many a time have I thought they put on the more airs, for the *number of cows* in their *Father's dairies*."[83]

If fear of labor was the consequence of excessive refinement, improvers postulated, perhaps it was also its cure. On the first page of the first issue of his Utica-based journal, the *Northern Farmer*, T. C. Miner announced an antidote to excessive gentility: his paper would "be exclusively filled with plain, common sense articles from the *real* farmers, who have *held the plow and swung the scythe*."[84] Fed by producerist politics and new narratives about the dignity of labor in the 1820s and 1830s, an alternative strain of improving rhetoric stressed plow holding and scythe swinging.[85]

Increasingly, public performances of labor, what might be called "conspicuous production," became a mark of authenticity, even among those with no need to work. When the flute-playing, Du Pont–marrying Herman Ten Eyck Foster died in 1869, a pillar of the state society, his memorialist lingered over "his working, where need was, with his own hands, and always where the labor was severe and ingenuity was required in its application, and in thus being the leader and instructor of his farm laborers." Foster's devotion to labor was proved in his death: he was crushed by falling ice while working in his icehouse and died of gangrene, "his place was the post of danger . . . he was always in the corner where the hard knocks came," his eulogist wrote.[86] In the same vein, accounts of body-damaging labor became proof of political authenticity. When former governor Silas Wright died of a heart attack in a field by his home (where he had retreated after his defeat by the Anti-Renter–Whig coalition), his eulogists worked to purify his memory by dwelling on the strenuous labor that had caused his death. "[Wright] was himself a principal laborer in all his agricultural operations," John A. Dix told a state fair audience, "plowing, mowing and harvesting, performing himself a full share of labor."[87]

But labor had cultural perils also. Its contradictions appeared strongly in the diary of Herman Coons, a farmer's son with literary ambitions. Harvesting on his father's farm, Coons showed his familiarity with standard praise for labor, describing, "the poetry of 'rocking the Cradle,'" praising the "exquisite relish [for food and drink] known only to the laboring class," and declaring, that "[harvesting] operates as a fine Panacea to dispel sedentary disorders. Hypochondria, vapors and blues ooze out at every pore in the form of sensible perspiration." However, Coons also admitted that the poetry of labor was "soon lost in the reality." When he left the farm to become a student at the State Normal School he did so, "judging it beneath my dignity to continue in the capacity of a common laborer."[88]

The growing presence of a small but permanent "laboring class" complicated the meaning of farm labor. Swinging the scythe was not enough to make someone a "real" farmer. On the 1850 census form, laboring men were often not marked as "farmer." Farmers, it was generally agreed, were heads of household, the directors of their family's labor, and sometimes of more. However, the acknowledged high price and scarcity of American free labor made it much more difficult for farmers to separate themselves from their workers. In an 1846 letter, J. R. Speed of Caroline complained that his foreman had been taken ill and that with hands hard to find at haying time he had done the work himself. "[I] never worked harder in my life," he wrote, "and never mean to work as hard again."[89] Writing to the *Cultivator* in 1848, J. S. Copeland attempted to separate himself from ordinary laborers, by describing his identity as split: "I have been my own director," he wrote, "manager and foreman, as well as a *laborer.*"[90] Agricultural journals fretted about the threat to status that labor posed and proposed remedies to labor's denigration. The *American Agriculturist* declared, "An American farmer should strive to be the *light* and *mind* of the country, as well as its bone and sinew—it is his exalted privilege—his destiny, otherwise he is but a mere *moving machine—a living automaton.*"[91]

Education, improvers hoped, could elevate the middling farmer above automata and justify their position of authority over laborers. Imitating the enormous increase in the number of medical and law schools established as antebellum doctors and lawyers sought credibility, improvers continually demanded funding for agricultural schools. Even without state aid, New York improvers set up an agricultural school in Columbia County in 1836; three in 1847 in Monroe, Dutchess, and Orange Counties; and finally, one in Ovid in the 1850s.[92] In agriculture as in medicine, education was to mark the professional from his fellows.

As improvers negotiated the ambiguities of middling, accounts of refinement or the value of labor did not congeal into coherent philosophies with bounded groups of adherents. In letters to journals, diary accounts, and public displays of goods, farmers jumped from one set of premises to another—sometimes trumpeting labor's virtues and other times fretting over its mindlessness, sometimes admiring neat buildings and sometimes obsessing over unnecessary parlors. Improving institutions and texts expressed these contradictions but did not resolve them. However, Herman Coons's nervousness about being seen as one of the "laboring class" reminds us to pay attention to another set of sometimes less-willing participants, whose work was also on display at the agricultural fair in Poughkeepsie but whose opinions and names were less likely to be credited, and whose status was much more tenuous.

Invisible Hands

In 1844, the fruit committee at the state fair awarded a diploma and one dollar to "Moses Humphrey, Poughkeepsie, a colored man, 80 years old, for fine specimens of Grapes." It was possible, in the 1840s in New York State, for an African American man, listed in the census as a "laborer," and (perhaps dubiously) as "illiterate," to compete with whites in a public forum.[93] Though this is the only explicit reference I have found, it may have been more common than it appears, since the race of winners was not generally listed and most entrants were never named. While rural and provincial Black New Yorkers are hard to find in the historiography, they were not hard to find in reality, either as minorities in places like Poughkeepsie or as majorities in places like "the Hills," an informal farming community of free Black people in Westchester, New York, that was preserved from white land speculation by the thinness of its cold soils.[94]

Moses Humphrey's prize was a rare moment of visibility breaking the norm of whiteness and relative wealth. It should remind us more broadly of the many hidden presences in improvement. Agricultural fairs and the pages of agricultural journals provided stages on which white men could perform publicly, but there were plenty of other people behind the scenes. As the century wore on, distant famines and political strife combined with domestic land bubbles to deepen divisions further, creating a small but growing class of seasonal wage laborers. By the 1850s perhaps one-third of the farm families in the Hudson Valley employed one non-kin servant in their household—some employed four or five throughout the year and dozens more at harvest time.[95] While white, native-born workers might experience agricultural labor as a transitional step toward tenancy or farm ownership, Irish and African American laborers, paid less and excluded from networks of neighborly credit, formed a durable bottom layer.[96] Although these did not approach in number the population of laborers in Great Britain, where laborers outnumbered farmers six to one, American farmers could still rely on the Irish immigrants, African Americans, and new arrivals from Connecticut who worked the fields of Amenia, or who stood back, we are told, when Herman Ten Eyck Foster took the "place of honor."[97]

Other hidden workers were family members of named improvers—sometimes younger men, but also women and children—who did the hauling, digging, and rock pulling needed to carry out many improving experiments. White women's fancy work, domestic manufactures, and fruits and flowers had their own sections at the fair and columns in the journals. But their field work, though reasonably common, surfaced only to provoke comment, as when twenty-year-old Della A. Roberts of Niagara County "having literally put her hand to the plough" defended the right of her family of sisters to do farm work "*in earnest*, and for *pay*" in the *New England Farmer*, in response to a scornful article in the *Rural New Yorker* during the Civil War.[98]

Of course, such workers were more than hands. The African Americans pruning trees and digging ditches in Andrew Jackson Downing's famous nursery were surely the source of some of the advice on grafting and insect deterrence that would fill his journal, the *Horticulturist*. Likewise, the insights on farm management published by Herman Ten Eyck Foster surely came in part from the laborers in his household, and visitors to John Holbert's prize-winning dairy in Chemung County would certainly have expected to speak to his wife, who likely ran it. Occasionally skill and knowledge could be parlayed into status. Myra B. Young Armstead has traced the career of James F. Brown, a Maryland fugitive from slavery who became gardener to wealthy farmers on the Hudson and began to move in the highest horticultural circles, experimenting and recording results and attending horticultural shows with his employer, Mary Anna Verplanck.[99] As we shall see, Sylvia Parmentier, descendant in her own right of a long botanical legacy, managed her Manhattan nursery after her husband's death, playing a key role in the launching of the mulberry bubble.

In general, however, the ideas and the labor of white women, young white men, and Irish and African American laborers of both sexes vanished into white male authorship in ways that we can only sometimes fleetingly see. When New Jersey inventor Ann B. Harned Manning's designs for a reaper and a clover cleaner were patented, the name on the patents was her husband's, a switch recorded by Manning's acquaintance the suffragist Matilda Gage.[100] At the Tompkins County Fair in in 1852, James McLallen made his self-crediting clear: "I got premium on Quinces and Butter," he wrote, "and Wm Hazely's Plowing."[101]

★

Seen in the light of the fractures, tensions, and uncertainties clustering around him, George Bancroft's speech at the 1844 Poughkeepsie Fair looks less like an agglomeration of well-worn clichés and more like an impossible balancing act. Many of his hearers were probably unhappy to hear the landlords who founded New York improving institutions invoked as guiding spirits—indeed the only named "farmers" in Bancroft's speech were Stephen van Rensselaer, Robert R. Livingston, and "the farmer of Westchester, the pure and spotless [John] Jay."[102] At the same time, Bancroft admired the "dignity of labor" in terms that Van Rensselaer and Livingston might have found remarkable. By stressing the independence of his imagined farmer, Bancroft tacitly acknowledged the collapse of the landlords' vision of an enlightened, profitable, and deferential tenantry.[103] His speech, in fact, was a little logically incoherent—to paper over rural difference, it had to be.

Even as Bancroft busily erased difference at the Poughkeepsie Fair, the displays around him performed it. Landlord families like the Wadsworths and the Rensselaers sent prizewinning cattle and grapes, though Rensselaer's grapes sat on the same table as those sent by the African American laborer Moses Humphrey. George Vail, formerly a wealthy merchant of Troy, demonstrated his entrée into elite breeding circles by winning the bull competition. Many of the finest and most elegant items, like "Catawba and Isabella grapes grown under glass, very large gooseberries, . . . a floral ornament seventeen feet high, shad from his fish ponds, &c." came from Robert Pell of Ulster, famous for the largest commercial apple orchard in the world—whose lavish life depended on his many laborers.[104] The exhibitions of the ladies' tent displayed changes in women's labor and shifts toward rural refinement among an aspiring new provincial elite. While Mrs. G. W. Henry's prizewinning entry into the flannel competition satisfied organizers' demands for domestic industry, in the manufacturing tent, the Middlesex Company of Lowell, Massachusetts, made clear that home textile production had become an exercise in nostalgia.[105] At the same time, marvels of fancy work like the wax imitation of fruit by Miss McDonald of Poughkeepsie and the gilt frames and divan by Miss Mary Sherwood of Fishkill forced organizers to acknowledge the growing strength of refinement by improvising new categories to include them.[106] As fairs grew larger and more significant, more and more meanings would crowd into them, slipping far beyond their organizers' control. In 1859, David Nelson marveled at balloon ascensions at his local fairs, enlivened in one case, by a puncture at "the hight [sic] of a mile" and in the other by a fire. (Surprisingly, no one was killed.)[107] Such displays helped draw agricultural fairs into the spectrum of public entertainments, from camp meetings and lectures, to circuses and menageries, available to antebellum farmers. To audiences they were spaces to shop, to drink, to wonder, or to ridicule.

As would have been clear to anyone looking across the fair, the single label "farmer" could refer to people across a highly visible, contested, and rapidly shifting rural hierarchy. It was these shifting and multiple meanings, hiding behind a pretended unity, that made agricultural improvement so attractive to politicians. Wielded skillfully, a farm could evoke the gentility of the British ruling classes, and, simultaneously, the republican virtue of the yeoman farmer. It could refer to labor or management or appeal to the interests of landlords or tenants. For middling farmers, it could demonstrate refinement in the face of accusations of coarseness or authenticity in the face of accusations of refined artificiality. It was from across this spectrum of identities that improving New Yorkers debated the meaning of the land and of the market and performed their identities, their worthiness, their probity, and their political legitimacy to each other.

PART TWO

Experiments

3

Experiments All for Worldly Gain

"This is indeed an age of excitement on worldly matters," twenty-two-year-old Alson Ward wrote in his diary in 1844, "new inventions are being made, experiments which all aspiring for worldly gain. This is right but how careful should we be that our aspirations are rightly directed. They should be for the good of others as well as ourselves and the glory of God our maker."[1] Like many devotional diary statements, Ward's was self-reflective—he and his family actively participated in the culture of profit-driven experiment that worried and excited him. At their mill and farm near Poughkeepsie, they not only raised cows, chickens, bees, pigs, hay, barley, oats, wheat, peaches, apples, locust seedlings, and watermelon; they also practiced moderate improvements—treating their soil with plaster, ashes, and stable manure and planting clover and timothy. They owned a thermometer, a water-powered corn-sheller, and a threshing machine, as well as a horse rake, a machine of their own design for "brushing peaches," and another for feeding hogs automatically.[2] That year they conducted their first explicit "experiment," harrowing in their cornfield "prostrating [the corn] to the earth, for experiment to see if it would injure or benefit it."[3] That winter, Ward and his father would conduct experiments on bees—building a hive in Alson's bedroom "for experimenting, &c."[4] Throughout the winter the Wards changed the bees' entrances and altered their food and mechanisms of feeding, lamenting, in the end, when they died.

Ward's claims to experiment were not unusual. New Yorkers from across the ranks of improvement were making experiments. Like Ward's, some of these seemed quite simple: one published note from "Experimenter" read, "I tried pulverized charcoal to keep the bugs from my melons and found that it did not a particle of good."[5] Others were more elaborate, taking over multiple fields, seasons, and articles in the agricultural journals. These improving genres of experimental reporting drew on those of natural science, developing forms of witnessing, prolixity, "naked" writing, and "invisible technicians."[6] However, they also developed their own features, most notably a complex relationship with place and an epistemological dependence on profit. Following the experimental culture of agricultural improvement on its own terms can illuminate improvers' understandings of nature, their relationship to markets, their social divisions and sometimes warring senses of credibility, and the ways in which they communicated (or did not) across the gulfs that divided them.

Experiments and Trials

Just as I did, many of the people reading this book probably learned a litany of features that constitute "an experiment" in school: the formulation of a hypothesis, the design of a test (including a control group), and the confirmation or falsification of the hypothesis, followed by the replication of experimental results. If we got far enough in our scientific education, we also learned the written structure through which to describe experiments. We learned to omit humans from these descriptions, studiously writing them in the passive voice. Our textbooks also reproduced the iconography of experiment: clean laboratory benches, white lighting, test tubes of different colors, and standing among them, a "scientist": someone with markers of credibility, like graduate degrees, a white coat, unworldliness, and glasses.[7] From this idealized viewpoint, antebellum descriptions of agricultural experiments may not look much like experiments at all. Muddy fields, subject to the vagaries of heat wave, rain, hard frost, and sudden irruptions of corn worm or blight are about as far from the clean laboratory as it is possible to get.[8] There were usually no recognizable "scientists" and few explicit hypotheses or control groups. The passive voice was largely absent, and human errors were frequently described.[9]

However, modern concepts of experiment would have been horribly out of place in the 1830s and 1840s. Even in the most elevated scientific circles, the "laboratory" was still a relatively novel space. Many people (not just the famous homebody Charles Darwin) still conducted experiments at home or, like Michael Faraday, translated experimentation into a sort of theater for general audiences.[10] Rather behind lawyers and doctors, scientists were developing a professional identity. Even the term "scientist" was only invented in 1834 (in Britain) and again in 1849 (in the United States) and was not widely used until later in the century.[11] The clear career path we now associate with science existed mostly in the imaginations of people who were starting to describe themselves as "chemists" or "geologists" and who could generally be found at secondary schools or working for the state geological surveys.

In the antebellum period, the word "experiment" was linked more firmly with its etymological cousin "experience" and was contrasted often with "theory," "book knowledge," and "speculation." Like other kinds of experiments, agricultural experiments were at once knowledge producing and knowledge communicating; they both revealed phenomena and persuaded others of the existence of those phenomena. We can get a better sense of this dual function from another word, one that was often used interchangeably with "experiment" in nineteenth-century improvement: "trial."

"Trial" may not now seem an epistemologically impressive word; "trial and error" is sometimes now used as a slightly insulting synonym for "tinkering." However, improving references to trials hark back to deeper interconnections between legal and scientific knowledge. Like experiments, legal trials are expected both to establish knowledge about the truth of an event and to convince others of that truth; they are both investigation and public performance. As Barbara Shapiro has shown, trial language grounded natural scientific traditions as far back as the seventeenth century.[12] Like other experiments, agricultural experiments borrowed a language of evidence, testimony, and (natural) law. They sometimes had witnesses and signed statements and, when performed at agricultural fairs, were presided over by "judges" and occasionally "juries."

The language of judicial knowledge making would have been familiar to many rural New Yorkers. Dreams of bucolic peace notwithstanding, rural life was litigious. Lawsuits for debt, libel, fraud, and damage caused by escaped pigs were common.[13] Trials, moreover, became a recognized form of public entertainment, both in the newspapers and in person.[14] During the winter, when work slowed down, John B. and Levi Weeks spent two days in nearby Ballston Spa to take in "a part of the testimony" of the murder trial of Joseph Glasser; Levi remarked afterward with apparent satisfaction, "a great many people in attendance, fine day. I took dinner at H. Merchants."[15]

Agricultural experiments could be trials of many different kinds of things: techniques, organisms, fertilizing substances or insect poisons, machines or accounting systems, landscapes or natural laws, principles of profit or particular visions of the future. They also tested the capacities and virtues of the person performing them. This expansiveness was not limited to agricultural experiment. To early Americans the language of experiment evoked social novelty as much as natural science. "The eyes of all Europe are turned to our experiment of self-government," New York senator Nathaniel Tallmadge announced to his colleagues in 1838, "and are confidently expecting a failure of the system, which we have predicted would regenerate the civilized world."[16] The members of the Oneida Community described the results of their "partial and temporary experiment" in free love, and the president of Kenyon College preached to New York audiences about the "experiment of the religion of Christ" that had been "varied sufficiently to put it to the fairest trial."[17] The Society of Friends, arguing against the removal of the Seneca, longed to be able to complete a "successful experiment" in "the complete civilization of our native red men," and the Auburn State Prison, "experimented," disastrously, with solitary confinement and enforced silence.[18]

Experiments also bolstered the expanding projects of American capitalists (in its 1840s sense, the holders of capital, usually bankers or investors). Transportation companies in particular described themselves as experimenting with new technologies, before expanding their use, as when Jasper Allaire's steam packet company begged for tax relief while their "experimental" ship carried paying passengers from New York to Charleston.[19] Economic experiments might also be performed at the scale of individual lives: Hannah Farnham Sawyer Lee's wildly popular novel *Three Experiments of Living* played out three economic scenarios in the life of one couple—who lived frugally in the first section, expansively in the second, and ruinously in the third.[20] Lee's work had many imitators, notably *The Fourth Experiment of Living: Living without Means* and *Three Experiments in Drinking*. She herself would write a sequel, *Living on Other People's Means*, in which a farmer's son, refusing to stay home and help his father with improving tasks, becomes president of the "Bubbleville Bank," expends their meager wealth on luxuries, and on his deserved death, leaves his parents in penury and his family farm to be half-cultivated by impoverished tenants.[21]

As in this last example, the language of financial experiment sometimes also implied folly. On one of the best known "hard times tokens," small coins issued by merchants during the Bank Wars and panics of the mid-1830s, a ship with "experiment" engraved along its side foundered, shattered by lightning and towering waves, while on its reverse a turtle labored under a box marked "treasury," wreathed by the words "executive experiment"—an attempt to pin the blame for financial disaster on the currency "experiments" of Andrew Jackson.[22]

However, while experiments were subject to accusations of volatility, fancifulness, and fraud, when successful they could also reveal the underlying structure of nature, demand particular forms of investment, justify or challenge social hierarchies, and elevate the experimenter.

Experimental Form and Experimental Parody

Experimental writing was various, encompassing meticulous articles sent in by leading improving lights but also private scribbled notes like those left by Edward Johnson, a farmer of Schodack, who in 1852 planted "as a matter of Experiment a paper of Drum Head Cabbage Seeds to see how soon I could get the plants and set them out."[23] Clearly, experimental description was not a perfectly disciplined form. However, if we dissect published descriptions of experiments (and the writings that made fun of them), the elements of a genre start to emerge.

Let's start with a not particularly distinguished or unusual experiment: a description sent by Jarvis W. Brewster to the *Cultivator* in 1835, the journal's second year. A self-identified improving farmer from Trenton, in Oneida County, Brewster admitted that he had only farmed exclusively for two years. However, he had made some experiments "in the growth of Indian corn, potatoes, and rutabaga turnip." Taken alone, his description of these feels like a bald recounting of details. First, Brewster described the land and its condition, "a ten-acre lot of stiff strong sward [grassy turf] that had not been ploughed for many years." Then he described his own actions, in great, if not stirring, detail—he measured off one acre for each crop, plowed, rolled, harrowed, and furrowed it "three feet apart from north to south." For his corn experiment, he manured the land at twenty loads to the acre, then planted the corn—giving his readers the date, and letting them know that his seed had been soaked in tar and water and plaster, and then put down four kernels per hill, one foot apart. The first row didn't come up (the seed had lain too long in the sun), so he replanted and it came up well. He described the weather, particularly the arrival of killing frosts in September. Finally, he gave an account of his profits. "94 bushels of corn at 6 shillings—$70 50" and "Expenses of $29.50" left an excellent profit of forty-one dollars. Here the trial was a test not of technique, but of place: "This section of country is celebrated for grass and grazing," Brewster wrote, "and most of our farmers have embarked in the dairy business, under a belief that the soil and climate is unfavorable to the growth of all kinds of grain excepting oats." Brewster was testing what he would have called the "adaptation" of Oneida County to the production of wheat, corn, and turnips, a concept that will be central to chapter 6.[24]

This choice of narrative elements was highly structured—common enough, in fact, that when Henry David Thoreau wrote his now semisanctified essay "Economy," he used them in a parody. As with the rest of "Economy," in his "experiment," Thoreau systematically inverted clichés of antebellum economic writing. Where Brewster explained that he had good soil covered with sward, Thoreau announced that he had a "sandy soil," "good for nothing but to raise cheeping squirrels on."[25] Brewster put on twenty loads of manure and plowed, harrowed, and dug—Thoreau "put no manure whatever on this land, not being the owner but merely a squatter . . . and I did not quite hoe it all at once." Brewster detailed his seed preparations and their cost; Thoreau boasted that he got his for free: "[seed] never costs anything to speak of, unless you plant more than enough." Where Brewster apologized for his lack of punctuality in planting—Thoreau airily failed at corn and turnips, planted "too late to come to anything," producing only beans and potatoes. Like Brewster, Thoreau finished with an account of his profit and loss—on his dreadful soil, with his free seed and cheerfully terrible practice, he made a little less than eight dollars. So that no one would miss his dig at improvement, Thoreau observed that he was "not in the least awed by many celebrated works on husbandry, Arthur Young among the rest." Then came the moral. With his tiny farm, relying on spades and human labor rather than plows and animal labor, shifting his fields rather than manuring them, he wrote, "I was more independent than any farmer in Concord, for I was not anchored to a house or farm, but could follow the bent of my genius."[26]

Of course, Thoreau was not specifically making fun of Brewster—an obscure figure, who had published nineteen years before. The appearance of something like a Brewster parody was possible because Thoreau was hugely outnumbered. Even those who didn't read the agricultural journals would have encountered texts structured like Brewster's as they filtered into almanacs and newspapers in the 1840s and 1850s.

What produced this parody-ready consistency? Thoreau was partly right to blame Arthur Young. As the most visible publicist of eighteenth-century British improvement, Young had consciously worked to shape the norms of improving experimentation. For decades, American improvers reprinted experiments from his four-volume *Course of Experimental Agriculture* (1771), detailing five years of experiments on his Sussex farm.[27] This work was clearly intended, not to invent agricultural experimentation, but to discipline an unruly genre. In "an age so fertile in book-making," Young complained, "which produces so many experimental husbandmen," the demand for agricultural books had even attracted "geniuses, in whom invention supplies the defect of land, feed, cattle, implements, and every requisite save pens and paper."[28]

In the antebellum United States, agricultural writing was also hard to discipline. Journal editors boasted of the number of their correspondents and were too anxious for copy to do much gatekeeping. Pushing indirectly for greater formality, American authors wrote admiringly about the reports coming from experimental farms established at Rothamsted in Britain in 1842 and at Möckern in Germany in 1850; agricultural journals and societies would consistently advocate for experimental farms of their own.[29] While their efforts largely failed until after the Civil War, the features of the genre that so irritated Thoreau spread into American practice through other routes. Perhaps most importantly, judging committees at county and state fairs began to require formal statements from competitors, recruiting thousands of people to Young-influenced styles of communication.

Breaking the genre of agricultural experiment down into its common features can help to fit it into broader patterns of experimentation and to see its particularities. Some of these features strongly resembled those of other experimental writing. Most immediately evident of these was the numbing level of detail, which Steven Shapin has called "prolixity." By adding layers of detail, Shapin argues, experimenters tried to give readers the impression that they were actually present, "virtual witnesses" able to attest to the matters of fact being described.[30]

Brewster's lack of speculation about causes was a similarly standard rhetorical gesture, one dear to the hearts of early nineteenth-century American men of science, who tended toward an extravagant suspicion of "theory building" and the "spirit of system" or "speculation." As Sophia Rosenfeld shows, during the eighteenth century, an emphasis on "facts" available to experience and a suspicion of "theory" had run through political and scientific discourse alike.[31] Echoing Royal Society experimental theorists of an earlier generation (and jurists of an even earlier one), antebellum improvers continued to deploy a "naked way of writing." When they ventured to suggest a cause or underlying law that might explain a phenomenon, they larded it with professions of self-conscious modesty and "theoretical innocence."[32] After postulating a theory that well-pulverized soil might hold moisture better because of the separation of the soil particles, for example, an author in the *New England Farmer* added hastily, "Such is my *theory*; but I am an enemy of theories, I always distrust them, I look only to facts."[33]

This ritual derogation of theory also appeared in a subgenre of accidental experiments, ostensibly performed entirely without expectations and therefore without preconceptions. Thus, "C," writing to the *American Agriculturist*, could report an experiment when "by accident, one row between two others was left without guano, and without any prejudice or partiality was cultivated in other respects," producing four times as much grain.[34] American improvers also followed Young in laboriously reproducing even those experiments "from which scarce any conclusions can be drawn," Young explained, "that my book might be the real transcript of my practice, and not a partial representation of experiments, picked and culled to serve the purposes of a favourite idea, or upon which to found a brilliant hypothesis."[35] Strengthening the impression that nothing had been excluded, experimenters often included realistic errors. But errors had a different status in improving texts. When Brewster confessed his errors—planting at the wrong time, on the wrong side of the hill—he was strengthening his case that Oneida County was good corn land, since even an incompetent farmer might make a profit.

To bolster their own testimony, and to give readers the sense that they were witnessing a real event, improvers often described witnesses around the scene of an experimental account. The most satisfying witnesses were skeptical neighbors, usually left anonymous, being treasured for their ignorance and subsequent conversion. One such story, appearing in the *Cultivator* of January 1853, dealt with the controversial practice of subsoil plowing, which was intended to bring up new nutrients from the deeper layers, but which threatened to unleash poisons and miasmatic exhalations. When a member of the Queen's County Agricultural Society attacked his land with a new Eagle D Plow, "the neighbors came round to see the folly, as they termed it, of the book-farmer" predicting that "he was killing the land in plowing up the yellow clay." The triumphant yields that followed naturally dissolved their skepticism.[36] Through such stories, improvers reassured each other in the face of opposition and stressed the power of their experiment over witnesses.

Though all experimental writing was prolix, the details of place had a distinct status in agricultural experimentation. Laboratory experiments have a peculiar relationship to place—by isolating particular natural phenomena in controlled environments, they are expected to produce natural laws that are universal, that is, to create placeless knowledge. Atoms smashed or mice drugged are supposed to respond in similar ways whether the laboratory is in London or Tokyo. Though historians have demonstrated the enormous social labor that makes the appearance of replication possible, local variability is, in theory, supposed to cast doubt on the validity of the result.[37] Fields, barns, and pastures, however, are inescapably variable.[38] The success of any particular new agricultural practice within farming depended on factors beyond any farmer's individual control: the fertility of the soil and the "tenacity" with which it dragged at the plow; the slopes and stumps that interfered with machinery; the exposure of hills to the north or south; or the average date of the last and first killing frosts. Markets also influenced experimental results; distance from towns, local or distant food fashions, the arrangement of roads and canals, and the local availability of credit and labor could all determine whether a particular technique was successful in any given location.

This posed a problem for improvers. If the effect of place was entirely fragmentary or atomistic, then global or even state improving projects would be impossible—techniques would not be able to move over distance. This helps to explain why improvers were so anxious to establish the boundaries of "agricultural regions" within which generalizations could be made, as we will see in chapter 6. It also cast the value of experimental farms into doubt. In New York, the most plausible push for a state-sponsored experimental farm was torpedoed by Jehiel Halsey and the skeptical agricultural committee of the 1834 state senate. "The diversity of soils, situations, and even climates, is so great in our 'Empire State,'" Halsey complained, that "it is evident that experiments tried upon a 'pattern farm,' located in any one part of it, cannot be made to indicate the proper system to be pursued on very many other farms in the State." According to Halsey, advice could not move between counties let alone countries. "Even if the whole scope of any one county were taken into view," he announced, "it would be impossible to found any system of farming upon it and make it generally applicable to the successful cultivation of farms in other counties."[39]

The failure of such pattern farms in New York and elsewhere has been seen sometimes as the failure of antebellum American agricultural science as a whole. But if the inescapability of place undermined the value of centralized experimental stations, it also made room for *decentralized* knowledge. Given differences of place, radically differing experimental results might be, not conflicting, but additive. When seeds that grew in one place did not grow in another, when principles of plowing that worked in valleys failed on slopes, these differences might be not errors, but information, revealing the qualities of places themselves. Agricultural journals repeated the Swiss botanist Augustin De Candolle's dictum that experiments should be repeated "on different soils and in different situations" and recorded and communicated in careful language.[40] Pointillistic pieces of information accompanied by references to place could, at least in theory, be assembled into a nonconflicting mosaic of results. Rather than aiming at placelessness, experimenters therefore worked to give a clear impression of their place using commonly understood cues, like Thoreau's sand-loving pines.

This does not mean that improvers always assembled these different reports of place. Agricultural societies or surveys sometimes worked to bring conflicting accounts together to make an authoritative map or survey. In general, however, readers encountered experiments as long threads of conversation over time. Newspaper-addicted Americans were used to performing the intellectual work necessary to assemble a picture from fragments, whether piecing together an understanding of the New Madrid earthquakes from widespread newspaper accounts or of a battle in the Mexican War from letters written home from various vantage points.[41] Even without attempting synopsis, moreover, they could make use of accounts that referred to their own place, or places with recognizable features. Rather than seeing this as a failure of experiment, we might instead see it as an element of another written tradition: "chorography"—the description of place that derives expertise from the local.[42]

While persuasively displaying places and techniques, experimenters also displayed themselves. Modern experimenters in the natural sciences often erase themselves by describing experiments in the passive voice, rendering their operations as seemingly self-evident functions of nature. Brewster, by contrast, carefully placed himself at the scene. His experiments are described in the active voice—"we plowed," "we put things up." Characteristically, who "we" was never became entirely clear, but it was Brewster himself, he hastened to say, who shook the new canister for seed over the field, while a boy followed him with a rake. This move, indicating one's own responsibility for delicate or skillful labor, was common—"I was obliged to hire a team and a man for the ploughing," Thoreau wrote, "though I held the plow myself."[43] Brewster, Foster and, mockingly, Thoreau all both performed labor and cast themselves as commanding of the labor of others. Such gestures might lack the drama of Herman Ten Eyck Foster's death in the icehouse, but they were part of the same culture of conspicuous production.

While evidencing their practicality, improvers in the 1830s and 1840s were also careful to display their book learning. Brewster's practices, the harrowing, the seed soaking, and measurement of distance between seeds were all recommended practices. Even in apologizing for his late planting, Brewster took a position within an existing debate about planting time: "I am decidedly in favor of early planting," he wrote, "I would never leave it later than 10th May. Urgent business calling me from home was the cause of my late planting at this time."[44] But Brewster was also suggesting a particular vision of time and thus a particular idea of regulated virtue. Punctuality was an easy virtue to display in experiments—cautionary tales were to be found in killing frosts missed or hay brought in too late, or by missing the tiny windows in which innocent green specks on the earth sprouted and sprawled into thrashing tangles of weeds. Just as entomologists could demonstrate the virtue of patient attention, and geologists claim masculine heroism, agriculturists could convert their experimental accounts into easy little sermons on capitalist virtues.[45] In "Economy," Thoreau was arguing against the need for industriousness because his targets, Brewster and his fellow improvers, were arguing so strenuously for it.

Harvesting Knowledge

As they wrote themselves into the field, experimenters wrote others out of it. While he himself is the only individual he mentions by name, Jarvis Brewster was clearly not alone in the fields. In the following census, Brewster listed four people as employed in agriculture in his household. Since the only other male in his household was under five, three of these may have been the adult women, listed only by their ages.[46] On crowded farms, other workers often vanished into the passive, the general "we," or the deceptive "I." Paid laborers appeared as costs in the final columns of accounts or the occasional glimpse offered by an error, as when another rutabaga experimenter complained that his seeds had been "drilled in by a bungler, who made the drills too distant."[47] This erasure of "invisible technicians" is common to other genres of experiment, stretching back to the instrument makers, butchers, laborers, and secretaries who made Robert Boyle's experiments possible and forward to the lab techs who do not figure in the crowded author lists of modern scientific publications.[48]

Adopting the genres of experiment allowed improving farmers to recast others' knowledge as "discoveries." While priding himself on a new technique for growing melons on a hill, for example, "D.T." acknowledged that "the honor of first discovery" might belong to his friend Richard M. Williams of Georgia, whose article had not yet reached the Northern agricultural papers. However, an extensive quotation from Williams's article makes it pretty clear that Williams's discovery had been handed to him by his enslaved foreman, described only as "Old Tom," who had asked for the seed, made the hills, directed the planting, and made Williams a profit of $3.75 per vine. (In sad contrast, the acre of melons planted according to Williams's own plan "on a level in the Yankee fashion" had cracked and rotted.) That both Williams and D.T. acknowledged Old Tom before assigning credit to Williams suggests that they expected readers to see an enslaved man not as a discoverer, but as a sort of knowledge resource. Further naturalizing Old Tom's role, D.T. pointed to a broader category of African American practice that could be mined for discoveries. "It appears then," Thomas wrote, "that this method has long been practiced by the negroes of Georgia."[49]

Other accounts suggest the persistence of knowledge made by people acknowledged even less frequently—New York improvers were well aware that the practice of intercropping corn and squash had been practiced first in New York by Haudenosaunee and Algonquian peoples—those who hadn't seen Haudenosaunee fields themselves might have seen the Pennsylvanian author John Lorain's often-quoted list of "savage practices" that included intercropping along with hilling corn, paring and burning the soil, and the use of seasonal signs in corn planting.[50] "C.B." of Phelps inadvertently revealed to the *Genesee Farmer* that he and his neighbors had borrowed these forms of intercropping. "In cultivating the pumpkin," C.B. wrote, "I have usually followed the example that has been set me by neighboring farmers, and planted the seed along with that of my corn." While the wide leaves of the pumpkin shaded out weeds in the space between the hills of corn (a space much wider than we would see in a cornfield today) C.B. wondered if it might not be worth experimenting to "plant them in fields separate from other crops" so that the vines they produced did not diminish the corn.[51] In testing the efficacy of intercropping, improvers moved Haudenosaunee knowledge into the same systems designed for credit-worthy novelties. Clearly, experimental culture was expected to extract knowledge from unnamed knowers.

Accounting for the Field

Of all the features of modern experimental knowledge, the one most alien to Ward, Brewster, and other improvers would have been the ideal of "disinterested" experiment—conducted not for money but for pure knowledge. This idea is an old one, part of the gentlemanly science laid out by the early experimenters of the Royal Society, where a gentleman's fortune rendered him immune to lying, by freeing him from the need for gain. We can see its vestiges, transformed and expanded, in the language of "pure" science that was given immense institutional power after World War II.[52] By contrast, improvers sometimes described personal financial interest as a source of reliable knowledge. Young's *Course of Experimental Agriculture* did this clearly. "As I embraced agriculture not as an amusement, but a business, and with a fortune that would not allow me to be indifferent about profit," he explained, "I fought after TRUTH, and tried a number of experiments merely to discover her; totally indifferent on which side I found her, and solicitous only to be convinced of the most profitable methods."[53] The relevant disinterest here was not in money but in theory—it was Young's own driving need for profit that was to lead to all-capitals truth.

Improving experimenters often indicated the centrality of profit by including a formal account, a statement of costs and revenue like those following both Brewster's experiment and Thoreau's parody. Such accounts were a key epistemological feature. Only clear calculations of profit, Young announced, could allow any kind of true assessment of the value of particular techniques. Profit was a better proof of success than yield, since a huge yield could be produced by expensive, and thus "ill-adapted," unnatural methods. A field might produce "vast crops of corn" and leave its owner bankrupt if the techniques were too expensive. "To assert that an acre in one method of husbandry pays two pounds clear of all expenses, and thirty shillings in another method, is stating a clear comparison," Young wrote, "but to say, that the one yielded thirty bushels, seed deducted, and another twenty, seed deducted, is, in comparison of the two methods, saying nothing to the purposes."[54]

By 1841 columns of accounts had become so common in experimental reporting that one reader complained. "Too many of our agricultural journals are filled with the short details of their raising corn and potatoes," he wrote, "reducing to the standard of dollars and cents the items of their labor—hire—manure—board—value of day's work by oxen—interest &c. &c., as a merchant taking account of stock, book debts, and liabilities to discern the result of a year's business, as if the cases were parallel." Labor on the land was not to be counted in dollars and cents; it was the farmer's "duty and his pleasure, as well as his privilege, to labour if his farm is in good heart."[55] This, however, was not the cry of a tradition-bound farmer lamenting the intrusion of the market—the author was a clerk, seemingly sick of accounting, who finished by asking how much a young man not used to hard labor might make starting out on a farm.

It was also an exception. From the late 1830s on, American improvers were devoted to accounting. "Few points," the *Cultivator* announced in 1837, "are more essential to success in any business than well-kept accounts; and these are as essential in farming as in other operations."[56] Given an appropriate accounting system, mused a later article, "What a clear sun-light would be sent into every dark corner of doubt; and the dim objects of twilight become clear and obvious in full glare of day."[57] Readers responded; agricultural journals overflowed with increasingly elaborate examples of specialized ledgers, daybooks, and labor books.[58] This was particularly significant at fairs, where wealthy farmers could cram cattle with wasteful amounts of feed, or boost yields by expending impracticable amounts of labor and manure, encouraging, as we have seen, the shift to Young's style of reporting.

Of course, accounting was growing outside of improvement as well. A growing population of clerks worked through accounting manuals at the business schools that proliferated from the 1830s onward and a sea of housekeeping-advice literature demanded similar, though unpaid, mathematical labor from well-off urban women.[59] Demand for accounting skills derived a painful edge from the violent fluctuations of the new national economy. Accounting would prevent "procrastination, 'the thief of time,'" while producing "active and painstaking habits" and inculcating "lessons of industry and prudence."[60] As New Yorkers watched businesses fail and fortunes evaporate, accounting promised (generally illusory) control over volatile circumstances through personal discipline.[61]

While agricultural accounting drew strength from these developments, it was not a sophisticated urban phenomenon pushed on an innumerate rural population.[62] Numeracy had left its mark on the language of Northern farmers—the commonly used verb "to calculate" extended beyond computation to mean "to think" or "to plan." ("To reckon" and "to figure," also mathematical terms, retain these meanings in colloquial speech.) More importantly, for many farmers accounting was already a part of daily life, as attested to by the thousands of account books extant in rural historical societies. While merchants and bankers at the upper level of the state society would have been familiar with the mysteries of double-entry bookkeeping, and improving slaveholders and their agents calculated plantation profits as a whole, Northern improvers built on the kinds of accounting systems that helped maintain good relationships in rural neighborhoods.

Middling farmers were used to recording, not farm profits, but personal exchanges.[63] Working in a cash-poor economy, they frequently exchanged goods and labor rather than money. Since debts often remained unresolved for long periods, farmers kept accounts to track their exchanges. The account keeper would write a neighbor's name at the top of each page followed by three columns. The first listed goods or tasks exchanged, and the second two, labeled "Dr." and "Cr." (for "Debtor" and "Creditor"), indicated the value and direction of exchange. Thus in 1844, when the farmer Joseph Hanson gave a bushel of beans to his neighbor George Derrickson, he marked a price in the "Dr." column under Derrickson's name; when Derrickson reciprocated with a bushel of oats, Hanson marked a price in the "Cr." column. At the end of the year, Hanson added up each column and, finding that he had received more than he had given out, paid the difference in cash.[64] Most farm accounts were intended to smooth over this moment, "settling up," which was often contentious. Salmon Bostwick commented wryly to his diary in 1826, "Settled with Daniel Johnson without a quarrel, which is a rare case."[65] Farmers' acts of "barter" were thus quantitatively tracked in a sort of shadow currency—they assigned monetary value to goods and labor in order to determine whether their exchanges were fairly balanced. Rather than describing profits, account books helped farmers manage quarrels and avoid court cases.[66] This form of accounting could also extend into intimate relationships; the Weeks brothers kept accounts with each other and with their father and aunt. Harvey Badger of Painted Post did the same with his children.[67]

Experimental writing converted these neighborly relations into relationships with farm fields. One accounting textbook suggested, typically, that farmers might "open an account with the field . . . making the field Dr. [Debtor] for the labor of plowing, dragging, sowing, harvesting, threshing, marketing, &c. for what it produces, the difference will show his gain or loss."[68] Farming in western New York, George T. Sprague took this advice to heart. "I commenced a plan this week of keeping Dr and Cr with the farm," he wrote. "The object is to ascertain its productive value (which depends wholly upon the manner in which it is tilled) & what crops are the most profitable the idea was suggested to me by reading the 'Rural' [probably the *Rural New Yorker*]. . . . The plan is to number each field and charging it with what ever is laid out on it and giving it credit for whatever is received."[69] Sprague's account books were not preserved, but others were. James McLallen, a former doctor farming near Ithaca between 1830 and 1860, for example, not only recorded accounts with fields and animals, but also coordinated his seasonal labor against thrice-daily thermometer and barometer readings.[70]

Populating the farm landscape with nonhuman debtors and creditors, these accounts made new knowledge. Accounting, improving authors argued, would allow the ideal agriculturist to take local variability into consideration, to "ascertain the adaptation of his farm to particular crops, or kinds of stock; to determine upon the relative values of each, and to vary them according to circumstances."[71] In providing a method of studying changing conditions, accounting was to make it possible for farmers to adapt the general advice of farm manuals to specific places. Moreover, by showing the judgment of buyers in the marketplace, buyers who presumably were mainly motivated by their own self-interest, the final prices marked in accounts were a form of interested witnessing.

Classes of Experiment

In laying bare the costs of experimentation, however, accounts also revealed the divisions in improvement. To both its detractors and some of its adherents, agricultural experimentation itself could be a visible luxury, offering wealthy experimenters the chance to modestly advertise capital invested, machines owned, labor managed, and land controlled. At the same time, public accounting also allowed wealthy improvers to separate their ostensibly virtuous experiments from the intentionally luxurious, less virtuous forms of production practiced by other wealthy New Yorkers. Forcing grapes and oranges in heated greenhouses or stuffing inefficient Alderney cattle until they produced their few drops of rich cream certainly bent the possibilities of nature, but they also derived their appeal from their conspicuous expense. Accounts demonstrating the profitability of an agricultural technique could convert spending into a risk undertaken for the benefit of poorer New Yorkers.

Given its costs and risks, some improvers suggested that experimentation ought to be the special task of wealthy urbanites and landlords. Here, they followed Young who had encouraged experimentation among the "nobility and gentry of large fortunes" as "they are the only people who can try experiments effectually;—it is a business much beyond the power of others."[72] American landlords, merchants, and bankers thirsted for such noblesse oblige. Such experiments, declared wealthy editor and political hopeful Jesse Buel, could be "of great value to the farming interest, and richly entitle those who make and publish them to the title of public benefactor."[73] The *Genesee Farmer* wrote similarly: "We would not advise farmers in middling circumstances to make expensive experiments nor adopt any novelty in husbandry on slight grounds, without being well convinced by testimony, observation or experience, of its beneficial effects," since, "a farmer, unless he be very rich indeed, cannot afford to be 'full of notions,' but must leave merchandize of that sort to the good citizens of the Metropolis."[74] The testimony here was to be the testimony of the rich.

Given the composition of his household, Brewster himself may not have been particularly wealthy, but engaging in profitable experiment allowed him to reach for the status of those who were. In his rutabaga experiment, for example, Brewster broke into a little sermonizing sidebar. Based on its low initial costs and reliable profits, he explained, rutabaga was particularly suited for the poor man, "who keeps but one cow and hires a tenement, with but one acre of land."[75]

For landlords, experiments that demonstrated the potential for profit had more urgent stakes. Alarmed by the westward drain of rural populations, and hoping to keep tenants at home and land values high, many were eager to prove that eastern farming could still be profitable *at all*.[76] However, they were not the only ones providing accounts to make claims about the capabilities of land. Writing from Chili in October of 1848, tenant Peter Tone sent the Wadsworths a sample account when complaining of the "verry [sic] poor remuneration" of wheat farming on Wadsworth land, particularly given the "Ravages of the Fly." After sending 250 bushels of wheat to the agent as well as paying $320 in rent, Tone was left with only ninety bushels, "falling short of the Amt. required to defray the Expenses attending it." Tone hoped to renegotiate his rent, threatening to leave agriculture or to move. "I hope I shall hear from you as soon as convenient on the subject and before making an engagement elsewhere," he wrote.[77] For landlords like the Wadsworths, accounts of profitable experimental practice promised to shift blame for poor profits from their lands to tenants themselves—*if correctly practiced*, they argued, farming could still be profitable. Accounts of profits here could promise a pathway to economic independence that rested on the labors of the poor, undermining any challenge to the larger structures of debt and tenancy that constrained them.

While landlords often mobilized experimental accounts to project certain kinds of political future, so too did the agrarian radicals of the Anti-Rent movement. An account of the profits of a ninety-acre farm in Monroe County became, in the columns of the *Albany Freeholder*, evidence that "a limitation in the quantity of land to be held would lead to great improvements in agriculture."[78] The calculations of radicals like Thomas Ainge Devyr, the *Albany Freeholder*'s editor, drew in turn on calculations by Irish and British Chartists. Feargus O'Connor's *Essay on the Management of Small Farms*, a sort of thought experiment in agrarian reform, laid out accounts of profit to suggest that Britain's laboring classes could pull themselves from the overstocked labor market by living on the bacon, potatoes, milk, eggs, vegetables, and honey from four acres, selling their surplus for a profit of £100 year. O'Connor in turn assembled his thought experiment from actual experimental accounts sent in to the *Leeds Mercury*.[79] Marshaled to demonstrate wealth, to avoid accusations of luxury, to describe the capacities of landscapes, to defend tenancy and to attack it, accounts were tools that could cut more than one way.

★

Antebellum agricultural experiment forms have not remained confined to the antebellum period. We can see one of their descendants in the modern field trial. Sociologist Christopher Henke shows us how modern field trials act as the intersection point between industrial agriculture and the scientific apparatus that supports it. More than just a knowledge-moving pipeline, they make knowledge in the fields of commercial farmers through the collaboration of agricultural extension agents, landowners, and field workers. Modern field trials resemble antebellum agricultural experiments in particular in their concern for what Henke calls "placiness," the simulation of the everyday. They also display what Henke's actors call a "pseudo-commercial" field, making the experiment look as much like market activity as possible. Henke places the origins of field trial in the early twentieth century, part of an attempt to convince farmers who were "wary of the 'book learning' of university researchers," of the efficacy of the science being produced by field experiment stations.[80] Writing about the experiment stations of the late nineteenth century, Charles Rosenberg makes a similar point about farmers' constant demands for profit reports.[81] Both Henke and Rosenberg imply that ordinary farmers participated in a culture to which experiment was alien.

But the experiences of antebellum improvers suggest that demands for profit reports, for trials in what might be called "real life conditions," and for simulations of commercial reality represent, not the adaptation of laboratory science to make something that antiscientific farmers could digest, but instead the continuity of older forms of improving scientific credibility. Improvers were used to experiments infused with money. They expressed expertise and performed practicality in experiments that placed them on the scene, and they communicated their successes with accounts, a form of scientific communication that stretched back to the learned agricultural practices of eighteenth-century Britain. Describing credit and debit relationships with fields and animals were a standard way of expressing improving claims.

Despite the absence of organized agricultural education in the first half of the nineteenth century, elements of the genres of agricultural experiment became, if not uniform, certainly widespread and broadly recognized. They were mobilized by participants in improving culture to demonstrate the value of particular techniques, the existence of natural tendencies to profit, and the importance of capitalist virtues. They became tools in conflicts between tenants and landlords and in the aspirations of middling farmers to modest virtuous display. However, for one group of improvers, they became a source of more direct profit. Machinery makers would bring improving experimental culture to a world stage in its most elaborate form: the machine trial. It was in this form that agricultural experiments returned dramatically and directly to their roots in trial culture.

4

Trying Machines

Even according to its supporters, the public trial of the Atkins's automaton in Geneva, New York, in 1852, was humiliating. In front of hundreds of people, the reaper ground to a halt halfway into the field. Its specially patented self-acting rake fell off, landed on the platform for holding the grain, and broke the operating lever, presumably with a resounding crash. As competing reapers pulled away, the Atkins's operators struggled to repair it, but by the time they had, the judging committee and much of the audience had gone home.[1]

Reaping machines that didn't work were not particularly unusual in the 1850s, but the pamphlet containing this story seems odd. First, this public nightmare was being further publicized, not by the Atkins's competitors, but by its maker, J. S. Wright. Second, Wright was forced to publish his account of the upstate New York failure three-and-a-half thousand miles away, in London. Wright himself was clearly shocked at having to do any such thing. He had arrived in London hoping to sell his patent (at a moment when American reaper patents were hot properties) only to find that news of the Geneva test had preceded him. Suddenly, experimental trials of machinery in New York State were internationally visible in a way that Americans were not used to.[2]

In fact, agricultural machinery trials like the one in Geneva were rapidly becoming the most public, expensive, theatrical, and carefully documented form of experiment in agricultural improvement, carried out at levels of precision unimaginable to Jarvis Brewster and to much greater acclaim. In the 1850s, they ran counter to the expected direction of knowledge movement, making the United States, rather than Britain, the source of machinery designs and laying the foundation for a century of global manufacturing dominance.[3]

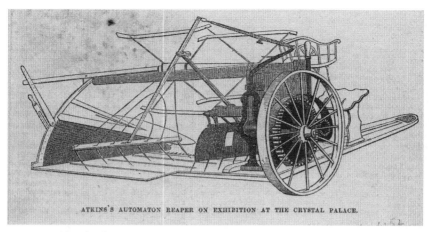

FIGURE 5. The Atkins's automaton. Note the raking arm, which replaced a human raker on an ordinary reaper. Courtesy of the Library of Congress.

Perhaps most importantly for this book, they were also a sign of agricultural improvement's turn to a system of commercial goods, spread by a network of manufacturers, warehouses, and seed stores, a network stronger, better funded, longer lasting, and more central to American agricultural improvement's growth than the network of landlords had ever been. As a more diverse coalition of New Yorkers took up improvement in the 1830s and 1840s, this commercial network bound them together; in doing so, it shifted participation in improvement toward acts of consumption and refocused improving knowledge making onto products.

We can see the rise of commercial networks of knowledge making most clearly by showing how improvers worked to make knowledge about machines. Not only were agricultural machines publicly tested, they were displayed in warehouses that were expected to act as museums and in museums that served the interests of warehouses. Such commercial spaces would become key to agricultural improvement, not least by printing and sponsoring the most famous agricultural journals. The hedges of documentation around machines would also rival the journals, providing an alternative publication route for the testimonies and experiments of improving farmers.[4] But machines would prove difficult to discipline and hard to know. The high stakes of machine trials in particular would ultimately turn trial-like experiments back into literal trials worth fortunes. By forcing a focus on patentable novelty, they would shift judges' attention past efficacy to investigate seemingly natural and commodified "principles" of design.

To open up this culture of machinery knowledge, we need to follow the

Atkins's automaton pamphlet back, certainly to the test in Geneva that it detailed but, first, to an earlier, even more storied reaper trial, the trial at the British Great Exhibition of 1851, which remains the Cinderella story of agricultural machinery.

Two Trials

The reaper test at the British Great Exhibition of 1851 is possibly the only generally famous moment in American agricultural machinery history. This is not because it was particularly well conducted—it was not. However, it occurred on a major stage, it lent itself to nationalist drama, and it reversed resented hierarchies of credibility and information flow. The media furor that followed it, combined with careful decades of mythmaking by its patent-suit-plagued victor, would make it legendary.

In the US, the best-known version of the story came from B. P. Johnson, the New York State Agricultural Society secretary, who was responsible for the US agricultural exhibits. In Johnson's telling, the trial had all the satisfying elements of a scornful neighbor story, but on a national scale. When he arrived at the exhibition, Johnson explained, he found the American exhibit sneered at. The agricultural implements in particular were mocked for resembling "the prints in agricultural works intended to represent plows that were used *several hundred years ago*."[5] The plow trials had been scheduled before the American implements and plowmen had even arrived. The reaper trials had not yet been scheduled, in part because the British had no practical reaping machines and did not expect the Americans to have any either.

This was not the result of disinterest. Reaping mattered desperately to British farmers, demanding an exhausting burst of labor wedged into a brief span of time. Grain and hay were at their best for a short window; they could only be harvested on fine days, to keep the harvested grasses from rotting, and were easiest in the morning hours. Wheat, the most culturally important grain, required extra labor. It was usually reaped rather than mown, requiring three skilled movements: cutting the plants at the base or middle of the stalk, gathering them in bunches, or "sheaves," and then tying a group of sheaves upright in a "stook," so that the fat seed heads could dry without spoiling. (Mown crops like hay could simply be cut and raked into heaps to dry.)[6]

While Britain had a much larger population of agricultural laborers than the United States, the wheat harvest pressed it past capacity. Long reliant on temporary urban harvesters, British farmers had become dependent on Irish migrant laborers, who often faced violence and intimidation from local agricultural laborers. By the time of the Great Exhibition, Irish migrant numbers had already peaked as Ireland recovered from the Great Famine.[7] British-born harvest labor remained unsteady, with boom-and-bust years depending on competition from trade and industry. In some years, even high wages could not attract enough workers to keep crops from spoiling in the field.[8] Adding to the pressure, the repeal of the Corn Laws had just dropped the price of wheat, a problem compounded by an agricultural depression in 1849–52.[9]

Certainly, reaping machines existed in the British imagination: both British and American improvers regularly referred to a sort of knife-pushing cart described by Pliny the Elder. British improvers had sporadically experimented with harvesting machines since the late eighteenth century.[10] However, most British farmers saw reaping machines as fanciful. The reaper trial, testing two American machines and one fairly theoretical British one (mainly there to ensure British representation), had been scheduled as a sideshow to a more significant event, the annual farm visiting day hosted by British improver John Joseph Mechi.

Conditions on the day of the trial were not encouraging. The field was soggy and the grain sodden and heavy—the first American machine, Obed Hussey's, clogged almost immediately in front of more than a hundred British landowners already primed to laugh. (A crowd of anxious laborers' faces reportedly brightened as it failed.) This only sweetened McCormick's triumph. Johnson gleefully described the general shock as the audience watched the McCormick machine trundle briskly down the field, shearing through the stalks of wheat much faster than human reapers could. He documented the gentlemen's cheers (but not the laborers' reactions), reproduced the special medal, and reprinted the British newspaper articles declaring that the reaper alone would pay for the whole ruinous expense of the exhibition.[11] Buoyed by nationalist fervor, the tale of the reaper spread into journals and commemorative prints. Although McCormick and his company would help make it legendary, at the time it was not clear that this was necessarily a McCormick-centered story. Many American reapers were arguably as effective as the McCormick. Suddenly an international market opened up, not for American reapers themselves, which were too cumbersome to transport, but for American reaper patent licenses. Hunting patent buyers is what Atkins's maker J. S. Wright was doing in London.

Though less famous now, the trial where the Atkins's automaton failed, the Geneva trial of 1852, was much larger and more elaborate than the one at the Great Exhibition. Organized by the New York State Agricultural Society as a nationwide trial, it was designed to correct the Great Exhibition's errors and to put to trial the much larger group of machine makers who had not

FIGURE 6. "The Great Exhibition of 1851." A print commemorating American successes at the exhibition. At the bottom, note the verse, "By Yankee Doodle too, you're beat downright in Agriculture / With his machine for reaping wheat, chaw'd up as by a vulture." In the background on the right, four perturbed Britons cluster by a reaping machine, the first saying, "Their Agricultural implements have taken all the prizes!" Courtesy of the Library of Congress.

sent anything to London. Competitors sent machines from as far away as Illinois, not for the prizes themselves—the top prize of twenty-five dollars was less than a quarter of the price of the machine that won it—but for the chance to be included in the report, to excerpt it in their advertising, and to have their successes reported through the web of agricultural journals linked to the state society.

Thousands of people attended the Geneva trial, watching as manufacturers and judges fussed over treadmills, chains, swinging knives, and different arrangements of horses and men before the machines went whirring into the fields. The society's committee of judges took every opportunity to exhibit good experimental form, conscious of their audience in print as well as in person. Their forty-four-page report carried prolixity to new levels, detailing the gentle slope of the meadow in the mower trial; the loose cobblestones covering the upper half and the patchiness of the lower half, made uneven where breakaway cattle had "poached" it; and the grass itself, a mixture of red top, fiorin, timothy, and aquatic grasses, tangled so as to "severely task the efforts of the most experienced mowers." These were minutiae that readers who had carried scythes themselves could have felt in their hands while reading. The society laid the fields out in carefully staked parallelograms of exactly two acres, inquired about local wage rates so that the (simulated) accounts included would be as close to accurate as possible, and placed a specimen of the clay loam from the test field in the society museum for future reference.[12]

From the society's perspective, the report achieved some of its purpose: the Geneva trial became the model for trials conducted around the US.[13] As Wright found, it also circulated overseas, though there it was not without competition—the judges noted that at the exact time of the Geneva trial, the Royal Agricultural Society was testing "*seventeen* specimens of reapers" (an American reaper won).[14] Reaper trials would become standard at the "Exhibitions of All Nations" that followed the Crystal Palace model.

This left Wright with a dilemma. Accounts of the collapsing reaper arm had to be counteracted if Wright was to sell the patent. First, he had to account for the automaton's poor performance in Geneva—he had plenty of reasons marshaled for its public collapse: the machine arrived late; it was assembled badly; and it was a windy day (apparently an unforeseen problem). However, the broader American culture of agricultural experiment and trial offered Wright tools beyond lame excuses. The rest of the Wright pamphlet shows the breadth of the system that American improvers already had for producing credible knowledge about machines. The moment Wright got his machine back from Geneva, he took it home, adjusted it, and set it working, then set up his own smaller trials. He invited his neighbors over as witnesses and got twelve to sign a document to certify that it did indeed work. Then he coaxed one of his neighbors to take the machine home and stage a trial in front of a new set of sixteen neighbors who also signed a statement. Both statements appeared in the pamphlet along with lists of prizes from state fairs, newspaper clippings, and reports of other trials.[15] To understand how neighbors were so used to this sort of thing, we need to look more broadly at the experimental culture of American machinery.

Experimenting with Machines

In both Britain and the US, inventors advertising new machines often referred to multiple phases of experiment and trial. First came a period of private experimenting, in which machine designs could be made workable—these were sometimes later translated into advertising as when J. H. Manny boasted in his circulars of having "incurred an expenditure of some thirty thousand dollars in experimenting" to produce a functioning reaper and mower.[16] Such private periods would later be used to justify the reaper's extraordinary cost.

The second phase was public demonstration trials: a single machine would be operated in front of an audience. Such trials were common enough in Britain by 1798 to inspire the popular comedy *Speed the Plow*, in which Sir Abel Handy, a ludicrous figure who had patented "a plan of cleaning rooms by a steam engine," plans a disastrous plowing competition to demonstrate the value of his "curricle plough" (funny because curricles were light, fashionable carriages) pulled by "Leicester horses" (funny because of improved Leicester cattle and sheep).[17] Like other kinds of experiment, machine trials sometimes served as entertainment and spectacle. In 1814, the British inventor and actor J. Dobbs ran a trial of his two harvesters onstage at his theater in Birmingham during the interval between the comedy and the farce. Dobbs himself played "the part of Robin Roughhead" working the machine "in an Artificial Field of Wheat planted as near as possible in the manner it grows." (Later reports claimed that the first trial was "completely successful"; the second was marred when Dobbs ran the machine into the scenery.)[18] Most trials, however, were more soberly conducted. In the later nineteenth century, McCormick's son would use the story of McCormick's first successful solo trial in 1831 to fashion a legally convenient past of lone invention.[19]

It was in the 1830s that the third phase, public competitions between machines produced by different makers like that at Geneva, took strong hold in the United States. The growth of these competitions came in part from the new competitive structure of American plow manufacturing. While British plowshares, which were made of wrought iron, could still be produced by village blacksmiths, the American switch to cast iron in the 1810s and 1820s made foundries into centers of plow construction. These were necessarily dotted across the landscape near available sources of fuel wood. Connected by new transport networks, these small foundries began to compete regionally and then nationally and internationally.[20] Cast-iron plows had become usual across the farming population more generally, made affordable by new manufacturing techniques that allowed plows to be made at many sizes for many budgets.[21] Centers of plow manufacture diversified into other agricultural implements and machines. Some, like seed drills, were at first limited to improving farmers. Others, like threshing machines, came to be almost universally used, supported by rental systems that allowed middling farmers to abandon the flail. In the early 1830s, harvesting machines still seemed fanciful; a friend wrote to one of the wealthy Wadsworths in 1831, "you have a machine for every purpose in the way of hay as mowing, spreading, raking and getting?" and teased him by prophesying next, "a machine for the purpose of perpetual digestion, then an eternity of segar [sic] in your mouth . . . Oh! For a machine to make fair weather and *love*."[22] By the Geneva trial, twenty years later, machines for "mowing, spreading, raking and getting" were real and plausible.

The landscape of manufacturers was well entrenched by midcentury. In 1855, the New York State Census alone counted 386 agricultural implement makers. While this number was dwarfed by that of New York's blacksmiths (16,948) and farmers (321,980), it was significantly larger than its population of professors (188), patent medicine makers (59), naturalists (10), or geologists (5).[23] Implement makers clustered at the north end of the Hudson, and around Buffalo and Rochester, but most counties had at least a few makers listed, and most county fairs could boast machinery made by local firms. In 1851, Levi Weeks remarked after visiting the fair in nearby Rensselaer County, "in the mechanized department, they go ahead of us."[24] One result of this distributed network was a great blossoming of designs. By 1850, the improving journalist (and later governor of Vermont) Frederick Holbrook observed that two hundred different plow patterns were manufactured. A similar expansion in reapers occurred within the next few years.[25] By 1855, in New York State alone, seven firms made about 2,500 reapers in nine factories, selling about half out of state.[26] This was part of a broader acceleration of the development of mechanisms; 19,661 patents were issued in the 1850s, compared to 5,516 in the 1840s, 5,077 in the 1830s, and only 2,697 in the 1820s.[27]

As implement makers and designs multiplied, trials pitting machines against each other became increasingly common and formalized. The American Institute, dominated by manufacturers, began plow competitions as early as 1835.[28] By 1844, only a few years after its founding, the state agricultural society not only tested plows, but also the different designs of dynamometers, devices that measured the friction of plows. The society also experimented with mechanical objectivity by replacing variably skilled plowmen with a robot: a sort of winch, "by which the plow was drawn with great steadiness and uniformity."[29] The robot did not appear again, but dynamometers became standard.

Like other forms of experiment, machinery trials were persuasive events, meant to overcome expected resistance and to encourage best practice. Seeing "by actual experiment the superiority of some [plows] over others," claimed the organizers of the American Institute plow trials, farmers "became satisfied of the great improvement in this valuable instrument of agriculture."[30] However, the diffusion of knowledge could not be separated from the profitable diffusion of machinery; persuasion about the value of implements was obviously advertising. Machine makers emblazoned copies of medals and references to their successful trials on their advertising materials and sometimes stamped them on the machines themselves. They also reprinted excerpts from trial reporting in thick pamphlets, documenting a series of triumphs or re-litigating an unfavorable result. Such pamphlets also included testimonials from individual farmers, who, not surprisingly, followed standard modes of experimental writing in showing how they had tested machines.[31] Through such texts, advertisers were becoming significant publishers of experimental results.

Commercial Networks and Knowledge Production

The rise of implement makers and other businesses dealing in improving goods did more than spark competitions; it also expanded the reach of agricultural improvement and created new powerful nodes in its networks. Not only did implement makers advertise with journals, at various points all the best-known agricultural journals issued directly from the same agricultural warehouses and seed stores that sold and sometimes built agricultural machinery. The *American Agriculturist* came from A. B. Allen's warehouse in New York; the *New Genesee Farmer* from the Rochester Seed Store; and the *Central New York Farmer*, from Comstock and Johnson's Agricultural Warehouse, in Rome.[32]

Though they might seem now to be a shameful conflict of interest, these commercial connections were far from secret. While Luther Tucker was both editing the *Cultivator* and running the Albany Agricultural Warehouse, he advertised using a woodcut of the warehouse with the *Cultivator*'s shingle hanging proudly from it. Originally the official organ of the state society, the *Cultivator* now issued from among the products it described.[33] Public criticism of such connections was likely muted by the dominance of warehouse-based journals. In response to a rare complaint, the *Maine Cultivator* valorized the connection in a piece that circulated nationally. How were editors to obtain "a good knowledge of agricultural mechanics," the *Maine Cultivator* inquired, "without a familiar acquaintance with agricultural warehouses, where may be examined the various implements and machines in use, and by testing such things, witnessing their practical operation and comparing their several advantages?"[34] It was editors unconnected with a store that, out of ignorance, "may advertise or puff a worthless article, and thus honestly aid in gulling the farmers."[35]

Like other editors, Tucker cast his warehouse as a site of learning. He declared two purposes for his first descriptive catalog: advertising was the second, the first was education—like a warehouse the catalog taught the value of labor-saving machines.[36] Tucker's successor at the Albany Warehouse, Horace Emery, took an even stronger line. He advertised the warehouse as "literally an Agricultural museum, where a farmer can well spend an hour or two, and feel well paid for his trouble in calling."[37]

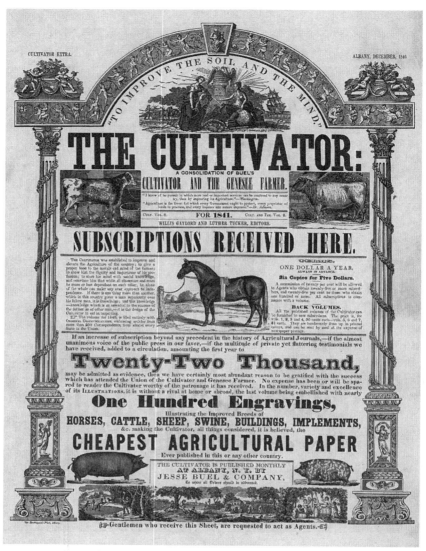

FIGURE 7. An 1857 advertisement for the *Cultivator* under Luther Tucker and Willis Gaylord's editorship. Note that engravings of goods and animals are part of the promised value—these were likely made as advertisements for Tucker's warehouse or supplied by breeders. Courtesy of the Albany Institute of History and Art.

FIGURE 8. An 1857 advertisement for the *Albany Agricultural Works* giving a sense of the scale of the Emery Brothers' national operation. Note the world's fair medals from New York and London reproduced on each side. Courtesy of the Albany Institute of History and Art.

Comparison with the state society's actual agricultural museum, a short walk from Emery's warehouse, gave this claim more weight. Displays at both the warehouse and the museum demonstrated the growing strength of improving commerce. Both were full of new seeds, available from seedsmen; new manures, available from fertilizer dealers; and new machinery, available from machinery manufacturers. The museum's walls were hung with portraits of expensive improved cattle, promoting the bloodlines sold by commercial breeders and importers like William H. Sotham.[38] Booksellers with growing lines in agricultural print—D. Appleton, Wiley and Putnam, and the new agricultural press Saxton and Miles—had provided books for the library, which jostled on the shelves with commercial catalogs from seedsmen and nurseries. Where the specimens of soils, stones, and plants in the neighboring "geological rooms" were inventories of natural wealth, the agricultural society's museum inventoried the products of cultivation and artifice. Alexander Walsh, the Albany County merchant who organized the agricultural museum, operated at this interconnection himself. His general store in Lansingburgh was nicknamed "Walsh's Museum" because of the variety of his goods—and he was practiced in displaying agricultural machinery. At Lansingburgh's agricultural fair in 1844, he assembled a twenty-three-foot-long, seventeen-foot-high pyramid out of agricultural machinery and implements, dotting the whole with "well-chosen mottoes" and titling it "THE FARMER'S COAT OF ARMS."[39]

American improvers had long characterized warehouses as a place where knowledge could be exchanged and produced. Back in 1818, Richard Peters of the Philadelphia Society for Promoting Agriculture had circulated a plan for a society-sponsored warehouse and manufactory. By drawing together the scattered products of American genius and the many neglected "valuable foreign instruments," Peters argued, the warehouse would, "become a highly useful *place of exhibition*." The sight of the latest machinery would teach buyers to understand and thus to desire them. As he collected machinery, Peters suggested, the society-appointed director could stamp the ones he approved with the society's mark—"emulation" would be "created in and forced upon" manufacturers, their excitement sparked by the exhibition and their practices surveilled and constrained by the pressure of expert judgment.[40] In mingling commercial and exhibition spaces, antebellum improvers also had international models. The museum at the Rochester Seed Store modeled itself explicitly on commercial museums operated by seed stores and nurserymen in Scotland, in Stirling, Edinburgh, and Perth.[41]

Warehouses remained important to improvers' plans. Funded by sales, commercial displays could reach places that museums funded by philanthropy and the state could not. As a new warehouse opened in Syracuse, a report to the state transactions noted, approvingly, "The great mass of farmers are ignorant of the improvements that are making in farm implements, and require to have their attention called to this advance."[42] By the 1850s, towns like Cortland, Auburn, and Rome boasted their own warehouses and nurseries; larger towns like Poughkeepsie might have more than one. Warehouses and seed stores also claimed an advantage over museums in that they could show what was and what was not selling—implying a faith in collective consumer knowledge that trumped the airy or corrupted speculations of state institutions. The *Maine Cultivator* suggested that farmers hoping to learn the truth about machines that had been falsely puffed in the press should "step into the agricultural warehouses and see such articles neglected and rusting."[43]

Beneath comments about shopping and discernment, of course, lay a naturalized and racialized notion of progress. The organizers of the society's museums promised to show "how fast improvements in agricultural implements have progressively been made, and in what the improvements consist."[44] To make the track of progressive development visible, the museum juxtaposed its latest devices with American plows, "of ancient and rough construction" from the early years of the century. These plows were familiar from many visitors' childhoods, as distant from them as a rotary phone might be from us, and still in use among poorer farmers around the state. Rendered as nostalgia, they made newer machinery seem a necessary accoutrement of the present. These evidences of rapid progress were further placed in self-congratulatory context, sarcastically compared to "a collection of agricultural and horticultural tools from India, showing how rapidly nations will advance in the arts under the influence of cool indifference," and to Chinese and Mexican plows from the 1840s and 1850s, "beside which a Michigan Premium Double Plow affords a gratifying contrast."[45]

Improvers believed that the mere sight of machines would prove shocking and educational. Machines at the warehouse and the museum taught quality and discernment and racialized notions of technological progress. But stilled and silent displays clearly could not match public trials for persuasion or knowledge production. When set working in public, however, machines proved harder to know than trial organizers had expected.

Knowing Machines

The Geneva test promised to address not just reapers and mowers, but all field machines and implements: seed drills, which planted seeds in rows, sometimes chased by a helping of the latest fertilizer; "cultivators," for weeding between rows; and threshing machines, which beat shocks of grain to separate seeds from chaff and straw. However, the trial centerpieces were reapers and mowers. While British observers focused mostly on reapers, to New Yorkers of the 1850s, facing new urban and industrial horse populations and burgeoning hay markets, mowers were perhaps even more exciting and led the trials.[46] As in Britain, both reapers and mowers promised to ease the tensions of the most time-pressed phases of the farmer's year—diaries from larger New York farms in this period often had long lists of tick marks in the back just to keep track of extra hands hired for the harvest. Control of the time of threshing also allowed farmers to respond to the ups and downs of the wheat market—at times when the price was low, they might leave their grain on the stem, but when it was high, the first to thresh might make significantly more than his slower neighbors.[47]

The judges at Geneva had a seemingly straightforward task—to compare the performances of machines. Dutifully they produced elaborate tables that attempted to capture machine performance. A table explaining the results of the mower trial recorded, for example, the "length of the stubble," "the condition of the grass when cut," and whether the machines had clogged, and then scored the overall quality, "assuming 50 as perfect, and 25 as perfect work of the scythe." (The machines scored between thirty-four and sixteen and a half.)[48]

However, the much-longer surrounding text makes clear that the measurement of performance was difficult. First, nature failed to be necessarily uniform—for example, even though they were taken from the same farm, some of the hundred sheaves of wheat counted out for the threshers had been nibbled by Hessian fly and wheat midge. Embarrassingly, even the same machine tested twice produced very different amounts of grain. Second, machines themselves were persnickety—some, like the Atkins's, arrived poorly adjusted or without anyone with the skill to operate them. Others lacked anywhere to attach a dynamometer, foiling any attempt to quantify their draft. Still others performed well during the trial but looked flimsy—this threatened not only their usefulness, but also their operators, who could be mangled if one of their rapidly whirring blades flew off later.[49]

These differences came under scrutiny in part because competitors themselves contested embarrassing results—like Wright, they often raised successes at other trials to attack a particular trial as unfair. Indeed, machinery trials became the only kind of agricultural experiment where formal replication was common. However, this same factor may have limited complaints. Whereas experiments like Brewster's could be fully local, deriving their authority partly from their connection to place, machinery makers needed replicability under a variety of conditions. Wet grain had doomed the Hussey reaper in Britain but heightened the triumph of the McCormick.

The status of skill was similarly ambiguous. On the one hand, as public displays of agricultural practice, trials offered an opportunity for conspicuous production: at a trial of mowing machines reported by the *Cortland County Whig*, the paper reported on machine "No. 3. Entered by L.L. Merrill and drawn by his noble sorrel grays."[50] Operator incompetence might also cause a machine to fail, as in the same trial, when machine No. 2, "owing to a want of skill in the driver, did not at first appear as well as its merits seemed to warrant."[51] At the same time, it was not in the interest of manufacturers to suggest that their machinery was difficult to operate or that particular successes were the result of skilled performance. The state society's failure to develop a machine replacement for drivers at earlier fairs meant that drivers were an unavoidable problem. They were more easily eliminated grammatically—at the Geneva trials, phrases like "the Ketchum Mower led off handsomely" attributed skill to the machine itself.[52]

Perhaps because of the variability of testing conditions, judges often attempted to acknowledge deeper qualities in the machine that might have been betrayed by circumstance. Thus, in the arrangement of teeth in Howe's cultivator, the Geneva judges saw "merits . . . which will probably lead to one of the most perfect in its class," even though the machine itself failed even to touch one-fourth of the land it was supposed to be cultivating and did "not effectually disturb the grass and roots" on the rest.[53] Even the Atkins's automaton was singled out for favorable notice for its automatic (if broken) arm.[54] To get at these points of apparent potential, judges and manufacturers often referred to mechanical "principles." Manny's mower for 1856, for example, promised a new divider that operated "upon an entirely new principle, and the only principle that will work in all kinds and conditions of grain or grass, and work perfectly through the worst condition of tangled clover, lodged or tangled grain, or even peas."[55] Principles would become the central feature of machines, both as seminatural objects and as legal ones.

Principles of Machinery

As described by manufacturers and other improvers, "principles" of machinery were at once physical features, ideas, and forms of patentable property. It was usually in these last two senses that they were most profitable to their inventors, who made their fortunes by licensing the right to produce their inventions to networks of small forges and machinery works, each claiming a territory. Thus, William Ketchum's mowers were sold across eight territories total and made by three different manufacturers in New York alone.[56] Rather than shipping unwieldy, fragile machines over long distances, inventors shipped designs, models, and "patterns" (wooden shapes used to guide builders).[57] Royalties for these rights could be substantial: McCormick received $6,750 of the $15,000 Manny and Co. made from his reaper in two years.[58] Possession of patterns in turn determined the status of warehouses and foundries. Mayher and Co.'s Foundry and Machine Shop, in New York City, advertised their "large collection of models and patterns," built up over twenty-five years.[59] The licensing of patented principles bound the decentralized landscape of production together. Recognizing this, historians have started to reshape their understanding of early American capitalism overall. Where once antebellum production seemed like a scattering of tiny firms, pale precursors of the seemingly real capitalism of postwar corporations, now, following these linkages, we see robust and complex networks, a many-sited system of production.[60]

At trials, therefore, principles were more than simply abstractions—the point of trials was not just to sell machines to farmers; it was also to entice observing manufacturers to pay for licensing agreements. This was taken to a remarkable degree in the international expositions. After winning the public trial at the Exposition Universelle in Paris in 1855, Manny and Co. claimed to have sold French rights to produce their reaper for 125,000 francs.[61]

What exactly constituted a principle, or patentable novelty, however, was debatable. Was it a new structure, a new way of arranging structures, or a new function? If, seeing a mechanism that disentangled stalks, a designer produced a different device that did the same thing, was this a patent infringement? Also, in order to count as an invention, did a mechanism have to work? If it did work, did it have to sell? The high-profile patent infringement suits that constantly embroiled the agricultural machinery industry, part of what Christopher Beauchamp calls "the first patent litigation explosion," meant that these questions were asked constantly.[62] Agricultural machinery drove a significant part of this explosion: Zorina Khan estimates that average awards in agriculture cases ($7,360) between 1790 and 1860 were about three times as high as those in manufacturing ($2,463).[63] More than a quarter of the suits in her 1850s sample were litigated in New York.[64] The list of lawyers involved in these trials shows their high profile—William Seward, the improving former governor and later secretary of state acted for McCormick in the 1850s; Edwin Stanton, soon to be secretary of war, acted for Manny (and employed and then snubbed the not-yet-famous Abraham Lincoln, who had been hired for his access to Illinois courts). Seward was proud enough of his defense of McCormick's patent rights that he had his final statement before the US Circuit Court published in a pamphlet, *The Reaper* (1854).[65]

Expecting to see intentional design in nature, improvers blurred the distinction between natural and artificial features. In defending McCormick's patent rights, William Seward suggested that the properties of mechanisms were features discoverable in nature and intended by nature for use: "Nature has furnished to us only seven mechanical powers," he argued, "with which we perform all the operations of human industry."[66] Seward suggested that a new combination of existing structures could constitute a new idea by making an analogy with the mechanical principles of the human body "Every machine is a combination of parts," he told the court, "the human frame, the most wonderful of machines, is a combination of parts, many of which are not peculiar to man, but belong to a large portion of the animal creation."[67] For Seward, machines that succeeded practically "lived," and those that failed in the market "died" as "abortions," gone "the way of all mechanical flesh."[68]

This bodily analogy would not have seemed unusual to antebellum Americans, already used to seeing the mechanical features of bodies as evidence of a divine designer. Moreover, agricultural machines were clearly imitating bodily movements that were still familiar to most people. Like other machines ending in "-er" (teller, computer), the "reaper" and "mower" began as kinds of people. Machines imitating the labor of human bodies had parts with bodily names—"teeth," "fingers," and "arms"—and Atkins's mower was even marketed as an "automaton": a mechanical imitation of living processes.[69] In experimentally uncovering the principles that made machines act like bodies, machine trials thus could be conceived of as uncovering hidden natural properties.

The practices of witnessing and reporting common to experimental trials in turn made it easy to fill courts with stacks of trial accounts and testimonials. Facing the "thousand pages of printed matter, of which 750 pages were the depositions of witnesses" and a "courtroom filled with models and drawings," the poor reporter for the 1857 Supreme Court case "McCormick v. Talcott et al." wrote, "The reporter despairs of giving any intelligible account of the argument in this case."[70] In patent litigation, experimental culture rejoined the judicial culture from which it had emerged.

*

More and more after 1840, agricultural improvement became the concern not only of farmers, "practical," "book," or otherwise, but of businesses; new agricultural practices increasingly centered on products. While the improving gentlemen of the 1810s and 1820s had imported machinery or animals from London, Paris, or Madrid, the spread of agricultural manufactories and warehouses, seed stores, fertilizer dealers, and local nurseries, as well as good roads and canals to move goods around, made the purchase of improving products easy and relatively normal. From the 1840s on, this world of products would fuel the expansion of improvement after decades of failures and would tie profit and knowledge making together in new ways.

As judges sniffed at freshly made butter, tasted apples, or felt the sharpness of plow blades, they tried the world of goods, observing in it the progress that they expected in the landscape. It was in such trials, forced by the jealousy of competitors (and in the case of machinery, the interest of courts) that reporting became most formal and detailed, returning the judicial language of experiment to actual judges of the law. Acknowledging this realm of experiment allows us to see the true scale of agricultural improvement, to see significant debates carried on, not just in journals but in catalogs and pamphlets and broadsides, and to see claims to expertise not just in nascent academic or state institutions but in businesses. As later chapters will show, more and more knowledge makers had a direct stake in the things they described. Pomologists sold apple trees, agricultural chemists peddled patent fertilizer, and agricultural geologists boosted land values.

But to conclude, let's close the loop between the kinds of experiments described by Alson Ward and Brewster in the last chapter and the elaborate reaper trials in this. In 1855, Levi and John B. Weeks, still farming with their father where the Mohawk flowed into the Hudson, decided to look for a mowing machine. After looking over different models in Albany, and inspecting the machinery at local fairs, they turned to their neighbors. On July 17 Levi went to a neighbor's and "saw a trial of the Manning mowing Machine, there were some 12 or 15 witnessed the operation." Though two of his neighbors promptly ordered Manning machines, Levi's response was cool: "it cut the grass very well but there is too much rigging and useless lumber I think about it. If Hallenbecks will do as good work as this I should prefer it as it's lighter and has less friction and draws easier and is less in price."[71] Arriving on August 3, the Hallenbeck agent also allowed trials, following his machine from Saratoga Springs out to the Weeks farm. After running it for an hour on the grass on the Weeks orchard, the agent pressed for a quick decision. The Weeks brothers felt rushed and annoyed. Declaring that they wanted "to test its powers and capabilities further," they listed flaws commonly described by judges at society trials: "It first is objectionable in that it cuts the small grass too high, 2d that it requires an uncommon swift walk for team—3—that it is difficult to get around and quite liable to clog with the mown grass." The agent backed down and left the machine for a longer trial. In the morning, the brothers shipped the mower back to Ballston Spa, having broken one of its knives on a bone in the grass "in consequence, I think," wrote Levi, "of a flaw there was in it."[72]

The Weeks brothers then turned back to the Manning machine (designed by Ann Harned Manning), which they bought after a second trial for $125.[73] (Levi paid his share with money borrowed from his aunt and sister.) Only a few weeks later, the Weeks brothers' machine became a spectacle of its own. "We had a good many visitors to witness the opperation of the machine," recorded Levi, "It done good work—and fully answerd our expections. We cut about 3 1/2 acres in about as many hours, fine day."[74] Within a

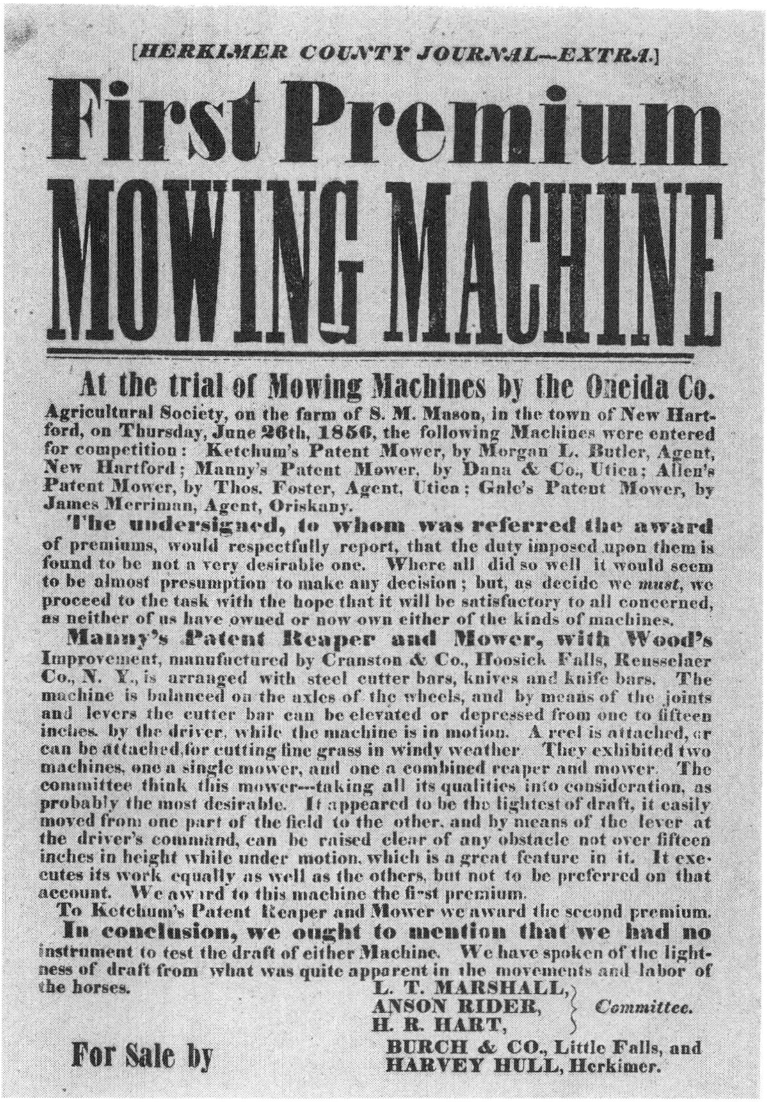

> [HERKIMER COUNTY JOURNAL—EXTRA.]
>
> # First Premium
> # MOWING MACHINE
>
> **At the trial of Mowing Machines by the Oneida Co.**
> Agricultural Society, on the farm of S. M. Mason, in the town of New Hartford, on Thursday, June 26th, 1856, the following Machines were entered for competition: Ketchum's Patent Mower, by Morgan L. Butler, Agent, New Hartford; Manny's Patent Mower, by Dana & Co., Utica; Allen's Patent Mower, by Thos. Foster, Agent, Utica; Gale's Patent Mower, by James Merriman, Agent, Oriskany.
>
> **The undersigned, to whom was referred the award** of premiums, would respectfully report, that the duty imposed upon them is found to be not a very desirable one. Where all did so well it would seem to be almost presumption to make any decision; but, as decide we *must*, we proceed to the task with the hope that it will be satisfactory to all concerned, as neither of us have owned or now own either of the kinds of machines.
>
> **Manny's Patent Reaper and Mower, with Wood's** Improvement, manufactured by Cranston & Co., Hoosick Falls, Rensselaer Co., N. Y., is arranged with steel cutter bars, knives and knife bars. The machine is balanced on the axles of the wheels, and by means of the joints and levers the cutter bar can be elevated or depressed from one to fifteen inches, by the driver, while the machine is in motion. A reel is attached, or can be attached, for cutting fine grass in windy weather. They exhibited two machines, one a single mower, and one a combined reaper and mower. The committee think this mower—taking all its qualities into consideration, as probably the most desirable. It appeared to be the lightest of draft, it easily moved from one part of the field to the other, and by means of the lever at the driver's command, can be raised clear of any obstacle not over fifteen inches in height while under motion, which is a great feature in it. It executes its work equally as well as the others, but not to be preferred on that account. We award to this machine the first premium.
>
> To Ketchum's Patent Reaper and Mower we award the second premium.
>
> **In conclusion, we ought to mention that we had no** instrument to test the draft of either Machine. We have spoken of the lightness of draft from what was quite apparent in the movements and labor of the horses.
>
> L. T. MARSHALL,
> ANSON RIDER, } Committee.
> H. R. HART,
>
> **For Sale by**   BURCH & CO., Little Falls, and HARVEY HULL, Herkimer.

FIGURE 9. An 1856 broadside advertising the results of a county society's mower trial. Courtesy of the Albany Institute of History and Art.

few weeks, a culture of test and experiment, first staged and financed by the sellers and then restaged and amplified by buyers, had spread Manning machines into the Weeks's neighborhood. Public machine trials had drawn their form from the private experimental forms suggested by people like Young. As public trials themselves became popular, private machine trials became an institutionalized form of advertising for networks of sales agents. But the Weeks brothers could use these forms of trial as well, using practices they had observed at the fairs and at their neighbors' farms to turn the tables on agents trying to sell goods. This gave them the dubious power of deciding themselves whether and when to enmesh themselves in debt by gambling on a machine.

PART THREE

Futures

5

Coining Foliage into Gold

In the summer of 1840, bruised by the Bank Wars and the shocks of the late 1830s, less than a year after resigning the presidency of the faltering Second Bank of the United States, Nicholas Biddle faced a further humiliation. According to the *New-Yorker*, he had to publicly quash a new and scurrilous rumor that he had been ruined, having sunk his personal fortune in "stocks, Morus multicaulis, and other profitable investments."[1] *Morus multicaulis* now seems like a strange specter to raise in a rumor of reckless speculation. It is not a stock; it is not a bond; it is a kind of tree, a mulberry. By 1840, however, it had also become a recognized emblem of the speculative fever that had characterized the late 1830s and had come so dramatically to its end in 1839.

Biddle had indeed invested in *multicaulis* in a serious way. He purchased a reported fifty thousand trees in December of 1838—visiting the Biddles in 1839, John Quincy Adams noted, "Chinese Morus Multicaulis; many thousand slips in pots." However, Biddle was far from the largest investor in what had become a national craze.[2] During the height of what became known as the "mulberry bubble," almost a million trees were advertised for sale in New York City alone. In Manhattan, dry goods merchants, druggists, attorneys, a bank president, a furrier, a clockmaker, and a tragedian were listed among the investors.[3] Markets sprang up in Manhattan, Philadelphia, Richmond, Boston, and Mobile, feeding a distribution network covering the eastern seaboard and extending west as far as Illinois. Dreaming of an American silk-growing industry to rival those of France, Italy, Turkey, or even China, investors trafficked in trees intended to feed an industrious but largely imaginary population of silkworms. Even during this wave, participants would describe it as a "fever" or a "mania." By November of 1839, however, *multicaulis*'s value would evaporate, leaving its adherents with fields upon fields of worthless stems. The mulberry's rise and fall showed the scope and variety of agricultural dreaming both in New York and around the nation.

How did an East Asian shrub, unknown to Americans until ten years before, come to be the focus of speculating attention, classed with stocks and bank paper? How did it accumulate enough concentrated belief to shape markets and ruin reputations? In part its rise came with the rise of improving institutions. As we have seen, the 1830s were a period of growth and precarity for agricultural improvement—commercial and print networks emerging from nurseries, warehouses, and seed stores were unfurling their tendrils. New York's landlords were at the height of their apparent wealth, speculating on western land before the collapse that would threaten their fortunes and their tenant relationships. Support for the New York State Agricultural Society was building among merchants, manufacturers, and middling farmers, though not enough, yet, to secure the state funding that would pay for fairs or support the county societies. The rapid rise of the *multicaulis* mulberry followed these new lines of growth, allowing us to see their interaction.

Multicaulis's brief, spectacular career also offers a remarkable window into the speculative booms and busts that rocked the volatile antebellum economy. As enormous tracts of indigenous land were seized and drawn into the market, as state governments embarked on colossal infrastructural developments dependent on infusions of foreign cash, as hundreds of new banks issued floods of increasingly dubious paper bills, and as antebellum Americans built themselves into a shaky edifice of credit relationships, conditions were ripe for sudden success or disaster.[4]

The fluctuations of the 1830s were mirrored in the world of agriculture by a much less studied series of agricultural crazes and manias. A commentator, "One of the Humbugged," listed some of them wryly in the *Genesee Farmer* in 1839, having been taken in by "the Merino fever, the Short horn speculation . . . the vine culture, the Italian and Siberian wheat controversy, [and] the Sugar beet business," as well as "the Immense Mulberry bubble."[5] Where bubbles and fevers have become central to the history of early American capitalism as a whole, agricultural manias have been sidelined in agricultural history, the exception that proves the rule of American farmers' presumed natural conservatism. But the *multicaulis* bubble directly linked the world of finance to the commercial and print networks of improvement. Its rise and fall were tethered to the larger economic boom and bust—its peak spanned the moment between the Panics of 1837 and 1839; it drew strength from the first and was defeated by the second. In the story of *Morus multicaulis*, the worlds of finance and agriculture connect in ways that illuminate both.

Multicaulis's rise built on a deeper history of silk fevers as well as on international networks of print and botany that brought *multicaulis* to America and gave it a public face. Multiple improving constituencies would fit *multicaulis* into their stories of national development or local reform, creating a coalition of promoters. But *multicaulis*'s particular bodily characteristics mattered as well, fitting it for absorption into the genres of storytelling, calculation, and description that silk speculators used to sketch plausible visions of the future. In an economy that was just beginning to be made visible through numbers, the *multicaulis* boom shows us new kinds of calculation and projection, new ways of imagining the future and new ways of deciding to believe in it. In this mode the quantification and labeling of living things led not to a stable, rationalized landscape, but to the dreams of fever and mania.

Global Stories of Silk

In telling stories about silk culture's American future, antebellum improvers looked first to its global past. Several tantalizingly successful episodes of silk culture hardened into legends that American silk promoters of the 1830s would tell and retell. At the height of the mulberry bubble, newspapers breathlessly rehearsed well-worn stories: the one about the pair of monks who had stolen mulberry seeds and silkworm eggs from China in the sixth century and carried them, hidden in their walking sticks, to the Emperor Justinian in Constantinople; the one about the Greek silk workers captured by Roger, king of Sicily in the twelfth century, who had carried the secrets of silk culture to Palermo; and the one about the culture that flourished throughout France under the benign influence of Henry IV in the 1550s and Louis XIV in the 1660s.[6] For material proof of these thriving ecosystems, Americans could look to a range of affordable goods; silk ribbons and sewing silks were available even in country stores by the 1830s.[7] Living in a landscape populated with transplanted European exotics like cattle, clover, and honeybees, it was easy to believe that silkworms could move too.

Americans were not alone in telling these stories or in attempting to re-create them. Alarmed by the drain of precious metals to China for botanical luxuries like tea and silk, and envying French and Italian domestic silk, governments in Britain, Russia, and the Germanic states had been actively promoting exotic plant cultivation for more than two hundred years, hoping to create self-sufficient territories by importing the species their countries lacked. Encouraged by a botany that saw even tropical organisms as gradually adaptable to northern climates, they nurtured many exotic species, hoping (sometimes with justification) that they would grow acclimated, and nourished dreams of Northern-grown tea, coconuts, cochineal, and sugar. British projectors had tried to grow silk repeatedly throughout the seventeenth century; Russian, Prussian, Saxon, and Swedish rulers had done the same in the 1740s and 1750s.[8] Such dreams helped stimulate the development of the international botanical networks of the eighteenth century and motivated their most famous builder, the Swedish botanist and taxonomist Carl Linnaeus.[9]

Linnaeus was particularly active in silk. With support from the Swedish state, he sent collectors to seek hardy mulberry trees that could be trained to survive Scandinavian winters, planting twenty-five thousand on the fringes of Stockholm and thirty-six thousand outside Lund. Supportive Swedish courtiers designed and wore Swedish-grown silk robes; Queen Lovisa Ulrika supervised the production of silk brocades at the summer palace; and Gustav III wore coronation robes of Swedish silk in 1772.[10] Though Linnaeus's mulberries ultimately withered, the community of botanists and the definitions of national wealth that had animated his efforts remained.

Growing outward from medical gardens and private collections in the seventeenth century, this network had been buttressed in the eighteenth century by state gardens like the Royal Gardens at Kew and the Jardin du Roi (later the Jardin des Plantes) at Versailles—which had, in turn, established competing subsystems of colonial botanical gardens.[11] Moving seeds and cuttings, these gardens determined colonial fortunes; for example, by saving the sugar plantations of the Caribbean from global cane epidemics with a supply of new varieties in the 1790s.[12]

Silk production had also anchored British colonists' earliest hopes for North American profit. Encouraged by reports of the "great Mulberry trees" seen near Native American homes and by the cocoons of wild American silkworms "as bigge as our ordinary walnuttes," settlers in Jamestown had immediately tried to produce silk.[13] They met with some success—when John Rolfe and his wife Rebecca, or Matoaka (known to modern readers as Pocahontas), arrived in England in 1616, they carried samples of Virginian silk with them.[14] James I supported these efforts enthusiastically, sending trees, silkworm eggs, "silkemen," and the first of a series of silk treatises aimed at Virginia.[15] The London Company, hearing reports of "vines and mulburitrees groweinge wilde in great quantities" also counted on profits from Virginian-grown silk and wine.[16] As it became apparent that silkworms fed on American mulberry trees produced poor silk, Virginia colonists imported a new kind: *Morus alba*, the white mulberry, a Chinese relative of *Morus multicaulis*, that was already heavily used by the silk industries in Italy and France. Starting in 1656, colonists were legally required to plant ten white mulberries every hundred acres; in 1662, they received a bounty in tobacco if they managed to set out fifty thousand. While the much greater profits of tobacco meant that few planters took up this second offer, enough imported white mulberry trees appeared in Jamestown to make them a handy way to demarcate property lines.[17] Following Bacon's Rebellion and William Berkeley's return to England, however, these efforts were abandoned.[18]

In the 1730s, the Trustees of the Georgia Colony would likewise pin their hopes for Georgia on silk culture: a mulberry leaf fills their original seal, adorned only with a silkworm egg and a lolling silkworm.[19] In their initial plan, the trustees required colonists to plant one hundred white mulberry trees, and "adventurers," coming to the colony on their own initiative, to plant one thousand.[20] Despite miasmatic swamps, and sporadic interest from colonists and colonial officials, Georgians also made some silk. In 1767, production peaked at 1,084 pounds of silk only to collapse with the withdrawal of government bounties.[21] Attempts at a North American silk culture surfaced several more times with little success; silk growing fizzled in South Carolina in the 1690s and the 1730s, in Philadelphia in the 1760s and 1770s, and, following the Revolution, in Ohio, Pennsylvania, Kentucky, and across much of New England.[22]

What had caused these successive failures? As they tried to argue once again that the American landscape could produce silk, American silk promoters had to wrestle publicly with this question. Their struggle to explain previous failures helps reveal the complexity of any attempt to move an ecosystem and to build it into an international market. As American silk manuals revealed, silk culture required, not only orchards of mulberry trees, but also billions of tiny, delicate livestock. Silkworms were subject to a distressing array of diseases, to fatal indigestion, rot, freezing, suffocation, and sudden death during thunderstorms. Thoroughly domesticated animals, dependent on human assistance, they required clean dry shelves, sufficient ventilation, and above all a consistent supply of mulberry leaves for about ten weeks, leaves that were both fresh and free of the moisture that would cause both leaves and worms to rot.[23]

Having survived the hazards of larval life, the worms would finally produce their valuable cocoons, but these presented their own problems. By boiling the cocoons before their occupants could chew their way free, silk growers produced a product that could be sold by the bushel. However, to reap the real profits of silk culture, growers had somehow to wind the fragile strands into thread thick enough to weave with. Since each worm produced a single fiber more than a mile long that was thicker at one end and tapered gradually to the other, winding three threads together to make an even "reeled silk" was not easy and not much like spinning cotton or wool. In Europe, silk was the basis of specialized local economies of skill.[24] While Americans tried to recreate these knowledge systems by importing experts from Armenia, France, and the Italian states, or by establishing public "filatures" where American growers could bring their raw silk to be reeled, the complexity of reeling combined with uneven political support kept these efforts from permanent success.[25]

Weak political support may have been the most decisive element in the ecosystem's failure. Like all agricultural organisms, American silkworms and mulberry trees had not only to be physiologically adapted to their places, but they had also to be adapted to market conditions. Though protected from competition with wild nature, Americans' mulberries and silkworms competed directly with their Chinese, French, and Italian counterparts through lines of shipping and international lists of prices current. Lacking a regime of skilled, cheap labor, colonial officials created safe havens for silkworms by giving bounties for trees planted and cocoons produced and reeled and by supporting raw silk prices. When these havens were dismantled by discouraged officials or forestalled by revolution, ecosystems of silk culture collapsed as well.

The combination of all these factors allowed silk promoters in the 1820s and 1830s to flexibly interpret silk's previous failures, turning each early episode into an object lesson in the chilling effect of insufficient state support, the disruptions of war, or, in the North, of the incapacity of enslaved people for delicate work. At the same time, they drew hope from old accounts and specimens of American silk that showed that, unlike Great Britain, the American landscape was at least capable of silk production. They were assisted in this by the lone surviving example: Connecticut, where in the 1820s, families and small businesses churned out silk sewing thread on a small but encouraging scale.[26]

A New Silk Coalition

When *Morus multicaulis* arrived in the United States in 1828, the newest wave of silk promotion was already under way, partly launched by some of the same New York landlords who had promoted the merino mania ten years before. Stephen Van Rensselaer III in particular had reported on the immense possibilities of silk production in the House of Representatives in 1826. Between 1826 and 1832, key members of improving networks generated a flurry of federal reports, which would be echoed in agricultural journals like the *New York Farmer* and by eight separate silk journals (and even more manuals) in the mid-1830s.[27]

Once again, visions of American silk drew strength from examples across the Atlantic; silk projects were under way in Sweden, St. Helena, Saxony, and Ireland, where in 1825 "the British, Irish, and Colonial Silk Company" had planted four hundred thousand white mulberries in County Cork.[28] As one might expect, Rensselaer lingered on the possibility of maintaining increased rents; the Irish project had succeeded so far, he noted, even though "a year's rent of land exceeds the price of the soil in many parts of our country."[29] Once again the white mulberry trees left behind by previous projectors became sources of potential seed for propagation, promoted by private projectors and the state.[30] Again these calls were supported by state and private bounties for early adopters—the state legislature offered a premium for raising white mulberry in 1825, and in 1828 the American Institute began to distribute white mulberry seeds free to anyone who would accept them.[31]

However, this latest interest also fed on factors particular to its cultural moment, most obviously the rapid rise of another American fiber. "Be assured," the *American Farmer* promised its readers, "the production of silk may be made as profitable here as the production of Cotton at the South."[32] In 1820, American cotton production had hit 730,000 bales, passing India to become the largest source of cotton in the world. Cotton was already the prime animating force of American trade, making up 40 percent of all exports, and driving the expansion of the domestic trade in enslaved people.[33] "Where fifty years ago, eight bales of cotton were produced, 1,200,000 are now produced," the *Journal of the American Institute* would point out in 1836. This rate of change, the editors held, altered the stories that could credibly be told about the future: "it is predicted that in a few years as great an amount of silk will be raised—why not?"[34] Stories of the role of the cotton gin in the expansion of cotton also seemed to imply that the mechanical difficulties of silk reeling could be overcome. As the *American Monthly Review* pointed out in 1832, "as in many other things the mechanical genius and enterprise of our country have outrun all foresight and prognostication."[35] With cotton's example shimmering before them, promoters in both the North and South projected a similar trajectory for silk.

Homegrown silk also promised other advantages. Weedy, hardy mulberries grew enthusiastically on the increasingly exhausted eastern soils that were now refusing to produce tobacco and wheat and promised to slow the flood of westward migration that alarmed landlords. Moreover, by providing Americans with a further textile industry, enthusiasts argued, silk would promote the union of the manufacturing, commercial, and agricultural interests so frequently invoked in antebellum political speech, particularly by pro-tariff manufacturing societies like the American Institute.[36] Similar narratives of united interest had propelled the Merino mania, promoted by Chancellor Robert Livingston almost two decades before.[37]

As silk weaving promised manufacturers new springs of wealth, silkworm feeding and silk reeling began to seem like appropriate substitutes for the home spinning that was increasingly romanticized by Northerners worried about women's factory labor.[38] According some Northern promoters, it was the intelligent labor of white women that would save a form of cultivation that they argued had foundered on slavery. "When speaking of some causes which in former years suspended and retarded the progress of the silkworm in the American colonies," wrote the Manhattan nurseryman Felix Pascalis, "we had occasion to contrast the ordinary labours of black slaves" with the "intelligence, delicacy, and unwearied attention required for little animals, which must be nursed and fed by the same hand during five or six weeks; cares . . . so easily but exclusively well discharged by females." Sensitive white female bodies, by Pascalis's account, were ideal for the delicate work of silk production since "[women's] outward senses can be better guides for them than thermometers, to judge of the temperature and pure air that are requisite."[39] Silkworms here became white child substitutes, best treated with maternal care in the home. The image of white women reeling silk in the home became one of the central motifs of Northern silk promotion, allowing a union of luxury and domestic industry that spoke to middling farmers' desires for and fears of refinement.

As light, largely indoor work, lacking heavy lifting or exposure to the elements, silk culture could also be made to promise solutions for "the almost useless" labor of all kinds of naggingly unprofitable populations. "The lame, the halt, the widow, and the orphan," described in the North were matched in the South by enslaved people too old or sick for field work.[40] Since it was not an established industry and could be practiced in confined spaces, silk growing could also be an employment for prisoners—one that wouldn't seem to threaten free working men with unfair competition. In 1835, the New York State Legislature would require the state farm at Sing-Sing Prison to grow mulberry trees for free distribution and to furnish white mulberry seed to keepers of the county poorhouses.[41]

As reports, articles, and advertisements moved through the organs of agricultural print, silk culture and mulberries assembled a broad coalition of support, from politicians seeking a task for prisoners, to manufacturers seeking a new source of raw materials, to agriculturists seeking a new source of profit and a sink for female, child, or coerced labor. Their arguments became the talking points for supporters of American industry, dozens of agricultural journals, a number of hopeful silk companies, and of course the network of nurserymen and agricultural warehouses who planned to supply the seeds, trees, and silkworm eggs. It was into this ferment that *multicaulis*, an unknown, short shrub with large, floppy leaves, dropped in 1828.

A Promising Stranger

One commentator would write at the height of the bubble that *Morus multicaulis* "came here as a stranger, and had its own character to substantiate."[42] In fact, it came with letters of introduction, riding international pathways of botanical exchange built to smooth its path. *Multicaulis* had been drawn into the botanical networks of western Europe by French botanist Georges Perrottet, who was introduced to it by Chinese merchants in the Philippines in 1821.[43] According to his later reports, Perrottet immediately noticed several things about the *multicaulis*. First, its leaves were very large, much larger than those of the white mulberry. Second, it was extraordinarily quick to grow—its roots threw up not a single slow-growing trunk, but "numerous small flexible stalks," each of which could reach a height of seven feet in a single season and could be harvested in the first year of growth. Third, it reproduced incredibly easily. Eight-inch chunks of its stems, cut and thrust into the ground, reliably struck root. Familiar with the white mulberry, and primed by his Chinese contacts' promises of silk, Perrottet did not see these points as the qualities of an invasive. He distributed trees in various climates, leaving specimens in French Guiana and Paris and experimenting on it himself in Senegal, and then published his official description in the papers of the Linnaean Society of Paris.[44] Over the next few years, these gardens produced a series of reports and living specimens, which circulated in France in the early 1820s. By the time it reached the United States, in fact, *multicaulis* was already a minor celebrity in France.

Only seven years after Perrottet's first encounter with *multicaulis* in the Pacific, his reports and descendants of his tree had arrived eight thousand miles away in Flushing, Long Island. This speed is not surprising. For decades, Flushing had been a major center of American botanical and horticultural activity, with close ties to European botany. Only a few years earlier, in 1824, the Flushing branch of the Linnaean Society of Paris had publicly gloried in its links to global botany by holding a gala celebration of Linnaeus's 115th birthday, launching "the new and elegant boat Linnaeus," in Manhattan, flying flags inscribed with Linnaeus's name, before proceeding to the most famous of the Flushing nurseries, the Linnaean Garden, where they read botanical papers, chanted poetry, and crowned a statue of Linnaeus ("locally made in imitation of a coin") with a chaplet of flowers, while Governor De-Witt Clinton offered up "thanks to the source of all light, for having devoted such a master spirit to the illumination of a benighted world" and celebrated the extension of "the empire of useful truths in Botany and Husbandry." Thomas Jefferson, frail in his eighties, sent his regrets, but cheerfully pointed out that his approaching death would allow him to express his admiration for Linnaeus more directly. "As [the prospect of attending] recedes from my view," he wrote, "another advances with steady and not distant steps, that of meeting the great naturalist himself, and of assuring him in person of the veneration and affection with which his memory is cultivated here."[45]

It was at the same Linnaean Garden that the *multicaulis* would first arrive, joining a living collection of more than four thousand species, assembled from four continents by the firm of William Prince and Sons, which consisted of William Prince, his son William Prince II, and *his* son William R. Prince. Prince and Sons was in a strong position to circulate both the tree and the texts that would make sense of it. Extended by their nearly one hundred agents, their plant and tree sales network stretched from Montreal to Mobile, over to the Great Lakes, and across the Atlantic.[46] In establishing a transatlantic trade in European fruit trees and North American exotics, they had made links with every well-known botanist in North America and several European botanical societies, with the members of the American horticultural societies, and with the editors of several newly founded agricultural journals. Moreover, they had the support of a group of French and Belgian emigrants with nurseries in Flushing and Manhattan, including the president of the Flushing branch of the Linnaean Society of Paris; the physician and nurseryman Felix Pascalis; and Andre and Sylvia Marie Parmentier, cousins and husband and wife, both scions of a famous family of nurserymen who had worked for Louis XVI.[47]

To the Princes and their colleagues, the *multicaulis*'s origin story meant a great deal. Its voyage from "the elevated regions of China" to the heat of Manila, to France to Perrottet's experimental groves in Senegal, illustrated its flexibility in encountering different climates. However, like French botanists, American nurserymen required more than a textual reputation. They began a phase of experiment and trial, producing reports and in public letters feeling out *multicaulis*'s place in the future of American silk. The most important of these reports, made by Felix Pascalis and known afterward as "The Pascalis Prediction," appeared in the *American Journal of Science* in 1830.[48] *Multicaulis*'s speedy growth and its dazzling reproductive properties, Pascalis explained, fitted it for a shining future of profit. "After the discovery of this plant, a doubt no longer exists that two crops of silk may be raised in a single season."⁴⁹ By 1831, Pascalis had sent samples of *Morus multicaulis* as far east as Cincinnati, and Sylvia Parmentier (now a widow) had more than a thousand trees for sale at a dollar apiece in her Brooklyn nursery. By 1832, Parmentier reported that she had repeated Pascalis's successful experiment with double cropping and had won a prize for her *multicaulis* collection at the fair of the New York Horticultural Society.⁵⁰

Through such texts and displays *multicaulis* became a recognizable name in agricultural and horticultural circles outside of New York. In the same year as Parmentier's prize, Boston's *New England Farmer* informed readers that it had acquired its own French source for the new tree: the famous French naturalist François André Michaux, author of the first major work on North American trees, the *North American Sylva* (1817–19).⁵¹ By the middle of the decade, when a growing number of writers on silk declared *multicaulis* the most promising of the mulberries, it received the full force of the renewed enthusiasm for silk.

Multicaulis and broader silk fever expanded together. In February of 1837, Connecticut representative Andrew Judson wrote a letter to then representative John Quincy Adams, which, read in the House, showed Congress that the Northeast, and to a lesser degree the South and the West, were deeply engaged in silk culture and the growth of the mulberry. In New York, silk growers' companies had been incorporated in Troy, Poughkeepsie, and Albany as well as in New York City; companies had also popped up in Pennsylvania, Delaware, Rhodes Island, Massachusetts, and Maryland. Nurseries of mulberries flourished in five towns in Maine; successful experiments had been conducted around New Hampshire and Vermont, and a silk company capitalized with $75,000 was stocking 250 acres with mulberries in Concord. Connecticut, where silk culture already had a foothold, boasted several working factories, and considerable nurseries in eight towns. The state legislatures of Vermont, Connecticut, and Massachusetts all paid bounties either on cocoons or on mulberries planted. In the South a few silk companies had appeared in Georgia, the Carolinas, Florida, and Alabama, and in the west companies appeared in Ohio, while trees had been planted in Indiana and Kentucky.⁵² Only nine years after the introduction of the first tree, American agricultural journals were routinely advertising *multicaulis* trees in tens of thousands.

Panic-Driven Growth

The Panic of 1837 followed closely on Judson's letter. Fearful of the instability that seemed to be emerging from the aftershocks of the American Bank Wars, the Bank of England's directors decided to stem the westward flow of precious metals. Refusing to discount bills of exchange drawn on American firms, they wiped out the value of transatlantic instruments of credit and set off a cascade of failures across the United States. Banks stopped payment, businesses folded, the value of land dropped, specie was almost unavailable, and paper money was increasingly useless.[53] All kinds of ventures failed. From their ruins, *multicaulis* grew.

Indeed, it was not until after the panic that *multicaulis* moved decisively into the world of city commerce, snatched up by men of business looking for financial rescue. In 1838 hundreds of advertisements for trees and buds appeared in commercial dailies like the New York *Journal of Commerce*. *Multicaulis* trees were sold at auction houses that specialized in real estate, or spices, or in one case, fine paintings. While nurserymen ran some of the private sales listed in advertisements, merchants in other trades also joined the market in large numbers. That same year Prince and other nurserymen sent agents to their French contacts to purchase tens of thousands of trees from *multicaulis* plantations near Versailles and opened plantations around the country to supply an anticipated demand.[54] In the fall of 1839, shipments of cuttings would come into Manhattan from all over the country—South Carolina, Georgia, New Jersey, northern Florida, and from the working silk farms in Connecticut and Massachusetts, which had de-emphasized actual silk in favor of the now more profitable mulberry trees. Prices skyrocketed—trees sold for twenty cents a foot, for two cents a bud, and then climbed higher. Prince's catalogs didn't even have prices printed, since he didn't want to rein in their advance.[55]

As the market for *multicaulis* heated up, nurserymen scoured European collections and their own stocks for new varieties that might achieve similar success: the Dandolo, the Alpine, the Canton, the Elata, the Rose of Lombardy, the Roman, the Pyramidalis, the Oriental, and the Lily-Leaved all joined *multicaulis* in nursery catalogs and on auction blocks.[56] The invariably colorful botanist Constantine Rafinesque jumped hopefully on this bandwagon. "Mr. Perotet [sic] has been much praised and probably well paid for introducing from afar the Morus multicaulis," he wrote in a manual on mulberry trees, "it remains to be seen if I ever will be praised or paid for proving that we have several equally valuable kinds, within our reach, growing wild within 120 miles of Philadelphia, which I can show in my Herbarium, and lead any one to the spot where they thrive upon rocks."[57]

Financing for the new volume of trade was not easily had in the specie-starved late 1830s. Speculators began to depend heavily on instruments of credit. Trees sold for mortgages and bank stock; sellers at one auction of

250,000 trees in Germantown, Pennsylvania, offered buyers who spent more than $1,000 up to six years of credit, at 6 percent per year, "to be secured by bond and mortgage on unincumbered real estate or other approved security."⁵⁸ Stuck in "losing concerns," explained the *Journal of Commerce* that same month, "people are willing to exchange any kind of property for trees."⁵⁹

While few records of their transactions remain, a dunning letter from the nurseryman William Kenrick to William R. Prince's son L. Bradford Prince shows that among nurserymen themselves, even more daring financial sleights of hand were in play. On Prince's death in 1869, Kenrick, by then impoverished, was still trying to collect $1,000 that Prince owed him from the mulberry days, which had been the subject of a lawsuit in Kenrick's favor in 1842. Like Prince, Kenrick engaged in deals on a new scale. Kenrick claimed that he had made "the extraordinary amount of 30,000" though he had laid out $20,000 of this fortune in advance for nurseries, land, and labor. Half of the money had come in a single enormous transaction with the Silk Company of Baltimore, which had bought fifty thousand trees in one order. Five acres of land near Richmond, Virginia, had brought Kenrick $15,000. Confidently, Kenrick dove further into mulberries, planting thirty acres and layering them "so as to increase them if possible to a million . . . by fall and then to sell half or 500,000 and to keep half." But, according to Kenrick, Prince and Sons were even more heavily involved—Kenrick had seen a plantation of trees "which I calculated would be worth . . . $75,000." This sight convinced him to enter into a risky kind of bargain with William R. Prince. He told Prince's son later that after breakfast, hospitality, and "some compliments," he had signed a draft for $1,000 payable to Prince's firm, indicating a transfer of funds that had never taken place, an imaginary debt that Prince promised Kenrick would never have to honor. By fixing his name to paper, and trusting in friendship, Kenrick had inflated Prince's paper assets by $1,000, a practice called "kiting." Promising to pay Kenrick later, Prince used his draft to pay a further large bill for his mounting nursery expenses. Nothing backed this bill at all. It was held aloft, "floated," only by the value of Kenrick and Prince's reputations, and by continued high mulberry prices.⁶⁰

Leaves, Stems, Numbers

What held prices up? Throughout this process, *multicaulis* had been heralded and supported by masses of text, from Perrottet's early reports, to the reports of federal officials and American nurserymen and women, to the broadsides, catalogs, silk manuals, agricultural journals, and newspaper articles that raved about it by decade's end. Across these texts, accounts of the *multicaulis*'s value gradually coalesced into several genres, which, appearing alone or in combination, were repeated with only slight variations. It was these genres of description that made the *multicaulis* a foundation on which fortunes could be gambled. By examining them, we can track the development of speculative visions around the *multicaulis* and, at the same time, see some of the standard ways of establishing credibility throughout the broader culture of improvement.

The most basic genre of *multicaulis* description concentrated on the desires and appetites of silkworms themselves. This meant examining the *multicaulis* leaf. In lyrical passages, silk promoters lingered over leaves, describing their tenderness, their luxuriance, "their glassy smoothness, and extreme succulency."[61] In the early reports out of Flushing, consumer testing backed up these flights of poetry—after experimenting with the *multicaulis*, Parmentier, Pascalis, and Prince circulated reports that worms had left lesser varieties "and devoured the new Chinese mulberry with avidity."[62] Worms that gorged on *multicaulis* grew almost twice as quickly as their less fortunate fellows, and produced cocoons "of a much larger size . . . the whiteness of snow and . . . a most beautiful shining appearance."[63] Following these reports, other nurserymen began to agree in their advertising that "the superiority of the foliage of this tree as food for the silk-worm over all others, has repeatedly been tested, and is proved beyond a doubt."[64]

Lending strength to these reports and experiments was the leaf's undeniable scale. Reaching lengths of fifteen inches long and widths of twelve, *multicaulis* leaves were easily achieved agricultural giants—they made impressive specimens whether exhibited at Niblo's Garden in Manhattan, at the offices of the agricultural journals, or at dozens of agricultural fairs.[65] For a group of agriculturists primed to perceive profit in four-inch white mulberry leaves, the sheer size of the *multicaulis* leaf, bending stems "almost to the ground with the weight of foliage," was easily interpretable as staggering wealth. This perception of value resolved into an iconic image that dominated broadsides, silk journal covers, and manuals in the later 1830s; in it, the modest leaves of the white and American mulberries lay at the center of the page, dwarfed by a looming *multicaulis* leaf that stretched from side to side of the page behind them.[66]

Expressed in this stark graphic, the dimensions of *multicaulis* leaves made their value clear. Expressed in pounds, they added a satisfying weight to a further calculating story: the estimation of the product of one acre that served as the coda for most silk manuals.[67] It is hard to give a real sense of the relish with which silk promoters embarked on these estimations, calculating over and over again just how many worms, cocoons, or pounds of silk could be

FIGURE 10. Cover of the *Family Visiter and Silk Culturist* showing how the *multicaulis* leaf dwarfs more familiar mulberries. Courtesy of the American Antiquarian Society.

raised from a tiny plot of land in the ten-week silk season or the enthusiasm with which they debated each other's methods of estimation and accounts of projected profit. Though simulated and real account book pages like those reproduced by Brewster and mocked by Thoreau were already a standard mode, in silk promotion they reached a new level; some authors, defeated by their numbers, averaged them or published different estimates in long tables of comparison.⁶⁸ Producing more pounds of leaves per acre, within easy reach of the ground, on whippy stems that could be stripped at a single pull, *multicaulis* infused old calculations with new life. Where early in the silk fever, Jonathan Cobb had claimed an already astonishing eighty-nine dollars per acre for the common white mulberry, in 1839 the *multicaulis* promoter John Clarke would claim a staggering $640.⁶⁹

As the value of money faltered after the Panic of 1837, such calculations grew more and more seductive. Even if they were not forged, paper bills, it became clear, were backed by an often-illusory stockpile of metal in untrustworthy banks that had, in any case, ceased payment.⁷⁰ By contrast, *multicaulis* leaves seemed to be backed by a tangible use value. The reams of verse that emerged from the bubble often linked *multicaulis* leaves directly to coins—"Each twig shall yield me coin" announced a perhaps satirical poem in the *Spirit of the Times*, "Till wealth shall make me bend."⁷¹ Similarly, the poem "to Farmers," printed in Jonathan Dennis's silk manual of 1839 told farmers aspiring to "wealth and ease" to stock their farms

> with mulberry trees
> The silk worms will their wealth unfold
> And coin their foliage into gold.⁷²

Just as earlier generations of silk promoters had offered silk culture as a way to stanch the flow of silver leaking to China, new silk promoters touted *multicaulis* as a better substitute for the mines of Mexico.⁷³ "On the subject of *specie, banks, hard coin, cash payments, mines,* and *Mexico,* we have read essays, lectures, pamphlets, and volumes without number," complained John Clarke, "Only set the silk worm to work; stop the enormous drain of specie abroad, by producing all at home; and we effect at once more than all the mines, more than all the ponderous tomes, . . . pole to pole, could accomplish in a century. The next lecture we intend to hear on the mystery of banking, mines, specie, and hard cash, shall be given by the silk worm."⁷⁴

The qualities of silk easily lent themselves to impressive economic discourse. Large numbers embellished all accounts of silk production—millions of worms, tens of thousands of trees, and millions of dollars, an annual cycle of mass birth and mass death. In an 1836 item, "Silk Worm Mortality," the *Journal of the American Institute* marveled at the "fourteen billions of silk worms [that] die victims to the production of silk consumed in England in one year."⁷⁵ Since huge numbers of silkworms were needed to produce relatively modest quantities of silk, pleasantly impressive numbers could be found even in small enterprises. When Clarke wrote that "a gentleman in Fryeburg fed last season 5000 worms, which produced the usual quantity of silk," he did not have to reveal that "the usual quantity" weighed just over one pound.[76]

The fetishization of the economic possibilities of animals and plants after the panic was not limited to the mulberry. In the same pages in which it advertised the mulberries, the resolutely urban *Journal of Commerce* published notices of other agricultural giants exulting over their exact and massive dimensions. But accounts of the "Mammoth Squash" in Manchester, Connecticut, or the giant hog, who had become the subject of betting on Long Island, could not capture the attention of investors in the same way.[77] The *multicaulis* had bodily qualities that lent themselves to economic projection. These came out in a further genre of speculative writing, which became dominant after the panic.

If the first vein of *multicaulis* description dwelt on the desires of silkworms, this second discussed it in terms of the desires of speculators. To do so, it focused on the *multicaulis*'s main reproductive organ: its stem. While common white mulberry trees grew from seed, like most fruit trees, *multicaulis* was propagated by cuttings or buds—sections of stem thrust into the ground or attached to the trunk of another tree, producing a copy of their parent. This method of reproduction had some advantages—for one thing, it was fast. While it took years for seeds to produce profitable white mulberry trees, the weed-like, prolific *multicaulis* sprouted usable leaves and feet of stem in a few months. A *multicaulis* bud or cutting was thus an investment that, theoretically at least, paid dividends of silk in the same year as its purchase.

More importantly for the purposes of the speculating market, *multicaulis* cuttings also rapidly produced more *multicaulis* cuttings—slender stalks, growing five or six feet could be chopped up at the end of a season to produce eight to ten new trees. This sectioning of the stem made it easy to imagine the *multicaulis*'s multiplication, a point that became the foundation of hundreds of stories of future profit. Gideon B. Smith, the editor of the *American Farmer*, and one of the first to receive a *multicaulis* tree from Prince, wrote one of the most influential of these, demonstrating that the ten trees bought in 1831 would be one million trees by 1835, trees that would moreover produce the foliage of ten million common white mulberry trees and that could be harvested for leaves every season for immediate profit.[78] The dollar spent on each original tree in 1831 (several times the cost of any fruit tree and the price of fifty white mulberry trees in 1833) could be justified, in that it would pay not for a single tree but for its innumerable offspring and for the vast trade in silk to be built upon them.[79] Of course, multiplication was not a quality restricted to the *multicaulis*. Full-grown white mulberry trees produced pounds of seed each year, but this fecundity meshed poorly with cycles of speculation—any mature tree could flood the market with millions of seeds, making seed relatively valueless, even as longer waits and smaller leaves made their ultimate product less desirable. The *multicaulis*'s yearly, orderly multiplication better paced it for the cycles of speculation than the lifecycles of less fashionable mulberries could. Growth would be fast, but not too fast.

During the mid-1830s, advertisements for shipments of *multicaulis* increasingly focused on their stems and buds. Auction notices announced how tall the trees in each lot were, whether they were trimmed or untrimmed, and whether they branched or not. Working backward from this information, the purchaser could reckon how many young trees could be produced from each purchased individual. Trees with lots of buds represented not individuals but countable crowds. In the Prince and Sons nursery catalog, each six-inch section of trunk was priced separately as was each root and each bud.[80] An advertisement for an auction of twenty thousand trees on Frederick Street emphasized this form of calculation, promising that its trees were "all securely packed in boxes, and the numbers of buds in each marked thereon—the counting having been done by sworn counters, every reliance can be placed on their correctness."[81]

As *multicaulis* speculators made the stem the object of desire, they defined trees and cuttings more and more precisely; an auction poster from Philadelphia declared that "every shoot of 12 inches in height . . . having root, or particle of root sufficient to grow it, shall be considered a tree."[82] Silk manuals of the late 1830s gave meticulous directions for purchasing cuttings and trees: no buds with unripened wood, none less than an eighth of an inch in diameter (too young) or more than one-half (too old), or standing too close together on the trunk (grown too close together). Trees were to be branched, if older than one year, since branches could be sold and their size was to correspond with their stated age. Their roots should be bright yellow and smooth, and their branches, if cut, should leave three buds behind to preserve the trunk. However, as Jonathan Dennis acknowledged, the realities of the market might mean that trees could not really be inspected. He suggested that "it is the best way to purchase by the foot, either with or without the branches," and insisted that in the winter, they should include only the old wood to the mature ends, not the tender shoots that would wither in the cold.[83] Thanks to such information, buyers knew to be impressed when William G. Harrison promised one hundred thousand *Morus multicaulis* trees "stout in stem and well branched"[84] or when, in their terms of sale, the Princes boasted of the scrupulous fairness of their cuts and measurements. "The imperfect wood at the ends of the shoots has been cut off," they explained, "which forms a point of far greater difference to purchasers than most of them are aware of, as in

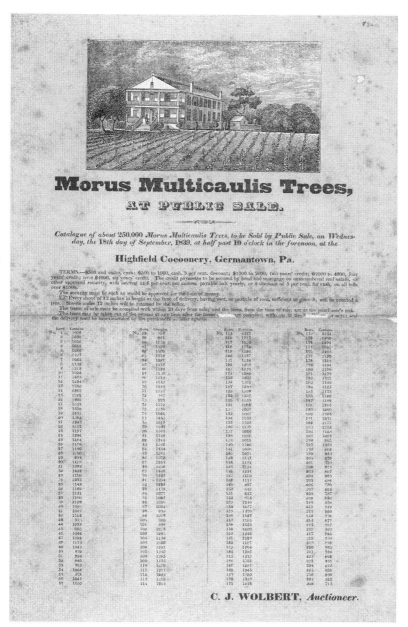

FIGURE 11. Broadside for a *Morus multicaulis* auction near Germantown, Pennsylvania. The figures at the bottom refer to numbered boxes, each of which contains the number of trees in the second column. "Trees" here are "shoots of 12 inches in height . . . having root or particle of root sufficient to grow it." Courtesy of the American Antiquarian Society.

most cases much immature wood is measured, or the buds on it counted and paid for."[85] Expertise in mulberries now meant the ability to measure the future in inches and to avoid being cheated in the process.

Speculating desires even changed the bodies of the trees themselves. Articles began to complain that to make the trees grow tall for the speculative market, dealers grew them in highly manured soil, producing long stems, so more cuttings, but a watery leaf, so less silk. As stems grew longer, saleable cuttings shrank. The two buds per cutting initially recommended dropped to one bud, and one-foot lengths became six inches.[86] As profits grew greater, nurserymen tinkered with buds and found that buds laid under glass or in hotbeds in winter repeatedly sprouted new shoots, each of which when cut off produced shoots of its own. This, the *Cultivator* informed its readers, allowed an increase of 120-fold and the "pretty sum of $3,000" between January and July.[87] It seems likely that the fashion for smaller and smaller cuttings began to stretch the endurance even of the *multicaulis*. Thousands of trees and cuttings purchased in the fall of 1838 did not grow in the summer of 1839, "ruined," one nurseryman (and defender of single-bud sales) postulated, "by oft changing hands, and removals hither and thither, and long continued exposure . . . imperfect and bad packing and drying, or the alternate freezing and thawing of their roots, and by careless management."[88]

Failures like these began to stain *multicaulis*'s reputation, lending support to Southern improvers' claims that *multicaulis* was not, after all, well adapted to the North. "There is no instance on record," a letter writer calling himself "Anti-Puff" pointed out in January in the *Farmers' Register* from Virginia, "of a tender plant becoming acclimated in such a brief period.—but in eight years, at most, the multicaulis, a native of the Phillippine [*sic*] Islands has been endowed with the constitution of the apple tree, by the potent spells of Mr. Prince."[89]

Nevertheless, the division of mulberry trees into agreed upon standards for feet and inches made possible a final genre of calculating story, one that became dominant in the winter of 1838. This was the price report, a form of text already common to all commodities sold at auction, in which commercial and agricultural papers published the prices at which mulberries had sold in their local markets or at notable auctions among their corresponding papers. As auctions became the common way of selling trees wholesale, and stem lengths and bud counts became an accepted way of valuing them, reports of prices expressed in feet of tree became comprehensible and comparable. Accounts of impressive auction sales, embellished with occasional prodigies—individual trees that sold for five dollars a bud, making their owners instant $40,000 fortunes—began to justify further auction sales.[90]

The significance of these reports in creating the markets they described can be seen in the efforts speculators made to manipulate them. In response to a mammoth sale reported in the silk-focused *Northampton Gazette*, the editors of the *Journal of Commerce* demanded in August of 1839, "Who are the buyers? Are they persons who wish to establish a price for their own commodities, as at the outset of the speculation last year, or are they the original speculators?"[91] High prices at the first auction of the year, maintained the *Journal*, had been artificially created—dealers had gone to the auction and bid high to set the price at which their own trees would be sold at their own auctions. This was the moment of pure speculation, when increasing prices became evidence of prices' future increase, without any reference to silk or leaves or sometimes even to trees themselves.

By 1838, the disconnect between the trees' use value and their exchange value—that is, the value of their leaves versus the value of their stems—was perfectly evident to all concerned. At thousands of dollars per acre, tree profits were far higher than even the most enthusiastic silk promoters' projections for actual silk. Worms and eggs languished on the market, and working silk manufactories in Massachusetts and Connecticut neglected existing cocooneries to sell buds. Some commentators viewed this separation with alarm. "One thing seems to be certain," Jesse Buel wrote in the *Cultivator* in the summer of 1839, "the public attention must soon be directed less to the *buds*, and more to the *leaves*, less to speculation in trees, and more to their use in the manufacture of silk—or . . . we shall ere long fail to realize all our golden dreams."[92]

Certainly, not even the most optimistic silk promoters expected tree prices to rise forever. "The greater part of the present prices of the multicaulis is undoubtedly a bubble," the *Family Visiter [sic] and Silk Culturist* informed its readers bluntly, "and that part must burst sooner or later, (when, we cannot predict)." But, they maintained, trees that contained generation upon generation of potential trees, and whose leaves gave them an intrinsic value, were not like other commodities. The multiplying powers of the tree would mean that even dropping prices could result in a profit. If demand for trees kept prices even to one-sixteenth of their present level for two years, the *Visiter* pointed out, investors would still realize a profit through the one-hundred-fold increase in their stock.[93] The balance between the leaf's putative intrinsic value and the calculability of its body allowed promoters to explain the bubble to themselves as they inflated it.

They were assisted in this by their projections of a further kind of desire, not of speculators, or of silkworms, but of American consumers. This desire, they argued, could be measured directly, since the tariff on imported silk meant that the federal government published figures for the estimated amount that Americans spent on imported silk. From the earliest federal reports, silk promoters had used this figure, which by the mid-1830s had reached $20 million, to postulate a reliable demand for silk. In the early part of the silk movement, this figure had been used to show the drain on American resources. By the end of the 1830s, promoters like Prince were using it to forecast the date of the end of the bubble, by predicting the number of mulberry generations needed to meet a fixed American demand. Working backward from $20 million, they calculated that approximately eighty million pounds of raw silk had been purchased. Taking into account the three thousand silkworms needed for each pound of silk, the one pound of leaves needed for forty worms, and the pounds of leaves that could come from each tree, such texts would work out a comfortably enormous number of trees necessary to meet the projected demand. In the case of William Prince, who had begun to add projected demands from "Mexico, Texas, Chili, Peru, and the whole chain of the West Indies Islands" to American desire for silk, this number would eventually reach five hundred million, an infinite sea of trees.[94]

Belief Falters

On the morning of October 4, 1839, two steamboats "thronged with passengers" left Manhattan heading for the Linnaean Garden. Waiting for them was a mulberry extravaganza—an auction on a scale new even to the Princes. While the crowd ate a cold collation provided by the management, more than two hundred thousand splendidly branched *multicaulis*, Alpine, Elata, and Expansa, sold for over $50,000. However, the *Journal of Commerce* acidly commented, "we learn from a gentleman who attended the sale, that the *number* of purchasers was not large." Near Philadelphia, another giant sale threw fault lines in the market into higher relief. The sale began by selling trees at twenty-seven and a half cents each, before the sellers—unwilling to part with trees for less than thirty cents—stopped the proceedings. "All present," the *Journal* added, "appearing to prefer 'the needful,' to multicaulis were mute, and dispersed."[95] Both these sales, the *Farmer's Register* would argue, had been staged in an attempt to prop up mulberry prices. Indeed, the *Register* held, false initial reports of the Philadelphia sale, "which we, (as well as most other persons,) were so credulous and foolish as to believe, (for a few days,) to have been a bona fide transaction," had been purposely placed in the *Germantown Telegraph*, so that it would unwittingly be republished by the *Farmer's Register* among other papers.

However, newspaper columns had proved difficult to control. "Two weeks after the publication," the *Register* maintained, "not an individual in the country remained deceived on the subject. This pretended sale was the last blast blown to swell the multicaulis bubble, and served to burst it effectually."[96] In Manhattan, reports of radically variable prices began to arrive. Trees sold in Batavia, New York, for fifty cents each; in Montgomery, Alabama, for one dollar each; and at Camden, New Jersey, for four to seven cents. On October 17, the *Journal of Commerce* published a notice of a sale where five thousand five-foot trees "of luxuriant growth" had sold at two cents each, and ten thousand had found no bidders at all.[97] The newspaper price reports that had bolstered the trade in trees had turned against it.

In part, the market was more and more undeniably weighed down with trees, which agricultural journal editors estimated in the tens of millions. Two days before the Linnaean Garden sale, the *Journal of Commerce* editors had commented with some sarcasm,

> We hope all who want mulberry trees of the genuine stamp, will make haste to purchase, no matter at what price as the variety is likely to become extinct! It is estimated that not above 100,000,000,000,000 have been raised in the United States the present season.[98]

The fields carpeted with *multicaulis* that had previously demonstrated silk culture's strength now undermined their own value. At the same time, broader economic changes had begun to undermine potential purchasers. Banks in the South and West, weighed down with untradeable state bonds, began to fail, and British bankers again began to tighten the credit they had extended to American merchants.[99] Once again paper money, and the instruments of credit with which *multicaulis* buyers had been making their purchases, dropped in value.

The collapse of confidence thereafter was rapid. By mid-November, the *New England Farmer* pronounced the bubble over. "The Morus multicaulis speculation is now at an end," the editors wrote, "it has fallen suddenly like a tremendous Colossus, and it now lies sprawling with a good many under it who are crushed by its fall.... The country in many parts is covered with the Multicaulis almost as thickly as wheat stubbles are with the Roman wormwood."[100] Within a few months of the end of the mulberry bubble, *multicaulis* was rapidly being expunged from the Northern landscape. Where speculators in the land boom were left with unprofitable tracts hanging heavily on their hands, *multicaulis* speculators had nothing. As debts came due over the next few years, the Prince family, holding piles of mortgages and IOUs from investors no longer willing to pay, sold most of the Linnaean Garden to a competitor; their correspondent William Kenrick retreated hastily to Europe.[101] Half-hearted attempts to establish a new market in silkworm eggs proved hopeless in this new economic climate. Far from being the axis around which a bright future spun, *multicaulis* ceased even to be a commodity. Like Roman wormwood, which we call ragweed, *multicaulis* trees were weeds.

★

Told after the fact, stories of speculative bubbles have a depressing sameness. Looked at from afar, the *multicaulis* bubble seems identical to the tulip mania, the South Sea bubble, or even the modern mortgage and dot-com booms. Antebellum Americans cast these ups and downs in medical terms, as fevers or manias; we now turn to the naturalizing power of the modern business cycle. Either way, the changeless up and down, the moral of folly and punishment is clear: prices are inflated by greed, then pop. This narrative arc makes such stories satisfying, a nice mixture of comedy and morality play, spiced maybe with a little schadenfreude.

While acknowledging their undeniable similarities, we should also acknowledge these bubbles' differences and see them as a window into a particular economic moment. They were not simply periodic fits of greed and madness, they were also astonishing moments of belief, moments when large numbers of people came to agree on a particular form of value and a particular kind of future. Investigating them in this light, we can learn much about the kinds of evidence that different groups require for belief and the different ways that they assign value and imagine futures. These vary enormously. For example, the tulip mania emerged from a fine art and specimens market. Tulips in seventeenth-century Holland were valued, like the fine paintings with which they were classed, for individual qualities of beauty and rarity. Their sudden spike in price, like the simultaneous spike in painting prices, came when a newly wealthy Dutch bourgeoisie began to participate in kinds of connoisseurship that had previously been reserved for a tiny class of enthusiasts.[102] By contrast the millions of *multicaulis* trees were theoretically identical and expected by a great range of improvers to found a new industrial order.

Speculation on living organisms was, moreover, necessarily shaped by the character of those organisms. In 1630s Holland, tulips were perfectly suited to a market based on the collection of rarities; beautiful and various, they came from far away, reproduced slowly, and occasionally broke out into unexpected and brilliant colors and shapes that could themselves be collected. Similarly, *Morus multicaulis* was ideally adapted to its speculative moment. Its giant leaves made its value recognizable, while its method of reproduction, yearly division into a manageable number of sections, each of which

FIGURE 12. "A huge budget of bubbles." From G. P. Burnham's *A History of the Hen Fever: A Humorous Record* (1855). Among the bubbles pictured: "Dwarf Pear Trees," "Paper Money," "Tulip," "Alderney [cattle]," and "Female Novelists." Earlier in the text, the author addresses the "shades of morus multicaulis victims! Shadows of defunct tulip-growers! Spirits of departed Merino sheep speculators! Ghosts of dead Berkshire Pig Fanciers!" Author's collection.

promised an annual return, made it a foundation on which speculators could build a structure of calculation and projection.

So what does the *multicaulis* bubble show us about its particular time and place? First, the *multicaulis* boom reveals a new facet of the moment of financial instability in which it appeared. Like the other agricultural fevers that accompanied it, Chinese tree corn, hops, Rohan potatoes, Durham cattle, and Berkshire pigs, the mulberry represents a more general turn to biological forms of value at a time when many other forms of economic value seemed untrustworthy. The Panics of 1837 and 1839, the ferocious Bank Wars of the mid-1830s, and the speculations of the whole decade before, not to mention the incredible flow of counterfeit and dubious bills, showed Americans with painful clarity that monetary value depended on fleeting social agreement. In this context agricultural animals and plants—not only self-evidently useful, but self-reproducing—came to seem almost magical sources of true wealth, a belief given force by the extraordinary profits being realized in cotton. This interest in the possibilities of living growth helps explain why the American Institute, an industrial organization, developed not only an interest in mulberries, but a whole agricultural auxiliary just at this period, and why even commercial papers carried reports of agricultural monsters. More importantly for this book as a whole, it helps explain why so many state legislatures, New York's included, began to support agricultural improvement societies during the depression that followed the panics.

Following the *multicaulis* bubble also reveals the mechanisms that improvers used to create credibility in this period. To become the focus of a bubble, *multicaulis* had to move through existing print and trade networks and through genres of writing and experiment already developed to create trust. By the time urban financiers were scrabbling for a savior, *multicaulis* already had an established economic story to seize. Speculative interest in its stem depended on early impressions of the value of its leaf, on trials of silkworm taste, on standard projections for the profit of one acre: forms of testing and proof common throughout improvement. Once the multiplying powers of the stem became the subject of speculation, they could be fitted into developing understandings of the shape of markets and of bubbles. These depended on new ways that Americans were beginning to imagine the economic future. The $20 million spent on foreign silk exemplified a kind of statistics emerging from international markets, familiar to us but still relatively novel to antebellum Americans, that seemed to reveal the desires, not just of single-city markets, but of whole nations. The sudden scale of the *multicaulis* bubble depended in part on this way of aggregating desire and on a developing conception of the nation as a single economic entity.

It may seem strange to have chosen an episode of failure to discuss economic projection and storytelling. Since the American silk empire remained imaginary, it is easy to dismiss the modes of description that conjured it up and the people who were moved by those modes of description. But episodes of failure show us things that success cannot. By focusing on a vision that dissolved, we can avoid attributing its construction to a kind of economic rationality only visible after the fact. The kinds of networks, stories, and visions that launched the *multicaulis* boom were not so different from the kinds of stories told about cotton, or about other new crops like hops or sugar beets. Like *multicaulis*, these plants moved through international botanical networks, carried forward by reports and descriptions and acre estimates created by self-interested promoters. Like *multicaulis* these crops benefited (if less dramatically) from reproductive qualities that allowed them to be reliably moved and multiplied. Like *multicaulis*, their acceptance came in part from perceptions of national demand based on international trade statistics. Now we see the new landscapes created around these organisms as the natural result of economic or environmental fate and belief in them as a clearheaded apprehension of reality. In fact, their promoters had to interpret similar signs of potential fortune in a similarly chaotic economic landscape. Examining the development of stories of the economic future that proved baseless allows us to step back and see the social relationships and forms of storytelling on which all accounting with the future is based.

Other fundamental features of the *multicaulis* market would long survive the bubble. Nurseries and markets in cuttings grew and became transformative, the search for value in nutriment would fuel new enthusiasms for agricultural chemistry and more and more precise calculations of the monetary value of food. The same urban failures that inflated and doomed *multicaulis* would help provoke the revival of the agricultural societies, finally rewarding the lobbying efforts of the state society with appropriations. Telling stories about the future remained at the root of agricultural improvement, and would continue to follow the same reporting pathways through state reports and the agricultural journals. The next chapter shows how these elements played out again in other kinds of stories about the future, a current of capitalist development depending not on a fabulously universal crop, but on local particularity, on geographic differences that were made to forecast divinely intended economic futures.

6

Divining Adaptation

There is a kind of map that I had several versions of as a child. Sometimes they were posters, sometimes they were puzzles, but they always covered each state with a piece of food: Idaho had its potato; Texas, its longhorn steer; California, its bunch of grapes; Wisconsin, its piece of cheese. Despite their cartoonishness, such maps are remarkable artifacts. They map a real living landscape, the actual habitat of billions of corn plants, tens of millions of cattle and orange trees. At the same time, they naturalize a landscape built on fragile, contingent social structures, a hotchpotch of indigenous American and global techniques and organisms assembled by imperial and commercial structures and spread over violently appropriated territory. The organisms pictured are commodities; most of them also reached their current habitats through trade networks—their ancestors purchased from catalogs and at auctions.

Making this landscape seem natural, making Wisconsin seem like "Dairyland" and California the natural home of French grapes, has taken an enormous amount of work. It is easiest to see this work by going back to a time before it seemed to be complete. In antebellum New York, the kinds of regional agricultural reputations displayed so confidently in my twentieth-century map were still markedly unstable, made so by a disorderly and rapidly shifting landscape. After the Erie Canal opened, first-growth forests were hacked down and replaced by fields and eroded soil from newly plowed hillsides silted up rivers. Old soils in the Hudson Valley rebelled at their former crops of wheat, even as newly uncovered layers of decomposed leaves in western New York sprouted what would become Genesee wheat, a new global good. At the same time, new roads, canals, and railways made previously loosely connected landscapes sharply relevant to each other. Farmers in western New

York, then Ohio, Illinois, and Michigan, undercut the grain prices of eastern farmers, while newly numerous mouths in New York City clamored for eggs, milk, fruit, and cheese.

In the midst of these changes, individual New Yorkers making individual decisions reorganized their state into regions of food like those on the map described above, occupying or obliterating older Haudenosaunee food landscapes. A wheat region formed around Rochester in the west, and the Mohawk Valley became a recognized cheese region. The farmers of the lowlands around the Hudson shifted to hay and beef, farmers in Otsego County turned more and more to hops, while farmers in the Catskills and the Southern Tier produced butter.[1] By the mid-1830s, even the way that farmers ate had changed. Flour from the Genesee Valley began to appear in Hudson Valley stores soon after the canal opened; by 1837, farmers in inland Maine bought flour from distant western farms and mills.[2]

We sometimes tell the story of regionalization in a way that implies a certain inexorability. Farmers, perceiving the different natural capacities of the landscape, this story goes, rearranged it to meet the demands of hungry markets. Among improvers, this sense of inexorability was stronger. They were convinced that they were treading a path laid out by unchanging natural laws. They expected that, like Great Britain, they would soon have regions of cheese, meat, hops, or fruit and that the place of those regions was predetermined. On the ground, however, the process of regionalization was chaotic and often painful. The landscape differed radically from its British models. Economic information was fragmentary, and visions of the future were both plentiful and hard to seize. While some ventures, like the hop boom, made instant insecure fortunes, others, like the Merinos, simmered for decades, boomed, and then withered in the face of new taxes. Claims of natural regional "adaptation" to particular crops were common, and the effort to persuasively predict future agricultural regions would become a major improving project during the 1830s and 1840s, one that would outlast the *multicaulis* bubble, even as it depended on some of the same kinds of storytelling.

Improvers perceived nature as functionally "adapted" to particular purposes, that is, as divinely and intentionally constructed for trade. Making claims about the destiny of particular regions required myriad acts of imagination, interpretation, persuasion, and disciplined performance.[3] To understand them, we will follow the struggles of one performer, Zadock Pratt, in his transparent bid for the reputation of Prattsville. Pratt's machinations reveal a wider culture of economic storytelling and show how the seemingly top-down visions of the state could be composed of a mosaic of booster claims.[4] At the same time, it shows how accounts of divinely intended regionalization could be used to conceal the labor and skill needed to create valued agricultural environments. In Pratt's case, they hid a landscape of market development built on women's expertise.

*

Perched on a hill in the Catskills, 140 miles north of New York City, Pratt's Rock is one of the stranger places in New York State. A sort of tiny precursor to Mount Rushmore, it was commissioned by the industrialist, agriculturalist, and Democratic congressman Zadock Pratt. Between 1843 and 1863, Pratt's carver cut a dozen emblems into the hillside. Each symbolized one of his accomplishments: among them, a fist clutching a scroll inscribed with the words "Bureau of Statistics," which Pratt had been instrumental in founding; a muscular arm with a hammer marking his identification with the mechanic he had once been and the men he had employed; an image of the kind of horse that he bred; the profile of his son, killed in the Civil War; and a deep hole that was to have been his tomb but which leaked and so is empty. Pratt's own face, in stony profile, dominates the top of the hill.

If Pratt's name is obscure now, he and many of his contemporaries never expected it to be. His rapid ascent from humble origins made him an excellent subject for homilies on the possibilities of the new economy, and his public identification as a workingman made him a splendid candidate for the Democratic Party.[5] That same ascent had lifted Pratt through the New York rural hierarchy. His father had been an evicted Livingston tenant farmer turned part-time tanner; Pratt himself would retire from industry having employed, he claimed, forty thousand men and paid them $2.5 million. When he launched himself into agricultural improvement in the 1840s, he simultaneously performed rural retirement, gained political exposure, and forwarded a pragmatic local boosterism. The boosterism is embodied in two of the carvings on Pratt's Rock: a hemlock tree and an inscription that read, "On the farm opposite, 224 pounds of butter were made from each cow from eight cows in one season."[6] These signs marked a shift in the surrounding mountains and a claim that Pratt was making about the nature of that shift.

The hemlock carving commemorated a vanished landscape, the source of most of Pratt's wealth. Back in 1824, thick hemlock forests had attracted Pratt to the Catskills. Their bark was rich in tannins, necessary for leather production, but since the bark lost its tannin if moved, it represented a kind of wealth that was rooted in place. Dozens of tanneries sprang up in dozens of new tanning towns, processing hides shipped up the Hudson from South America. It was mostly the workers in Pratt's three-hundred-vat tannery who populated Prattsville, renamed for Pratt in the 1830s. By the early 1840s, though, the

FIGURE 13. Endicott and Co. print of Prattsville. Note Zadock Pratt himself, seated under the hemlock in the foreground. Courtesy of the American Antiquarian Society.

Catskills had started to run out of hemlocks—Pratt himself claimed to have deforested ten thousand acres. Surrounded by stumpy, hilly land, Prattsville had evidently outlived its purpose.

This was a problem for Pratt. While his tanneries could move to the Adirondacks, he was president of the town bank; he had paid for half the building of the local academy and half of each of the local churches. He owned the newspaper and represented the region in Congress.[7] To preserve his town, Pratt began to encourage what he described as the next natural stage of its progression—you can see what that was by looking at a print that he probably commissioned from Endicott and Co. in 1848. Pratt himself appears reclining under a hemlock in the bottom corner, the author of the scene. In the center we see Prattsville itself, picturesquely seated among hills denuded of hemlock and unrealistically cleaned of their unsightly stumps. In the foreground is Prattsville's imagined future—pastures of horses and cattle. The inscription on Pratt's Rock was intended as a more specific prophecy and proof. The future of Prattsville was butter.

In claiming adaptation to butter, Pratt aimed not at a general quality, but at a particularly valuable rival: "Goshen" or "Orange County Butter," by then famed for its "freshness and delicious flavor," its stability in hot weather, and its ability to command twice the price of "western butter."[8] It was easy to see why Orange County might be a good target for developmental fantasies. Orange County bank notes, known as "butter money," were tinted butter yellow and traded at par in New York City, unlike notes from other country banks.⁹ A comic tourist guide to New York City written in 1828 claimed that when New Yorkers had grown suspicious of country bank money, the Goshen bank directors "did incontinently determine to starve the good citizens of New York into swallowing their notes by cutting off their supplies of Goshen butter." In the consequent "horrible scarcity," New Yorkers, "actually reduced to the necessity of substituting Philadelphia butter," capitulated, accepting country notes now colored a triumphant butter yellow. Farmers in Orange County had profited early from good land on the Hudson and from marketing innovations devised with the commission house of Van Auken and Cook.¹⁰ Looking back on his boyhood, dairyman K. P. McGlincy recollected that it was "almost an impossibility to sell a pound of butter in the New York market that was made west." Like many dairymen, McGlincy put his western New York butter in Orange County buckets.¹¹

Pratt took aim at Orange County's butter reputation by working his way into agricultural improvement, in particular into the system of county fairs. Since he supplied the funds for the county prizes, Pratt probably had little difficulty getting himself elected president of the Greene County Agricultural Society in 1844 and then chairman of the state fair butter committee in 1846. Through public experiments, speeches, articles, and exhibits at agricultural fairs, he would argue for decades that Prattsville was part of a region "peculiarly adapted" to butter production. In an article for the patent office's annual yearbook, he would sketch the borders of butter land with great specificity: "That belt of territory varying from twenty-five to fifty miles in width, which begins with Orange County, near the city of the New York, and extends from the Hudson River in a northwestwardly direction, perhaps one hundred or one hundred and twenty-five miles, into the heart of the state."¹²

Examining the ways that Pratt worked to build a reputation for butter can help us denaturalize the commodity map—making clear that shifts in crops were not straightforward responses to markets or climates but the results of human choices, constrained, certainly, by climate, soils, labor conditions, and markets, but also driven by particular persuasive visions of an agricultural future. To do so, we must first explore the meaning of the word that centered much of improving thought: "adaptation."

Adaptation and the Function of Landscape

"Adaptation," the closeness of fit between structure and purpose, is a word so frequently used in improving texts that it becomes almost invisible. Often, it refers to human acts of design: using the laws of mechanics, plows could be better adapted to the soils they were to overturn, and animals and plants could be better adapted to their places through selection, hybridization, and acclimatization. As with the "principles" of agricultural machinery, adaptation had a double meaning. Seen in wild organisms—in the fitness of flattened teeth to grinding grass or hooked seeds to the fur of unwary animals—it offered evidence of a different designer: the divine watchmaker most famously described in the work of the natural theologian William Paley. Paley's work had been easily absorbed into the postmillennial evangelicalism that dominated antebellum American Protestantism, infusing scientific and religious writing with references to the "sermons" that could be read in stones or in the "book of nature."[13] Pratt's common turn of phrase, "adapted to pasturage," referred to this kind of design, an unquestioned structure underpinning the material world.

Just as American naturalists saw the wing of the bat or the human eye as structures demonstrating a purposeful design, improving New Yorkers expected to uncover providential intention built into their landscape. The most frequently trumpeted evidence of this in antebellum New York was the extraordinary crossways break in the mountains that became the Erie Canal.[14] Canal projector Gideon Hawley exclaimed in his first promotional letter, "It appears as if the Author of nature, in forming Lake Erie with its large head of water into a reservoir; and the limestone ridge into an inclined plane, had in prospect a large canal to connect the Atlantic and continental seas: to be completed, at some period, by the ingenuity and industry of man."[15] New York was a half-built landscape awaiting the builders of canals. It was in this standard vein that Governor William Seward would argue in 1839 that Nature had herself demanded three railroad lines. "The policy of our state is so legibly written upon its surface," he told the state legislature, "that to err in reading or to be slothful in pursuing it, is equally unpardonable."[16] Claims of destiny were not uncontested; Seward's "natural" railroads led to the home counties of his political allies (as the Albany papers pointed out).[17] However, as battles for canals and railroads saturated political rhetoric and threw different parts of New York into competition, a wider array of New Yorkers became used to manipulating the language of providential regional destiny.

Adaptation of particular places to particular functions was an expected feature of both wild and cultivated landscapes. Naturalists' books—"floras" and "faunas"—described bounded nations of organisms, created for a particular place and adapted to it and to each other.[18] Improvers drew on a related concept. "Most plants and animals," the *Dairyman's Manual* explained in 1839, "have their natural zone beyond which they deteriorate or do not live. The potato, for instance, deteriorates south of latitude 40."[19] Such ideas were not confined to textbooks; an 1837 advertiser hoping to sell two farms "adapted to wheat" in Chemung County worked with the same assumptions.[20]

The idea of different agricultural regions echoed on a larger scale the landscape of mixed farming that farmers already knew well. Given a mixture of slope and bottom land, and a range of exposures to the sun, farmers could create the dozens of kinds of places needed for local food production—a stand of sugar maples for early profit, a gentle slope for grain fields, a drained gravely space for an orchard, a fertile lowland meadow where hay could be cut, and steeper slopes with fast-moving springs where sheep, cows, and children could be raised free of disease. Surveyors assessing land for the Holland Land Company in the 1790s had kept this in mind, looking automatically for undulating "diversified" land.[21] Canals and railroads seemed to offer a chance to make the same moves on a grander scale, to arrange an entire state like a mixed farm.

To perceive an inherently purposeful specialized landscape was also to see a landscape intended for commerce. Trade came from difference; the different functions of landscape were intended to create a uniting market. The British agricultural geologist and chemist James F. W. Johnston expressed a common sentiment in writing, "All study of natural history, and of physical geography, shows that the Deity intended that one part of the world should minister to the wants of another, and that they should mutually interchange commodities and productions." For Johnston, writing in a tradition stretching back through Adam Smith and David Hume to Plato and Plutarch, this became an argument for free trade. "Perfect freedom of commercial intercourse," he wrote, "is consistent with, and pointed to, by all the arrangements and productions of soils, climates, and seasons."[22] Some American improvers, by contrast, suggested that the different landscapes in the United States would bind the Union together as an internally self-supported system. "Each district of our country," the author of the *Dairyman's Manual* commented, "seems adapted to some peculiar culture, rendering each dependent upon the others, as if to unite us closer in the bonds of fellowship and good feeling."[23] Differentiation could thus justify economic nationalism and free trade at once.

Unlike the flora and fauna of natural history, however, the adapted landscapes of improvement were incomplete. Improvers did not simply observe natural laws; they were also to carry them out. In a speech to the Greene County Agricultural Society, Zadock Pratt told his audience that the mission of improvement was to discover ways to "produce all the results that the Creator ever designed to put within our reach."²⁴ Where naturalists described existing places, canal projectors and agricultural improvers looked at one landscape and saw another. "Giving a stretch to the mind, into the womb of futurity," Elkanah Watson wrote in a tract that would help provoke the national canal boom, "I saw those fertile regions ... overspread with millions of free men. Blessed with various climates,—enjoying every variety of soil,—and commanding the boldest inland navigation on this globe; clouded with sails, directing their course towards canals, alive with boats, passing and repassing."²⁵ Likewise, gazing at the Adirondack Mountains, the improving scientist Ebenezer Emmons imagined the "herds of cattle and flocks of sheep" that "may one day give life and animation where the silence of the day is broken only by the rustling of the wind through an unbroken forest."²⁶

Americans were not unusual in projecting elaborate, invisible, future landscapes over quite different real places. As European empires expanded and European landholders appropriated commons and "wastelands" at home, such acts of imagination had become commonplaces of the global improving project. Thus, in 1840, Charles Bruce, superintendent of tea on the Assam frontier, had assured the Agricultural and Horticultural Society of India that Assam would soon rival China in the production of global luxuries. "The whole of the country is capable of being turned into a vast Tea garden," he promised, "the soil being excellent and well adapted for the growth of tea."²⁷

Almost invariably, accounts of the hidden economic function of landscape also justified the subjection, removal, or extermination of people not included in the projected future. Charles Bruce's vision of a tea garden on the Assam frontier depended on violent annexation.²⁸ Likewise, when hopeful New Yorkers called their towns things like "Mount Merino," "Wheatland," or "Butter Hill," they knew themselves to be overwriting a landscape very recently occupied by Haudenosaunee and Algonquian peoples. Mount Merino, named for the five hundred sheep sent there during the Merino mania was a hopeful retelling of Oriskany Creek, which the Oneida named for the nettles that grew there.²⁹ The continued resistance of the Seneca in western New York kept the state legislature actively scrambling to extinguish Indian land rights and constantly renaming and claiming their land.³⁰ More broadly, assumptions about the connection between particular places and kinds of bodies ran deep into antebellum American and imperial British accounts of white adaptation to "temperate" climes (and the consequent inevitability of settler societies), of Black adaptation to labor in agonizing heat, or the ill-suitedness of Native Americans to survive in their own land.³¹ When Pratt described cattle of European descent as "adapted" to the Catskills, he was implicitly making claims about other bodies of European descent as well.

Seeing Adaptation

How were the functions of landscape to be perceived? The very earliest settlers in western New York had an enormous advantage in this respect, immediately appropriating Haudenosaunee knowledge and Haudenosaunee towns and pathways.[32] The well-drained bottomlands identified as the best land by the Haudenosaunee were the first to go to new settlers: Mohawk Valley settlers planted wheat fields at the Mohawk's "Niskayuna" or "Extensive Corn Flats."[33] Settlers building the city of Aurora on the site of the Indian town they had called "Peachtown" knew something of its capacities, and names like Ga-Jik-ha'no, "place of salt" in Tuscarora, had provoked a successful developer's salt rush to the lands around Onondaga Lake, in the 1790s.[34]

For later settlers, however, the landscape was more opaque. Retrospectively, regionalization seems to be a simple matter of slotting crops into place based on climate and soil, but what should go where? While improvers expected to follow a British path, British observers themselves found the American landscape hard to read—a confusing tangle of forest interspersed with clearings, burned stumps, and roots and the occasional comforting field. Moreover, in the United States, where labor was scarce, different soils had different capacities. "The minute attention given to each atom of earth in Europe, requires a more numerous population than ours," the Rensselaer Institute geologist Amos Eaton wrote. "These, and other circumstances," he claimed, "render the European treatises on agriculture of little use to our practical farmers."[35]

Even for Americans, the "real qualities" of a landscape were elusive. At ground level, for example, climate was localized and plastic. As farmers cleared forests, layers of shade trees falling away allowed the sun to bake soil it had never touched before. As they approached the Great Lakes, farmers encountered unfamiliar seasons and new blanketing layers of snow. Their early disorientation was heightened briefly by the June snowstorms of the "Year without a Summer," 1816–17, and more lastingly by the ebbing away of the Little Ice Age.[36] Such transformations were easily integrated into learned theories of climate change that encouraged Americans to believe that cultivation would have a warming, softening effect on the landscape, revealing new possibilities over time.[37] Seemingly more solid, soils shifted as well. The layers of black mold left by decades of decaying leaves might be skimmed away by erosion in only few years. "It is when that coat of manure is gone, and the land worn out by constant cropping," argued the early geologist William Maclure, "that the soil shews its fertility."[38] Existing crop cover could be deceptive—looking at wheat growing in the southern Genesee valley, improving chemist

Ebenezer Emmons could condemn it as not "the natural crop."[39] It became easier to believe in a natural sequence of crops, perhaps, when wheat culture rose and fell as fast as it did in the Mohawk Valley, collapsing in only a few decades.[40]

New Yorkers also could see that internal improvements seemed to rearrange the capacities of cultivated nature. Having come of age as canals and railroads sprawled across New York, farmer Isaac Carr would write urgently to his son, settling on a new farm in Wisconsin in 1857: "Can you see anything that seems likely to make it Become anything in the future is there likely to Be A RR there or is it in the Center of the town or is it near the County Seat or is there anything Else that will be likely to make it Become anything."[41] As Carr knew, regardless of the soil, such features could determine whether an agricultural landscape lived or died. Closer to the city, indeed, proximity to urban markets justified expansion into lands that seemed, at first glance, impossible. If the unwelcoming sand flats of Long Island could bloom and fruit with a regular infusion of city horse manure and dried night soil, perhaps other landscapes were flexible too. While "tanlords," like Pratt, looked to repurpose their cutover, improvers like Jesse Buel stretched their farms to meadows and plains that early surveyors had read as barren. Iron share ploughs, cutting more easily through sticky clays and new fertilizers added to hungry sands extended the range of soils available to farming. Even unhealthy bottomland, thick with mosquitoes and squishy with stagnant, potentially miasmatic water could be drained or ditched to grow hay.

Moreover, the shifting array of available species, breeds, and varieties—the sorghum, sugar beets, silk, Berkshire pigs, and Durham cattle promoted by the agricultural journals—invested each soil with a sense of alternative possibility that some improvers found overwhelming. Pratt publicly wrestled with this last aspect of adaptation. At the Greene County Fair in 1845, he argued against the fashionable importation of "foreign" breeds and varieties. "Is it not better, as a general principle," he asked his audiences, "both as to animals and vegetables, to choose and improve the best of such as already are adapted to our climate and soil?" He argued that breeds and varieties should be produced gradually and locally from specimens grown "upon our own or on a neighboring farm" so that when moved, they would not be forced "by the irreversible laws of soil and climate to change their character and adapt themselves to their new locality." Tobacco lost its scent and corn its prolific nature if moved, he pointed out, "As well might you attempt to transplant the beautiful hemlock of our mountains, where the God of Nature placed them, to regions designed for the live oak and the olive, as to neglect the peculiar varieties of grain that our own regions produce in perfection and cultivate those whose distinctive properties are the result of a different soil and climate."[42] Pratt's address suggested a sort of democratic agricultural science, in which each farmer had access to the materials of success without resort to the expensive imported animals being displayed in the cattle fancy. The article made the rounds of the agricultural journals and was reprinted entirely by the Anti-Renter *Albany Freeholder*. Its popularity reveals that while adaptation was a universal concern, its meaning was not fixed.

The Gender of Good Butter

To match crops to the landscape, farm families had to adapt themselves as well. A shift to a new regional agriculture was not a light matter. Concentrating on a smaller number of species could fundamentally restructure family relationships and family labor. This was particularly true in the case of butter, as Nancy Osterud, Sally McMurry, and Joan Jensen have shown. Since "butter and egg money" often went to women, increased butter production sometimes gave women in dairying districts new grounds on which to negotiate their status at home. At the same time, a greater volume of raw milk in the heat of the summer meant a more intense daily race against spoilage and an enormous amount of labor poured into the work of churning. Dairies elaborated their technology in response. Where at the beginning of the century, women had made butter for home consumption by beating cream in a bowl with a spoon, by 1850 some were using massive churns, holding up to fifty gallons of cream and run by treadmills powered by sheep, dogs, or calves. But sheep and dogs could not skim cream or wash hundreds of milk pans or work out the last skim milk from the butter and pack it in firkins.[43] Farm women complained as dairying occupied a longer and longer piece of each day and tried, in letters to the agricultural journals, to get milking recast as men's work (a move that men resisted).[44] Dairy work also involved fewer tasks suitable for children than home fabric production had, intensifying reliance on older daughters and hired help.[45] Confirming this, Martin Bruegel has found dairymaids signing labor contracts in the 1820s, before similar contracts for male laborers. Indeed, improvers saw particular potential for profit in dairy because wages were held down by gendered expectations.[46]

Though women's work has been better studied in this instance, shifting to commercial butter production also altered men's work.[47] It meant putting more of the farm into pasture and meadow. It meant choosing grasses and clovers that could be dried easily for winter and produce sweet and copious milk—New Yorkers claimed to be able to taste the impurities that came when cattle were fed on turnips or garlicky grass and the freshness that came with the first grasses of May. It often meant building new structures—a larger barn for winter fodder and shelter or a dairy house. Keeping cattle close enough to the house for regular milkings meant hauling tons of manure away later. And feeding them certainly meant a much greater pulse of labor in the haying season, requiring the help of neighbors, laborers, or machines, and an anxious winter watching the hay supply diminish. It might mean finding new animals, particularly those with Durham or Devon blood, that were "thriftier," giving more milk. Such cows might be judged by the "lactometer," a rack of glass tubes into which a sample of milk from each animal could be poured—the thickness of the cream line in each the next morning marked the value of each animal. Improving farmers also introduced an array of new feeding machines—hay presses and "masticators" that chewed farm refuse to make it palatable to cattle.

Despite the profound shifts involved, many farmers hoped to become part of a butter region. Easier to move than fruit or vegetables or hay, butter offered access to a growing cash market paid by a growing army of contractors, encouraged by Orange County's success. Increasingly it was reported as providing a more stable income than grain crops. "In the grazing counties," declared Gordon's *Gazetteer*, "the buildings are generally of a better character, than in the grain growing districts."[48] Given such promises, a reputation as "a grazing county" was perhaps worth fighting for.

A Butter Battle

The Orange County butter debate of 1847 was precipitated by an insult. A naval inspector, receiving a test batch of butter from Chemung County, condescendingly dismissed it as "excellent butter for Northern New York" and lamented "the inability of any butter to stand the test of foreign climates and of time, that was not made in Orange County." B. P. Johnson, the secretary of the state society, treated the situation as an emergency. At stake was not only the navy contract for sixty thousand pounds of butter a year but also the expanding global market for hot-weather butter in India, the West Indies, China, and the American South.[49] In as many venues as Johnson could command—the state society's *Transactions*, the *Cultivator*, the *Genesee Farmer*, the *American Agriculturist*, and finally the US Patent Office *Annual Report*—he informed the navy that "the region peculiarly adapted to the production of good butter in this State" extended beyond Orange County's borders.[50]

"It is not believed that there is any such peculiarity connected with Orange County," Johnson would write, in lines that would undoubtedly have pleased Pratt, "as to give it pre-eminence over other counties in the Catskill

Mountain range, and some other localities in the State."[51] In fact, Johnson was likely in conversation with Pratt, who had been his fellow judge of the butter competition the previous year and with whom he maintained a continuing connection; the year that Pratt was society president, Johnson was the key speaker at the Greene County Fair.[52]

To make his case, Johnson called for testimonials: letters from dairy farmers and butter dealers. These showed that Pratt was not alone in shooting for Orange County's reputation. From Otsego County, J. W. Ball wrote, for example, "If our navy lack for Orange County butter, let them pay the Otsego dairymen navy prices and I will guarantee them butter that will keep the world over." Likewise, Joseph E. Bloomfield promised that Oswego County would make "as good 'Navy butter' as Orange county" as soon as the Harlem and Hudson Railway was finished.[53]

However, the letters also reveal that the boundaries of "Orange County" had, for market purposes, already spread far beyond the county line. "I think that not one-third of the butter sold in market as 'Orange County,' is made in that locality," wrote one western dealer.[54] In fact, the source of "Orange County" butter's valuable qualities was perhaps not Orange County. Butter dealers usually offered up two explanations. The first was the familiar argument about local adaptation: "Much of the southern tier of counties, and also of the central and northern portions of the State of New York," wrote Hawley, a butter dealer from the Southern Tier, "will, when well cultivated, produce the variety of grasses necessary to give butter the peculiar flavor and aroma of Orange County." However, in almost the same breath, dealers argued that Orange County butter was not a place but a technique. "The term, Orange County butter, seems to be misunderstood," Hawley wrote, "It does not mean (as I understand it) the locality where made, but a peculiar method of manufacture."[55]

Where improvers described the butter landscape as a physical place, dealers saw it as a set of relationships with expert women. Butter came to dealers in eighty-pound firkins branded with men's names, but dealers got them through annually renewed contracts usually negotiated with dairywomen.[56] They bargained over points of technique: oversalting; cream left to stand too long or put in pans that were not "perfectly clean"; butter with stray hairs or the faintest taste of dung; butter allowed to touch the lid of the barrel, packed in the wrong kind of wood, or gone rancid from poor storage. As such, quality was embodied in Orange County dairywomen themselves and could be expected to spread as they migrated across the Southern Tier. "A Minisink [Orange County] dairy woman that I know," another dealer wrote "could make a dairy . . . in some western spot, that would be a facsimile in eating and keeping, with that she now makes in that place."[57] Hawley claimed that another Minisink woman had accomplished this feat already. The sixty firkins from her new dairy in Broome County were "the best she ever made."[58]

As Orange County dairywomen moved west, their reputations moved with them; what seems like fraudulent mislabeling of western butter to us may have seemed like the acknowledgment of skill to them. Eager to break into the growing global markets for hot-weather butter, dealers wanted high production volume. To expand the skilled buttermaking region, they promoted good practice, as in the circular "to the Dairywomen of New York," appearing in agricultural journals in 1838. Dealers also actively worked to disrupt entirely place-based narratives of value.[59] Responding to Johnson, Hawley told a clearly well-worn anecdote: dining with "a well-known gourmand in New York" in 1847, Hawley commented on the excellence of the butter. His host responded that "such butter could not be made out of Orange County." The firkin being brought up, Hawley triumphantly pointed to the brand "John Holbert, (premium)" the mark of a prize-winning Chemung County dairy.[60]

Land speculators and local boosters, however, concerned themselves with the male spaces of dairy production, searching for the edge of the "dairy district."[61] In the *Dairyman's Manual*, William Townsend had seen "the district of country along the north lines of Pennsylvania and New Jersey, embracing the northern borders of the Mohawk valley, and stretching from Lake Erie into New England," as a space "destined to become . . . the great dairy district of the Union, nay, of the American hemisphere."[62] Writing from Oswego, Joseph E. Bloomfield maintained, "The true Dairy region of the United States is confined mainly to the streams and side hills of the several spurs of the Alleghany mountains that drain into the Atlantic."[63] In a rapidly fluctuating land market, boosters hoped that a reputation as good dairy land could translate into durable land values; looking back on the late 1840s, X. A. Willard remembered that in Herkimer County, a cheese district at the edge of the Adirondack Mountains, dairy farm prices had risen to fifty to sixty dollars per acre. "The dairy industry was esteemed the best that farmers could follow," he recalled, and "as the dairy districts were then supposed to be of quite limited extent, the dairy farmers of Central New-York not unfrequently plumed themselves upon having about 'the whole of a good thing.'"[64] Such claims could also be used in the endless squabble for internal improvement funding or to entice investors in railroad stock. Reporting to the state society from "the Luxurious Valley of the Chemung," E. C. Frost and A. J. Wynkoop described lands "admirably adapted to grazing." However, Chemung's promise required, "the completition [*sic*] of the New York and Erie Railroad."[65] To railroads, dairy development was nontrivial; by 1845, two-fifths of the partially finished New York and Erie Railroad's income came from carrying cooled milk.[66]

In short, Pratt's effort to argue that his place was adapted to butter had become familiar—lots of people were trying to pry open the Orange County reputation. However, butter was not literally written in the rocks of Greene County until Pratt's carver cut it there. Indeed, Pratt would later admit, it was not immediately apparent that Prattsville's rocky hills were good for anything at all. The stranger "whose ideas of all good farming are indissolubly associated with fields of smooth or gently varied surface, clad in unbroken herbage, or tilled to garden cleanliness," Pratt noted, "will find such anticipations most rudely shocked."[67] Moreover, the hills around Prattsville still supported the survivors of a previous vision of the agricultural future; between 1821 and 1835, a second Merino sheep boom meant that in 1835 sheep outnumbered cattle five to two.[68]

Like many New Yorkers, Pratt found himself arguing for the value of a conventionally marginal space. To do so, he relied on forms of proof that matter, not because they were innovative, but because they were standard: tree knowledge, geological knowledge, and publicly performed experiment.

Trees

While dreaming of future farms and cities, what New Yorkers actually saw, mostly, was huge trees.[69] The modern forests of upstate New York, full of teenage trees from the twentieth century, give a poor idea of this past landscape. Working on his brother's farm on the shores of Lake Ontario, Herman Coons reported, "an impenetrable forest in primeval state" full of "timber of gigantic growth; principally Beach [*sic*] and Hemlock of which species I have measured trees 15 ft in circumferance [*sic*] and proportionally tall."[70] For many, "farm making," cutting new farms from this forest, was both their main source of income and the main sink of labor. While bulky lumber could not be moved without river access, trees could be burned to make potash, a blessedly light and easily transported commodity used to make soap, glass, and gunpowder in tree-scarce western Europe.[71] Even after trees were cut, their place was still marked. Passing through the Hudson Valley in 1816, British traveler Francis Hall saw "many forests whose leafless trunks, blackened with fire, rose above the underwood, like lonely columns, while their flat-wreathed roots lay scattered about, not unlike the capitals of Egyptian architecture."[72]

It is perhaps not surprising then that to Pratt the hemlock tree carved in Pratt's Rock stood as evidence for the butter land beneath. In a report to the patent office, he invoked "the old tanners" who maintained, "that of all this region it is the hemlock lands which prove the best for butter-making and are capable of imparting the sweetest and richest flavor."[73] Such claims had deep roots. During the eighteenth century, Americans had produced a general interpretive structure for valuing land using trees; oaks marking the best land and stunted pines the worst with gradations in between.[74] When agents advertising twenty thousand acres of West Virginia land to the readers of the *Albany Freeholder* promised that it was "covered with a luxurious growth of Walnuts, Oaks, Hickories, Sugar, Paw Paw, &c.," they expected readers to understand that oak promised good soil; hickory and pawpaw added the promise of local subsistence; and sugar maple, sweetness and a quickly marketable good.[75]

New Yorkers deployed this expertise as a matter of course. Passing the Manitou Islands on a trip to Lake Michigan in 1845, Alson Ward dismissed them as barren, since "there is nothing but pines on the shore," though signs of a wooden railroad running back from the beach gave him hopes that the land was better in the interior.[76] Writing home about his search for a farm in Michigan in the 1820s, Seneca County farmer Lyman Chandler worked on the same plan. "Nature has divided the land into three distinct kinds," he told his parents. Chandler hunted for oak openings "verry [sic] moderately uneven, dry and pleasant to build on," which had the best water and were "better adapted to wheat than grass," but avoided "black walnuts and Witewood [sic]," which were "generally not good to build on account of poor water."[77]

Using trees to evaluate land also had precedent in learned tradition. Virgil's *Georgics* gave comprehensive lists of Roman soils and the trees that grew in them; readers could learn to find meager clay (at least in Italy) where "wild olive-shoots o'erspread the ground."[78] Classical familiarity may have eased the movement of tree knowledge into formal texts. The surveyors of the Holland Purchase in the 1790s had used it in their ratings of first-, second-, and third-rate land; many improvers did the same.[79] At the first state society meeting, the Pennsylvanian physician William Darlington integrated evaluation methods based on trees (rendered formal with Latin botanical names) with improving soil treatments. "The soils indicated by a natural growth of black oak, (quercus tinctoria,) walnut (jugulans nigra,) and poplar (liriodendron)," he informed the assembled members by letter, "are generally most signally benefited by the use of lime."[80] As the state society began to include questions about tree cover in its report forms, and its members mentioned trees in their accounts of experiments, tree knowledge became a standard element of claims like Pratt's.

Pratt was not exactly a settler, but he did know trees; he would cast his deforestation of Greene County as an act at once of harvest and of farm making; "Having thus done his best to harvest and convert into form for human use the crop sown by nature over these mountains and valleys," wrote someone who was likely Pratt himself, "it seemed to [Pratt] almost a matter of duty to replace the hemlock, if possible, with other crops; to render the land not less productive in cultivation than it had been as a wilderness."[81] Here, Pratt was behaving conventionally. However, there was a reason he referred to old tanners rather than old farmers. Hemlock never appeared as a marker of good land in advertisements or letters home; it marked the lands considered to be poorest, indicating steep slopes and shallow soils.[82] (In 1855, perhaps to strengthen his claims to a successor landscape, Pratt planted the "walnuts, chestnuts, beechnuts, butternuts, black walnuts . . . hard-maple . . . and locust" that would mark his land as good.)[83] This tendency in hemlock may come from sensitivity to disruption. Burned or cut, hemlocks tend to be replaced by maples or beech. They survive in poor soils because such soils are rarely cleared for culture. Hemlocks' presence was perhaps a relic of indigenous land valuation; they may have marked bad land because burning and cultivation had already weeded them from the good.[84]

If hemlock marked Pratt's land as bad, the expansion of dairy was also changing the meaning of previously "bad" land. Since most nineteenth-century farmers chose to graze their animals on their poorest land, reimagining cutover hemlock as perfect pasturage was not difficult.[85] Rocky outcrops, it could be argued, wore at cattle's hooves, keeping them from curving agonizingly. Slopes meant cleaner, faster-moving water, fewer flies, and well-drained soil, preventing foot rot. Pratt was far from the only person redescribing marginal land. Edward Peck, promoting land sales on rapidly developing but infertile Long Island argued that if lime was applied, "it is a fact, that in many parts of the country those lands called oak barrens [covered in "scrub" oak] and neglected for a time, have been found to be the best wheat lands."[86] However, a whole farm of hemlock stumps on a steep slope was something New Yorkers needed to grow used to. To be convincing, Pratt needed other kinds of evidence.

Rocks

To buttress their claims, Pratt and Johnson both dwelled on some knowledge more recently minted by state geologist William Mather. "All the country containing the Catskill division of rock . . . is admirably adapted for grazing," Pratt quoted enthusiastically from the state geological survey, "both for cattle and sheep, and the finest sweet grass and cold springs, offer as great facilities for making excellent butter, as the world affords."[87] To Pratt's readers, geology would have been a familiar way of framing land values; accounts of experiments already often opened with geological commentary. So did more general texts: Gordon's *Gazetteer* of 1836 explained that "the transition soils of Orange County . . . are, as the lime or slate prevails, adapted to winter grains, or grass and summer crops."[88] In fact, by the 1830s, geology had become fashionable, even sexy. Geological lectures were billed on the lyceum circuit, geological specimens adorned parlors, and Hudson River school painter Thomas Cole littered paintings with erratic boulders, representing catastrophic geological change.[89]

If tree knowledge emerged from farm making, geological knowledge welled up from internal improvements. In the 1820s and 1830s, diggers were riddling the state with holes—marl pits, quarries, mines, wells, and privies exposed strata and turned up fossils. Transecting the state, cuts made for canals and railroads would become the foundations upon which a new class of professionalizing geologists built their maps. Many learned their profession on the Rensselaer School Flotilla, a string of canal boats turned into museums and classrooms, gently towed through the strata revealed by the Erie Canal.[90]

These same cuttings made strata and soils readily available to more casual viewing and specimen collection. Sent to Lockport to collect specimens at a new canal cut in 1841, professional collector John Smith complained to his employer, survey geologist James Hall, that "everything like a specemin [sic] is picked up by the workmen and offered for sale." (Worse, he mourned, the temperance-minded Irish laborers could no longer be bribed with "the price of a quart now and then.")[91] This enthusiasm frustrated professionalizing scientists: in 1846, Jacob Bailey complained to Asa Gray, about "the small fry who will consume hours in describing the thickness, composition &c. of every layer of a clay bank near their village."[92] In the same pamphlet where he argued for the misunderstood virtues of scrub oak barrens, Edward Peck also had to reassure readers used to reading soil quality through railroad cuttings. "In passing through on the Railroad . . . the impression left on the mind to an ordinary observer, is erroneous," he wrote. The excavations were too deep, passing through "the proper coverings of the Island" and into deeper layers, "so that the whole impression left on the mind is made by sand and gravel." In fact, he assured readers, "the geological structure of this portion of the Island is almost precisely the same as that in the vicinity of East New York."[93]

As Mather's report shows, geological knowledge had also started to come from the state. Following in the footsteps of the Rensselaer-sponsored surveys of the 1820s, in 1836 the state legislature began to sponsor geology at a scale that, members of the British Parliament would comment, was more suited to a nation.⁹⁴ The survey's ambitions appear in the recipient list for the thirty volumes of the natural history survey, including not only a glittering group of international scientific societies, but also Queen Victoria and the emperors of China, Austria, and Brazil.⁹⁵ In fact, New York geologists were well placed to make international waves. Like Erie Canal builders, they occupied a strategic spot. In New York a particularly complete series of strata had been flipped sideways and crushed together, offering a diversity then rivaled only by the British Isles.⁹⁶ Like the British, New York geologists leveraged this to become a center of geological taxonomy. In the state survey, Welsh tribal names like "Silurian" and "Cambrian" gave way to North American place names: the Champlain Group, the Helderberg Series, and the Erie Group.⁹⁷ Where the British Geological Survey employed a single geologist until 1839, New York State appropriations supported nine geologists, four draftsmen, a zoologist, a botanist, and two biologists' draftsmen.⁹⁸ Gradually extended to a natural history and agricultural survey and to the geological, natural history, and agricultural museums established near the statehouse, these efforts would ultimately cost the state of New York more than a million dollars.⁹⁹

Geology's powerful links to agricultural improvement became most palpable in the geological rooms at the state house, where the state agricultural society met before the legislature gave them their own rooms across the hall.¹⁰⁰ Improving texts regularly quoted the surveys; journals followed their progress and supported their funding, and improvers took up geology as an allied pursuit. Leading society figure Henry S. Randall typically developed a fossil collection and an assortment of correspondents, including leading figures at the geological survey.¹⁰¹ Geological science could also double as genteel ornament. Under the title "Making Farm Life Attractive," the *American Agriculturist* suggested that farmers make "a cabinet collection of every kind of rock on his land" and "a collection of fruit drawings in watercolors."¹⁰² New York geologists needed improving support particularly badly, the sequence of New York strata ruled out the mineral that New Yorkers wanted most—the anthracite coal booming in Pennsylvania.¹⁰³ To placate disappointed legislators, geologists pointed hastily to sources of agricultural value: peat, gypsum, lime, marl, and most importantly, soil.

Except when looking at a canyon wall or canal cutting, geologists experience strata not as vertically arranged layers of rock, but as areas of land—territories where strata come to the surface.¹⁰⁴ Different strata broke down into different soils, creating, geologists postulated, different zones of adaptation.¹⁰⁵ The first geological map of the United States, produced by William Maclure in 1807, had promised in its title to explain the "Nature and Fertility of Soils, by the Decomposition of the Different Classes of Rocks."[106] The Catskills had fallen into the "Primary" category: hard, crystalline granites from which clear, healthful water poured in abundance, but which broke down into sandy soils too slowly to create the navigable rivers needed for trade.

The decomposition of rocks into soils helped fit geology into a natural theological framework. Thomas Dick's much-reprinted *Christian Philosopher* traced the decomposition of rocks, broken down by ice, divided by lichens, and occupied by mosses and heath until at length "a mould is formed . . . capable of rewarding the labors of the cultivator."[107] In his geological textbook, Edward Hitchcock called this same process "a bright exhibition of benevolent design."[108] Like other progressive laws, this process could be forwarded by human action. According to Maclure, plows and harrows acted "in aid of nature's operations to reduce the particles of earth to a state more fit for vegetable production."[109] Surveying Rensselaer County, Amos Eaton had noted approvingly that a few farmers drove their new cast-iron plowshares over rocky knolls when plowing, "to break up at every ploughing a little of exposed rock," creating good soil, they claimed, within a few years.[110]

By the time of Pratt's push, maps of adaptation had become more complicated, partly because geologists had identified more strata, but partly because soils themselves moved in unexpected ways, something that geologists had begun to map with greater intensity. During the 1830s and 1840s, geologists fiercely debated "diluvial" phenomena—great erratic boulders hundreds of miles from their home strata, heaps of gravel, and long scratches in rock faces, all aligned north to south. Clearly something, perhaps vast glaciers or floods topped by boulder-carrying icebergs, had pushed soils south, mingling whole districts in massive eddies and ripples, now stilled. Fascinated by these colossal patterns, geologists described their function: softening the landscape for tillage, producing "a vast amount of new soil."[111] With the rise of uniformitarianism in the 1830s, geologists began to scrutinize these processes; moving sand dunes, mudflats, and floodplains seemed to hold the key to past geological change.

Soils were shifting quickly. Mather, the geologist Pratt quoted, reported that the sandbars of the Hudson had been growing noticeably for the past four years, that piers into the river were lengthened by feet every few seasons, and that the inhabitants of the town of Hudson had had to cut a mile-long canal in the growing island of mud near their town to let the ferry through. Mather knew that this was the effect of agriculture. "Every shower moves more or less earthy materials in the streams," he explained, "far more, than when the soil was covered by its native forests."[112] As dairying, tillage, and tannery operations climbed previously forested hills, eroded soil poured downhill and choked the river.[113]

Though hill farmers probably interpreted the draining away of their soils differently, erosion was not necessarily seen as damage. In the *Christian Philosopher*, erosion suggested an encouraging anti-Malthusian future. "By these operations the quantity of habitable surface is constantly increased," Dick suggested. "Precipitous cliffs are generally made gentle slopes, lakes are filled up, and islands are formed at the mouths of great rivers; so that as the world grows older, its capacity for containing an increased number of inhabitants is gradually enlarging."[114] Accordingly, Mather watched the growth of mudflats optimistically. "[They] are sensibly increasing," he noted, "and will, at some future time, make valuable and productive lands. Many of them are now employed for hay and pasturage, and others are rapidly becoming adapted for such uses."[115] In deforesting the hills and loosening the soils, poor hill tenants and rich tanners were carrying out a progressive law, enriching more established lands in the valleys.

Following Pratt's enthusiastic deforestation and the consequent erosion, geology was pretty evident to the eye in Prattsville. "It is a country which nature has favored, to say the least, with a due proportion of mountains and rocks," Pratt commented.[116] By inspecting "the vast collection of mineralogical specimens of every size and form, cropping out . . . or scattered in loose masses, just where the primal convulsions of mother Earth may have left them," discerning visitors could confirm for themselves that Mather's touted "Catskill division" lay beneath Prattsville.[117] But this too was not rock-solid evidence. Accounts of good grazing land, even geologically informed accounts, varied. Descriptions of Prattsville's granitic soil bore little resemblance to, for example, the "strong tenacious clays" that John Claudius Loudon praised in the agricultural journals at roughly the same time.[118] Even in the geological surveys, William Mather used other evidence to suggest butter's real place: "A large proportion of the butter sold under the name of Goshen butter . . . is made in the mountain region of Delaware, Ulster, Sullivan, and Greene counties," he would explain. What some New Yorkers read as counterfeiting here attested to a deeper claim of truth.[119]

Experiment

For Pratt, too, the product of the land was key evidence, carved into Pratt's Rock: "On the farm opposite, 224 pounds of butter were made from each cow from eight cows in one season."[120] Pratt had originally meant his 320-acre farm to breed a special bark-hauling horse, but as the tanneries declined, it became an experimental stage on which an unnamed farmer, his wife, and two hired girls, a man, and a boy worked to demonstrate Prattsville's dairy future. It was well staged; perched just across the creek from Prattsville itself, on thirty-five acres of flat meadowland at the base of the mountains, sloping upward to 320 acres of precipitous hemlock cutover of the type he hoped to reclaim. Though he began to describe himself in the census as a farmer, Pratt did not need the money—though the farm netted $450 a year, Pratt was worth more than $100,000.[121] The farm was not a source of income—it was an argument.

Here again Pratt was acting more or less conventionally. When William Marshall had mapped "Agricultural Districts" in his influential *Rural Economy of the West of England* (1796), he had declared demonstrations of superior practice in his own mid-page, all-capitals phrasing, "A FIRM BASIS, ON WHICH TO RAISE FUTURE IMPROVEMENTS." Feats of farming, showing the capacity of land had become a standard genre of experiment, as described in chapter 3.[122] Like Brewster, Pratt had to model desirable practices without making them look too expensive, laborious, or dependent on the expert knowledge of his workers. This last point was particularly important in butter making, so clearly reliant on a sophisticated body of women's knowledge.[123]

To manage this tension, Pratt de-emphasized the kinds of women's practices that butter dealers had identified as key. His dairy used "ordinary barrel dash churns," though powered by water, and the butter itself was worked "in the ordinary manner." Rather than churning the milk, a practice that Hawley had insisted on, which required much larger churns and made less butter, the cream alone was churned, "which we think better than churning the milk and cream together." The "we" in this instance was likely a rare active reference to people Pratt generally referred to as "my farmer and his wife." But much of the credit went to structural modifications—the cemented floor and good ventilation of the milk house, the hollow logs that carried in springwater and carried out waste buttermilk to the piggery below the dairy—and, of course, to features of place: "good, pure, cold spring water, which is very essential in making good butter."[124]

If Pratt's farm was to be recognized beyond the boundaries of Greene County, it needed publicity. The Endicott print, with its smooth mountains, fat cows, and hemlock-shaded proprietor, carried Pratt's vision of his farm to a broader audience, as did the accounts circulating in the agricultural press in the late 1840s. In the late 1850s, Pratt began to enter the state society's farm competition, ensuring that his farm would be described in full by a society report, made once again by B. P. Johnson.[125] By the early 1860s, Pratt's farm had reached the patent office report, the most widely circulated federal document of the nineteenth century.[126]

In his position as local patron and agricultural society president, Pratt was in a good position to amplify his message through local fairs. In his first report to the state society, he remarked, predictably, that "the show of cattle and horses was particularly large, on account of the peculiar nature of the county, and its adaptation to the raising of stock."[127] However, Pratt had himself created this impression, by asking the society to distribute premiums that he had funded so as to excite "emulation in the making of butter and cheese as well as for encouraging attention to the best breeds of cattle suited to our Highland regions." The *Prattsville Advocate* likewise circulated encouraging statistics, describing the 560 butter wagons, each carrying twenty firkins, each worth twelve dollars, which had passed through Prattsville in two days.[128]

Pratt also demonstrated adaptation through his animals: he self-consciously selected "native" animals descended from those brought by seventeenth-century settlers, rather than buying expensive imported cattle popular in the upper reaches of the state society. Doing so allowed him to argue that he was demonstrating the capacity of the land for butter, not the capacity of improved breeds for butter production, and, at the same time, to demonstrate the value of natives, which he saw as better adapted to the landscape of their birth. More easily circulated than land and more concrete than images, these animals too became forms of material proof. In 1844, Pratt's dairy cow won third prize at the Poughkeepsie Fair in the native cattle category.[129] Perhaps the same animal, "a native milch cow," which Pratt called "Lady of the Mountain," and the *Prattsville Advocate* called "the most perfect animal ever exhibited," appeared at the Greene County Fair the next year. Pratt told the county judges he wanted no prize; he meant only "to show what the mountain towns could do."[130]

As in most improving experiments, the most easily circulated proofs were not images or animals but accounts. It was the account of production, after all, that Pratt had carved into the mountain and that he would use in a series of triumphant tables appearing throughout improving print, which would join a swarm of similar circulating estimates. Devoid of everything but Pratt's name and his place, these figures would continue to testify to Prattsville's value.[131] While evidence of superior practice often resembled the tables and longer columns of debit and credit that Pratt had circulated, Pratt's carved boast was actually a sort of genre of evidence in itself; in the 1840s, the state society had sponsored competitions for butter production specifying a number of cows in a particular period of time. In 1848, John Holbert, whose Chemung County butter had discomfited the New York gourmand, had won one too, making 248 pounds of butter from five cows over thirty days in the late spring of 1847.[132] Both Pratt's and Holbert's dairy feats were recognizable as "experiments in butter making."

It is within this same genre of superlative demonstration that we might understand another subgenre of agricultural giants: the thousand-pound "mammoth" cheeses and butter pyramids that were famously sent around the nation by Col. Thomas Meacham, an Oswego County dairyman, in the mid-1830s. Sent to President Jackson, Daniel Webster, and the "enterprising Citizens of the great and flourishing City of New York," Meacham's dairy monsters joined a line of proofs of producerist capacity (huge loaves of bread, massive calves' hindquarters, more and more mammoth cheeses) in American political culture.[133] However, they clearly operated on other levels as well. Though Meacham's cheeses were emblazoned with political mottoes, only their shocking dimensions made it into print. "Mammoth Cheeses from a Mammoth Dairy," said the *Cultivator* headline in 1835. While the *Cultivator* didn't entirely approve of Meacham, the society's own competitions resembled his monsters more than a little. Meacham's cheese, like Pratt's year's supply of butter, was an agricultural giant.

Prattsville in the Future

The *Prattsville News* published a poem in 1861 that went like this:

> How changed the face of nature here
> Since forty years ago,
> The mountains thick with hemlock trees
> The Creek did flow below.
>
> Instead of these, fine pastures green adorn the mountain's side,
> And cattle, sheep, and horses graze
> And lambs and colts beside—
>
> And who has wrought this rural change
> From mountain top to flat?
> That far famed persevering man
> The Hon. Z. Pratt.[134]

The 1865 state census shows that Prattsville had certainly changed over the previous twenty years. More than eight thousand acres were either in pasture or meadow, only eleven hundred were sowed with crops. The population of sheep, the previous future of the Catskills, had dropped from 2,572 to 944. Taking their place, the population of milk cows had risen from 727 to

961. In 1845, these cattle had produced cream enough to make 40,342 pounds of butter; by 1864, they produced 108,850. Each cow in 1835 had produced 55 pounds of butter per cow; now they made 114, a doubling that represented changes in stock quality, feed, and milking frequency, and certainly an enormous increase in women's labor. These shifts mirrored changes in the rest of Greene County; by 1864 it produced about 1.3 million pounds of butter.[135]

Pratt's hopes and prophecies notwithstanding, Greene County hadn't managed to fully inhabit Orange County's identity; Orange County farmers made 2.4 million pounds of butter and valued their farms at fifty-three dollars per acre. While Orange County had ten dairy and milk dealers in 1855 (New York City had 579), Greene County still had none. However, the butter landscape had certainly shifted—Prattsville was now on the edge of a new, much larger district—the entire Southern Tier of the state had been recast as butter land. Moreover, despite the departure of the local manufacturing base, formerly its farmers' main market, agricultural land values rose in Prattsville, climbing to about thirty-one dollars in 1865 from $16.69 ten years before (in 1845, the state census didn't record land values). This was almost certainly due to the spike in food prices caused by the Civil War, but it's worth noticing that Prattsville land prices, which had lagged behind Greene County as a whole in 1855 had advanced a dollar beyond the average in 1865.[136]

Other efforts to pry open the reputation of the Orange County butter region had also been partially successful. "The Orange County Milk and Butter Depot," establishing a warehouse in 1848, announced publicly, "Pure milk and choice butter is punctually furnished directly from the best dairies of the several counties through which the New York and Erie Railroad passes."[137] By 1867, Thomas Devoe, expert butcher and grocer, would explain in his manual for consumers, that, "the counties of Orange, Chemung, and Cortland (in New York) have the reputation of producing the best qualities and largest quantities [of butter]."[138]

How much can these changes be attributed to Pratt? Not nearly as much as he or the *Prattsville News* suggested, of course. As with most cases of promotion, it is hard to draw a clear line between cause and effect—the men and women who worked Prattsville's dairy farms made both the butter and the decisions. It's clear, though, that in agricultural print, Pratt's accounts of Greene County, amplified by B. P. Johnson, had become standard. They had also leached into the everyday texts that New Yorkers bought to manage trade and buy land. French's *Gazetteer* for 1860 told a story that Pratt himself could have written: "Villages of considerable magnitude, with churches, schools, stores, and taverns rose up in the wilderness as if by magic." Though the hemlock had disappeared, "The result of all this was to facilitate the occupation of the lands in the mountain towns and in many cases to carry cultivation to the summits of the most lofty ranges thereby opening one of the finest dairy and wool growing regions in the State."[139]

Pratt's production figures had a stranger fate. Immediately after Pratt's death, a few articles appeared referring to them as his "tall writing on buttermaking," evidences of his eccentricity as clear as his many wives (five, sequential) and his passion for stone inscriptions.[140] However, by 1881, the *Ohio Farmer* was suggesting that while some of Pratt's figures were "open to correction," they were actually too low, "a fair test of what every progressive farmer in the country ought to be able to show on an average of years."[141] Passing from "tall" to standard, Pratt had been surpassed by the future he had predicted. Like Pratt's figures, butter production in Greene County was durable but not permanent. Like almost all Catskills farming, it faltered and failed during the 1920s and 1930s.[142] The same thing, indeed, would happen in Orange County, where "Butter Hill," under its new, romantic name "Storm King," would become a slightly ironic symbol of wilderness.[143]

★

The children's puzzle that began this chapter had a hopeful ancestor in another map, produced by Ebenezer Emmons. Emmons had been commissioned to add an agricultural successor to the state geological survey, the five volumes of the *Agriculture of New York*. To crown the first volume, on agricultural geology, Emmons developed a map that reduced New York's multiplicity to six hand-colored districts, each indicating where a particular crop—wheat, corn, dairy—might be most profitably grown. To define them, Emmons layered together the same forms of evidence that Pratt had used. Like Pratt, Emmons drew on a settler knowledge of trees; his plates foregrounded, for example, "the superb elms," indicating "the deep and rich clay soil peculiar to this district ... soils that are rich in potash."[144] Each district was also linked to a particular system of geological strata: the cold hard primaries of the "Highlands" or the sea sands of the "Atlantic Division." Like Pratt, Emmons also linked trees and rocks to practice, though he had a broader store of sources to choose from. Like William Marshall, he interviewed farmers to determine their opinion of the best form of culture within the broad outlines of his districts, examined specimens of grain submitted for prizes, and assembled tables of fair premiums as awarded by region to show where prizes could reveal the capacities of the landscape.[145] Emmons's survey also hid women in the passive voice. The best butter was made from cows and slopes, not from exhaustively cleaned dairies or the secrets of temperature, churning, and salting. Women's skill was too mobile to be a useful form of value. Relegating

FIGURE 14. Agricultural map of the state of New York, 1846, produced by Ebenezer Emmons for the agricultural volumes of the natural history survey. The different patches, hand-colored in the original, represent areas of land "adapted" to different purposes. Author's collection.

them to the backdrop made it much easier to create a naturalized, apparently unpopulated map of regional agriculture.

Its recognizable credentials make this map look neutral, authoritative—the centralized vision of a powerful state. On closer examination, however, this impression fades. When we find Emmons writing in the main text "the butter which is made from cows feeding upon the rather steep slopes of the Catskill range, either of Greene or Delaware counties, is probably superior to any in the State," it is hard not to see Zadock Pratt shouting up through the page—particularly since Emmons knew Pratt's ally Johnson well; he worked next to him at the agricultural rooms and depended on him for specimens and texts.[146] Other claims within the texts have been similarly constructed; the apparent god's-eye view of the final map only blurs the bids for local reputation from which it was aggregated without interfering with their central message.

Perhaps more importantly, New Yorkers' view of their landscape was not filtered through the colors of Emmons's map. Though the *Agriculture of New York* circulated to libraries, agricultural societies, and major improving figures, most improving New Yorkers would likely never have seen it. What most New Yorkers did see were little bids for reputation like Pratt's, publicly stated, performed, and published—fragmentary and contradictory bids for attention—told and retold at fairs, printed on premium barrels, and embodied in cattle. The map is a brief snapshot, a frozen frame in a constant debate. Though it is ubiquitous and slower moving, the same is true of my puzzle.

Of course, assigning different regions to different organisms only mattered once new populations were established. To do this required more than just a reputation. While Pratt had wanted to create a local breed of dairy cow from the settler cows already in the Catskills, most improvers turned not to local farm populations but to catalogs and auctions that funneled organisms across the ocean or around the country. But first they had to fix names to bodies, to make organisms that could move through catalogs and catalogs that could move organisms. This is the subject of the next chapter.

PART FOUR

Values

7

Truth in Fruit

Buried in the state society's *Transactions* for 1842 is a short and irritable essay called "Hints on Describing Fruit." Its author, John J. Thomas, was in his early thirties, had just followed his father into the nursery business in Macedon, New York, and was beginning to make a name for himself as an improving author.[1] In the essay, Thomas bemoaned the state of American fruit culture. Despite the new interest in horticulture, he complained, "a good fruit garden . . . is at the present moment a great rarity in most parts of our country." Thomas attributed this neglect to "the difficulties in the introduction of the best varieties."[2]

According to Thomas, these difficulties were not physical, but textual. Fruit had not only to be beautiful, edible, and useful, it had to also be truthful. Thomas complained that "the numerous errors in the names of fruits" made it hard to procure "those which are genuine," a problem compounded by "the multiplication of new varieties differing but slightly from old and celebrated ones," and by "the meagreness, looseness, and inaccuracy of nearly all books of descriptions which have yet been published."[3] Even worse, he argued, the circulation of fruit trees around the nation fundamentally challenged the act of description itself. Movement changed the characteristics used to identify trees; changes in soil and climate made the same variety, grown from grafts of the same tree, produce unrecognizable fruits. Thus, Thomas pointed out, "the Virgalieu, which in most parts of New-York is decidedly one of the finest varieties, is pronounced by Kenrick in the neighborhood of Boston to be 'an outcast, intolerable even to sight.'"[4]

While Thomas fretted, the kinds of orchards he called a "rarity"—filled with named, grafted fruit—were spreading quickly, displacing a landscape of semiferal seedling trees. But Thomas was not wrong. Clearly the networks of print and plant distribution that constituted mid-nineteenth-century fruit culture sat uneasily together even as they depended on each other. These two issues—problems of identification and the instability of fruit trees as they moved in space—are the constant threads of this chapter. Examining improving debates about fruit descriptions and fruit reputation and exploring the landscape of texts and living organisms that they helped to create can help us see the changing contours of a category central to the relationship between natural knowledge and commerce: the variety.

Historians of science have devoted considerable attention to the commercial meaning of a broader taxonomical category: the species—examining the ways that species were created and their identities stabilized and tracing the rise of a global network of botanical gardens.[5] Within this context, the variety is a weak category, subordinated to the species.[6] However, for gardeners and farmers, species names like cow, horse, and apple were insufficiently specific; it was in "varieties" and "breeds" that the characteristics significant to production and ornament appeared.[7] In agriculture, moreover, varieties were not simply subcategories of species, but fundamentally different entities. At the species level, populations of varying individuals had to be described in a way that made them seem like members of a coherent entity. By contrast, populations of fruit varieties were entirely created from a single individual—chosen, cultivated, disseminated, and maintained through human networks of exchange. As such they had a peculiar relationship to print culture. Through catalogs, pomological manuals, agricultural journals, and advertisements, printed descriptions of new varieties circulated more rapidly than trees did. By establishing advance reputations, encouraging fads, and constructing "celebrity," these texts smoothed the way for new landscapes.[8] This movement of fruit trees was part of a much larger shift: the wave of introduced or created varieties of plants and animals that swept across the recently appropriated lands of the new United States. Recent scholarship shows that such "biological innovations" were crucial to the expanding American economy; just as new varieties of cotton allowed cotton culture to stretch into the black belt, new varieties of wheat made it possible for American farmers to multiply the American wheat crop by eight.[9]

Fruit varieties were thus not just philosophical categories. Chopped into cuttings that we would now call clones, they became an easily shipped, varied, and beautiful product, one that tapped into the ambitions of refined rural New Yorkers and moved easily through the commercial networks that connected them. However, businesses selling living things—seeds, cuttings, and living plants and animals—drew on and differed sharply from the markets in other refined goods. Classic histories of consumer culture have shown that botanical images—flowered silks and cottons, painted and carved plants on china and furniture—were common markers of refinement as indeed were gardens. However, taking seriously the production and distribution of living things in our narratives of rural consumer desire, purchase, and display has the potential to radically shift our sense of the geography and power structures of consumer culture. Not only did tree cuttings and seeds offer a much wider array of colors, scents, and forms than their painted and embroidered imitations did, they moved in very different ways. Even if the copying was not perfect, trees could be cloned long before ornamented furniture could be mass produced, and they could be shipped to places where furniture could not easily be sent. They could also escape from the standard systems of distribution in other ways—the purchaser of a clock could not become a distributor of clock cuttings, but the purchaser of a Royal George peach tree could. In places where silks might be hard to come by, flowers might be easier. On the other hand, living goods could not be improvised. While American provincial workshops might independently make elegant chairs out of local wood, the fruit tree trade required the maintenance of the social links through which reproductive material passed.[10]

Just as manufacturers and warehouse owners claimed to have the most valid knowledge of experimental machines, the loudest and eventually strongest claims to pomological knowledge came from the sellers of fruit trees: nurserymen. Teeming with millions of potential varieties, the North American fruit landscape of the early nineteenth century presented a different kind of market and a different object of study than European landscapes of the same period did. In stabilizing varietal names, pomologists altered not only their texts, but also the organisms their texts described. They replaced a wild profusion of trees with a regimented set of named varieties, which, if imperfectly controlled, was also radically simplified and newly subject to the desires of consumers. The ways that pomologists chose to do this mirrored strategies developed to make sense of the volatile antebellum marketplace. The tools of varietal credibility that pomologists developed—books of descriptions, fruit profiles, and varietal lists—strongly resembled tools developed for detecting false banknotes, assaying personal reputations, and reading the faces of strangers. However, even as the variety became a stable market category, it slipped from the control of aspiring experts into the unpracticed hands of eaters of fruit. Examining this category can help us see the particularity of improving knowledge and its fundamental differences from natural history and show how the commercial print networks of improvement worked to produce the landscapes they tried to describe.

"The Artificial Productions of Culture":
Varieties and Productive Taxonomy

Early nineteenth century agriculturists agreed that variability was a special quality granted to domesticated species. "By what means the first tendency to change their nature was given to domesticated plants," the British horticulturist John Lindley noted, "we are entirely ignorant."[11] This unaccountable but provident variability not only explained the widely differing kinds of dogs, cattle, fruit and flowers, but also made domesticated animals and plants gloriously subject to progressive development.[12] Through the variability of their offspring, domesticated animals and plants could be split and manipulated into new taxonomic categories—"breeds" in the case of animals and "varieties" in the case of plants. Much more than species, these were the meaningful categories for agriculturists. It was by shifting to late planting Mediterranean wheat, for example, that farmers in upstate New York and New England were able to sidestep the life cycle of the midge that devastated wheat crops in the 1820s and 1830s.[13] However, the *variety* was a changeable category, and the particular reproductive characteristics of fruit trees gave the word a very specific form.

Fruit seeds, and apple seeds in particular, produce highly variable offspring. Fruit tree seedlings can and do differ from their parents and each other, producing fruit of different colors, flavors, and shapes, which appear at different times of year and keep for different periods. This instability meant that fruit trees could not be propagated by seed; a fine tree could produce thousands of bizarre or useless offspring. However, as nineteenth-century fruit experts pointed out, it also meant that fruit trees offered an astonishing wealth of forms and flavors. Early catalogs describe striped apples; gray, egg-shaped apples; "Twenty Ounce" apples; and the "Surprise Apple," "yellow outside, and red to the core within."[14] Among hundreds of seedlings, the fruit grower might find a few valuable or unusual characteristics—some simple like sweetness, richness, thriftiness of form, fast growth, or early fruiting; others more esoteric like a taste of anise or oranges or a translucent or rosy flesh. To the novice, Andrew Jackson Downing wrote, planting seedlings "appears . . . a lottery, in which there are too many blanks to the prizes."[15]

A moment combining human luck and human judgment qualified a tree to receive a name and a description, and thereby to become a variety. In improving writing, these moments were cast as discoveries; thus William Prince, for example, boasted of having "discovered" the Sine Qua Non apple in a Flushing field.[16] However, as the horticultural writer Walter Elder noted in the 1840s, the act of horticultural discovery fundamentally differed from the collection of botanical specimens: "The botanist considers a plant with a double flower a monster—the florist considers it a beauty.... Species are the hobby of the botanist—variation the hobby of the florist."[17] Where naturalists sought representative individuals, fruit growers hunted "monsters" and "sports" and dealt in productive oddities. As fruit enthusiasts themselves pointed out, varieties were "the artificial productions of culture."[18]

As we can guess from Thomas's vehemence, the moment of naming and description was crucial—discovery offered the fruit tree a new mode of human-mediated reproduction. Since fruit seedlings did not resemble their parents, varieties were propagated asexually. Pieces of branch, or "scions," were cut from the original tree and either placed directly in the ground to root themselves or grafted to the root system of a different tree, sometimes of a different species like quince. Once established, these grafts branched, flowered, and fruited and could be split again. Since they had only a single original ancestor, fruit varieties sidestepped the years of selection needed to create seed varieties like Mediterranean wheat. Moreover, fruit trees' immense variability meant that new varieties might spring up in any seedling orchard, producing a continuous stream of novelties.

Nineteenth-century horticulturists and botanists agreed that these novelties were necessary. Just like individuals, varieties appeared to have life spans—after a few centuries they began to succumb to disease and rot. Since new scions literally stemmed from an original tree, some horticulturalists theorized that this decay was merely the aging of the original tree. In 1797, the British horticulturist Thomas Andrew Knight had suggested that varieties could be preserved by pollarding or coppicing the original tree as soon as its valuable qualities were identified—by extending the life of the original, these pruning techniques would also communicate longevity to its scions.[19] The Belgian pomologist Jean Baptiste Van Mons suggested a more familial (and republican) theory—varietal decline resembled the decline and weakening of aristocratic lineages, as a result of excessive cultivation.[20] Regardless of the cause, degeneration justified the continual search for new varieties and the creation of new celebrated plant forms.

When Thomas referred to "genuine" varieties in "Hints on Describing Fruits," he therefore meant, not trees that displayed valuable characteristics, but trees that had a demonstrable relationship to a single original. Unlike a species, the fruit tree variety was not a group of organisms found in a landscape, but a network of propagation spreading out from a single initial point. Also, unlike a species, the fruit tree variety reproduced solely through socially produced networks of exchange and trade, maintained its identity through techniques of naming and description, and lived and died entirely according to the dictates of desiring orchardists and consumers.

Grafted Fruit in North America

Of course, the practice of fruit grafting was already millennia old by the time Thomas wrote his essay—nineteenth-century authors often referred with wonder (and some skepticism) to a tree described in Pliny the Elder's *Natural History*, on which berries, grapes, figs, pears, pomegranates, and several kinds of apples had reportedly been grafted together.[21] However, the surfeit of grafted varieties that Thomas described was relatively new to North America.[22] This did not mean that New Yorkers lacked fruit. The first colonists had come bearing seeds as well as a few scions and trees from Europe. These trees rapidly spread beyond the colonists' advance—General Sullivan's 1779 expedition against the Haudenosaunee observed and then destroyed vast orchards bending with peaches and apples. Pouring into the state in the 1790s, settlers from New England brought fresh infusions of seeds. New York's landlords and developers institutionalized fruit growing by making the planting of orchards one of the improvements required by their leases. The settlers' practice of foddering pigs and cattle on windfall fruit and then allowing them to range in the woods sent fruit seeds beyond the bounds of settler and indigenous orchards—apple, plum, and peach trees sprouted in swamps, fields, and forests on their own.[23]

Settlers, Haudenosaunee people, landlords, and animals had created a layered American landscape of fruit by the 1820s. However, most of their fruits differed sharply from the varieties that Thomas hoped to regulate. A Hudson Valley lease from the period reveals a typical planting practice. The lessee, it declared, should, "the first year, strew apple seed or pomace [the refuse of cider pressing] upon a patch of land for said Farm, for a nursery, well prepared for that purpose, of at least fifty feet square, to the intent that within six years be planted a regular orchard of one hundred apple trees at least."[24] Fruits grown from such seedlings were variable and often inedible and were frequently made into hard cider, perry, or brandy. It was this kind of apple, not the tamer grafted kind, that was propagated by the itinerant Swedenborgian nurseryman John Chapman, otherwise known as Johnny Appleseed, whose legend was later sanitized by temperance-minded biographers.[25]

It was not until the 1820s and 1830s that grafted fruit began regularly to appear on tables and in markets, growing in habitats that spanned New York's rural hierarchy. First of these was the elite gardening culture that formed first among landlords and then among wealthy urbanites practicing rural retirement.[26] As well as alluring colors, shapes, and flavors, fine fruit varieties promised tangible and affordable links to the elite British landscapes that many improvers dreamed of. As editor of the *Western Farmer and Gardener* in the mid-1840s, Henry Ward Beecher expected his readers to share his fantasy of "imaginary visits to the Chiswick Garden [and] the more than oriental magnificence of the Duke of Devonshire's grounds at Chatsworth." Few (if any) Americans could afford the labor that the Duke of Devonshire's servants reportedly devoted to his "monster" Royal George peach tree, which extended over a hundred feet of trellis in its own greenhouse and produced 8,727 peaches in a year. For thirty-five cents, though, they could buy a piece of it and plant it at home.[27] As Andrew Jackson Downing wrote in the 1840s, "Fine fruit is the flower of commodities."[28]

At the same time, American orchards offered a more patriotic, homegrown luxury, an example of the sort of improvisational provincial gentility described by Cathy Kelly and David Jaffee.[29] A correspondent to the *Genesee Farmer* lightly mocked the novice gardener who "had just read an account of an extraordinary '*seedling cherry*,' produced by Mr. A., in one part of the country; a wonderful seedling apple, by Mr. B., in another, a no less remarkable pear, by Mr. C., somewhere else."[30] At a relatively low price, aspiring orchardists could begin to assemble rarities into "good fruit gardens." The agricultural fairs rewarded these comprehensive collectors; early exhibition reports often consisted only of numbers and names of varieties displayed. Through such accounts, published in the horticultural journals, readers could watch new fruits move through the system and perhaps decide what to buy.

As trees became a consumer good, so did fruit. Sprouting haphazardly from roadsides and home orchards, seedling fruit had been fair game for passersby, but in the 1840s, a series of state laws hastily enacted against "fruit theft" showed the new importance of markets in fine fruit.[31] Markets themselves became literally concrete: where in 1837, New York City had two fruit shops, eleven years later it had seventy-one.[32] Encouraged by the deregulation of public markets in the 1840s, a new class of urban grocers altered the way that fruit was presented and sold. Public markets had previously been segmented by time: prices dropped during the day, and battered afternoon fruits were sold to the poor. The new groceries were segmented by class; where street vendors served the poor from barrels, groceries in upscale neighborhoods competed to have the most elaborate displays—and new and attractive fruit varieties became key to their strategy.[33] Distressed by the brandy distilleries that depended on wild apples, temperance advocates also supported this shift to fresh grafted fruit varieties. The same kinds of anxious reformers who wrote articles entitled "What Should We Do with Our Apple Trees?" (and signed themselves "BURN THEM") suggested that a taste for fresh fruits precluded a taste for alcohol.³⁴

As fine fruit orchards became evidence of rural gentility, a new landscape of commercial orchards satisfied the new urban demand for grafted fruit. In 1851, Cincinnati alone consumed 24,414 barrels of apples, worth about $100,000, from the new orchards of western New York and Michigan.³⁵ Commercial orchards reversed the direction of the transatlantic luxury trade. Even before the Revolution, London markets had sold a few Long Island–grown Newtown Pippins, shipped packed in sand, though they were "too expensive for common eating."³⁶ After the removal of British duties on apples in the mid-1830s, American exports doubled within a decade—Baldwins from Boston and Albemarle Pippins from Virginia jostled with Long Island fruit in Covent Garden.³⁷ Overseas taste for American apples made possible places like the Pell Orchard of Esopus, New York, which claimed to be the largest commercial orchard in the world, producing about 13 percent of American apple exports in the 1840s.³⁸

As collectors sought rarities, middling farmers sought refinement, and "orchardists" sought urban markets, nurserymen, the sellers of trees, profited. During the first half of the century, the few dozen nurseries clustered around the major cities of the East Coast would grow to more than a thousand, selling between fifteen and twenty million trees annually.³⁹ Positioned between western and East Coast markets, New York nurseries became particularly dominant, with nationally known centers on Long Island, on the Hudson, and in Rochester. Their reach was broad—when the proprietor of the Llewelling Nursery arrived in Oregon by wagon in 1847, he carried trees from the nurseries of Ellwanger and Barry in Rochester and A. J. Downing on the Hudson.⁴⁰

By 1845 the Hudson Valley nurseryman Andrew Jackson Downing could boast exultantly that "the planting of fruit-trees in one of the newest States numbers nearly a quarter of a million in a single year," that "there are more peaches exposed in the markets of New York, annually, than are raised in all France," and that "American apples, in large quantities, command double prices in European markets."⁴¹ Perhaps more revealingly, in 1850, the New York farmer Lester Doubleday distinguished in his diary between "grafted" apples and "pig" apples, implying that, as advised by temperance advocates and agricultural improvers, he was using his seedling apples solely for stock feed.⁴²

If in the 1820s Americans had lacked grafted varieties, by the 1840s they were awash in them. This sea of varieties created a problem. Varieties passed "into each other by insensible shades; or often differ so slightly and by such variable characters," Thomas complained, "that it becomes exceedingly difficult to discriminate."[43] Miniscule gradations in the depth of the eye (the declivity at the base of the apple), the shade and pattern of mottling or striping, the grain of the flesh, and its acidity or sweetness determined the dividing lines between kinds. Andrew Jackson Downing slyly expressed the subtleties and frustrations of identification in a poem, "To the Doctor on His Passion for the 'Duchess of Oldenburgh'" that he scribbled to his friend, the physician William Darlington sometime in the early 1830s.

> Dear Doctor, I write you this little effusion,
> On learning you're still in that fatal delusion
> Of thinking the object you love is a Duchess,
> When 'tis only a milkmaid you hold in your clutches;
> Why 'tis certainly plain as the spots on the sun,
> That the creature is only a fine Dutch Mignonne
> She is Dutch—there is surely no question of that,—
> She's so large and so ruddy—so plump and so fat;
> And that she's a Mignonne—a beauty—most moving
> Is equally proved by your desperate loving,
> But that she's a Duchess I flatly deny
> There's such a broad twinkle about her deep eye;
> And glance at the russety hue of her skin—
> A lady—a noble—would think it a sin!
> Ah no, my dear Doctor, upon my own honor
> I must send you a dose of the true Bella Donna![44]

For orchardists like Darlington, apples were foci of passion and in Downing's writing, a sexualized longing for apple and female flesh.[45] However, the poem also demonstrates the virtuosity of apple identification—breadth of eye and color were signs that Darlington had failed to read—signs as "plain as the spots on the sun," which, like sunspots, were invisible to the unaided or at least inexpert eye.[46] It is easy to see why in his own book, *The Fruit Culturist* (1846), Thomas recommended that fruit growers mark the names of varieties on the bodies of the fruits themselves, by indenting their skins with a pencil.[47] The poem speaks to the weight that Downing, Darlington, and their contemporaries attached to names and naming, and to their need to reveal as milkmaids fruits that had passed for duchesses. In correcting his friend, Downing also asserted his right to give genuine names, a right claimed increasingly stridently by nurserymen like Downing and Thomas. They claimed, that is, to be pomologists.

Nurserymen Become Pomologists

In Britain and on the continent, "pomology," or the study of fruit, was centered in institutions, in botanical gardens, or in experimental orchards. By contrast, American pomology was created and circulated largely by the emergent commercial network of nurserymen that had so industriously promoted the *multicaulis* mulberry. Their claims to expertise emerged from botanical connections forged during America's colonial period.[48] In New York, the Princes, of *Morus multicaulis* fame, had first built up their business in the second half of the eighteenth century by selling American specimens to eager British collectors (by the Revolution, it is popularly thought, the gardens had become so well known that both sides purposefully spared them).[49] As early as 1823, their catalog shows that the shift to such grafted fruit trees "as have acquired a well merited celebrity" had already begun. By the 1830s, the Princes had begun to self-identify as pomologists. Like other nurserymen, they performed the main functions of pomology: they grew fruit trees, they tested them, they distributed them, and, perhaps most importantly, they wrote about them.

Like other major American nurserymen, the Prince family drew plants and credibility from the same global network of botanical gardens that had supplied them with mulberries. However, the Princes had made a particularly valuable British contact when the second William Prince became a corresponding member of the London Horticultural Society (LHS), an association of wealthy Whiggish fruit enthusiasts that had established a new fruit-testing garden at Chiswick. Since the LHS sent free specimens to corresponding nurserymen for testing and distribution, Prince suddenly had access to the 3,825 fruit varieties whose names would fill Chiswick's first catalog in 1823.[50]

Chiswick was not alone in casting itself as a center of nomenclature. Accurate naming was the central practice of both botany and horticulture. Without an accompanying name and description, plants could not be exchanged through far-flung networks of correspondence, collections could not be compared, and value could not be determined. It was for developing a system for naming new species of plant that Linnaeus was celebrated around the globe, and it was naming and describing valuable new species and varieties that allowed aspirants entry into botanical circles.

Within this transatlantic culture, "good fruit gardens" were comprehensive collections of accurately named fruit. To create them, nurserymen had to build a dense web of exchange relationships. Andrew Jackson Downing's Rosselet de Meester pear came from the experiments of Jean-Baptiste Van Mons at the University of Louvain, in Belgium; his Thompson apple came from Chiswick Garden; and his Downton Pippin from Britain's Thomas

Andrew Knight. However, this stream of specimens also required an exchange. Here, American nurserymen had an advantage. Where Van Mons and Knight struggled to breed new varieties, American nurserymen had a thick stack of lottery tickets in the seedling landscape. Fruits like the Cranberry Pippin—"a strikingly beautiful apple" that Downing "found growing on a farm near Hudson, N.Y."—eased nurserymen into international circuits of specimen exchange.[51]

Though American nurserymen started on the margins of this system, we should not assume they were permanently peripheral. Like many major nurserymen, Downing characterized his orchard as a site of knowledge production analogous to Chiswick, a place where varieties were tested and judged. "Little by little I have summoned [varieties] into my pleasant and quiet court," he wrote, "tested them as far as possible, and endeavoured to pass the most impartial judgment upon them."[52] The absence of an American state-sponsored botanical garden or privately sponsored testing garden such as Chiswick during the early decades of the nineteenth century gave greater weight to such claims. American nurserymen also benefited from the strength of local markets. By the 1830s, British fruit was in decline—undermined by the waning of cider and perry as working-class drinks, by medical tracts blaming cholera outbreaks on the eating of fresh fruit, and, in 1838, by the removal of almost all duties on imported fruit.[53] Moreover, even as American seedling orchards turned out hundreds of new varieties, British varieties had begun to suffer from inexplicable ailments.[54] While British fruit culture contracted, American fruit culture expanded.

Printing the Orchard: The Genre of Varietal Description

Participation in both pomology and fruit tree markets meant that nurserymen also became authors. They contributed broadly to agricultural print, but their key genre was fat books containing hundreds upon hundreds of varietal descriptions.[55] Since most people first encountered new varieties through text before they encountered them in the flesh, it was through texts like William R. Prince's *Pomological Manual* (1831) and Robert Manning's *Book of Fruits* (1838) that varieties became public and collectable.[56] Such books marked nurserymen as experts worthy of the attention—and specimens—of a global audience of botanists and horticulturists, demonstrating the scholarly accuracy and judgment of the author and their familiarity with other works. The dense interweaving of fruit texts appears in their citations; lists of works referenced, sometimes many pages long, *preceded* their tables of contents. While the practice of pomology depended on the possession of a well-stocked library, it also depended on the possession of an orchard. The credit of Prince's *Pomological Manual* rested in part on the twenty acres and four thousand species and varieties at the Linnaean Garden—its descriptions referred to (and advertised) varieties available in the Prince *Annual Catalogue*, a free list of varieties Prince had "tested" and was hoping to sell.[57] Descriptive catalogs gave nurserymen a space both to advertise the valuable qualities of fruit and to demonstrate that they had the knowledge to sell genuine varieties.

Some American entries in this genre matched the splendor of their European counterparts. Ebenezer Emmons's agricultural survey, for example, devoted a volume and a half out of five to colored images of fruit and fruit descriptions and, in its ambition and expense, resembled lavishly illustrated British works like *Pomona Londoniensis*, objects of desire in themselves that were beginning to reach a middle-class British market for botanical works in the 1830s.[58] However, a larger number of American works were cheaply designed for rapid national distribution. Priced at $1.50 a volume, William R. Prince's books issued from a carefully spaced network of printers, in New York City and Boston and the growing fruit region of western New York, but also in expected markets in the slave states and in Cincinnati, which in 1831 was a forty-one-year-old city of about twenty-seven thousand people.[59]

To reach these broader audiences, works like Prince's volume depended not on images but on precision of language to capture the identity of a variety.[60] Descriptions were intended to guide and educate the aspiring pomologist's eye (and usually his tongue as well) by directing his attention to a series of significant characters and providing a set of terms to denote them. Prince's descriptions of the "Mouthwater Pear" began with shape, "an exact pyramidal form," and then proportion, "its height thirty-three lines and its greatest diameter twenty-six, tapering very much towards the stalk." These proportions could alter, Prince explained, and the pear would assume "a turbinate form . . . twenty-eight or twenty-nine lines in each direction." Prince described the place of the eye, the length of the stem, and the color: "even at the period of maturity . . . an uniform shade of rather dark green . . . one may observe a grayish streak running lengthwise of the fruit." He then considered flesh texture, "rather firm, but melting," and then flavor, "pleasant . . . with some sweetness and richness." Season of ripening, thriftiness, and color of foliage followed. Then, emphasizing the extent to which accurate descriptions were credit among pomologists, Prince carefully corrected an error of identification appearing in two other texts that, Prince held, erroneously separated the elongated and turbinate forms.[61]

On first glance, it might be tempting to distinguish between botanical and commercial functions here—to describe color and shape as natural historical characteristics and taste and smell as commercial ones. In fact, the divide is not clear. Since books of description cataloged real, saleable orchards, all identifying descriptions were also advertisements. They acted as banked copy and were quoted in advertisements for new varieties or reproduced in the puff pieces on new fruits that circulated in the journals. Conversely, the characteristics of taste were significant identifiers of fruit, just as color and shape were saleable qualities. Even textual corrections were selling points, helping buyers of fruit and fruit trees to certify their purchases as genuine articles. However, as with all commercial goods, claims of authenticity created problems.

Confused and Counterfeit Fruit

The burning concentration with which antebellum nurserymen wrote about fruit identities, assembled authoritative sources, and established testing orchards, bespoke not the triumph of rational certainty, but a struggle against chaos, confusion, and deception. Once varieties became part of orchard practice, names and fruit multiplied together in a disorderly tangle. Following a practice of using European names for American organisms, for example, the new varieties springing out of American orchards often received old names or, even more confusingly, any one of the same set of praise names—William Prince may have discovered *the* Sine Qua Non, but "Sine Qua Nons" appeared everywhere. In a move that contradicted pomological definitions, moreover, the seedlings of European varieties often received the names of their parents and then were circulated in good faith.[62] Worse, multiple names were often carried over from Europe with new varieties to which new names might be applied locally, producing great lists of possible names for a single tree. Quoting Lindley, the New Yorker Michael Floy complained that a customer who ordered the

> Beurre Dore, Beurre d'Anjou, Beurre d'Or, Beurre d'Ambleuse, Beurre d'Amboise, Poire d'Amboise, Isambert, Red Beurre', Beurre du Roi, and Golden Beurre, White Doyenne', Doyenne Blanc, Beurre Blanc, Bonneante, Saint Michael, Carlisle, Citron de Septembre, Kaiserbirne, Poire a Courte Queue, Poire de Limon, Poire de Neige, Poire de Seigneur, Poire Monsieur, Valencia, and White Beurre

would be disappointed to realize that he was the possessor of only two kinds of pear.[63]

This confusion of names and the sheer number of varieties made space for more intentional deception. Once variety names began to define the value of fruits in the 1830s and 1840s, they became worth faking.[64] Advocates of the newly fashionable "Northern Spy" apple complained in 1845 that it "like all other popular fruits, is *counterfeited* by the men and boys who sell apples around the streets, and on the corners; every apple they can find, that in any way resembles the 'Northern Spy,' is so called by them."[65] These deceptions, bad enough with fruit itself, were far worse for those who bought scions, which did not display their valuable characteristics until years of labor had been spent on them.[66] "Many persons," warned the second William Prince, "apt to purchase trees without regard to any point but their cheapness . . . not unfrequently, after the toil and expense of years, find them, when they arrive at bearing, absolutely worthless."[67] The life cycle of fruit trees left both buyers and sellers vulnerable—the first to fraud, and the second to damaged reputations.

Nurserymen often blamed failed fruit on transient figures who increasingly filled out the network of plant distribution—tree peddlers and traveling "grafters" and "budders." The *Genesee Farmer* complained of farmers' willingness to rely on "some irresponsible, peddling grafter" who would throw different varieties together "promiscuously, without mark or label."[68] The Princes, ostensibly to fend off imitation by false nurserymen, used advertisements for their *Pomological Manual* as a certifying stamp for their receipts.[69] As in American market society more generally, where peddlers served as "a lightning rod for the anxieties of a developing market society," transient tree workers were easy targets.[70] However, they worried pomologists because they were influential. The farmer Moses Eames noted in his diary in April 1831 the arrival of "two men come to graft apple trees" on his father's farm, and the next day recorded, "they do 364 grafts."[71] Budders and grafters rapidly spread informal grafted orchards beyond expert nurserymen's control.

Though less frequently mentioned, informal channels of exchange posed even worse problems for pomologists. Scions regularly circulated among neighbors in the grafting season (from late February to early April), losing their names and gaining new ones along the way. Agricultural journals and societies further confused matters by distributing specimens themselves. The effects of such practices can be seen in the *Genesee Farmer*'s distribution of the "Williamson apple" in 1835. Though articles referred to the apple as the Williamson, after the owner of the first known tree, it was locally known as the "Land Office Apple" because the original tree had grown for forty years near the land office of the Pultney estate. However, it was agreed locally that Williamson had brought the tree with him from England as a scion, meaning that the phrase "Land Office" obliterated a past European name and that identical specimens might have recently been reimported. Through such movements, fruit identities became uncertain.[72]

To combat such uncertainty, nurserymen turned back to the power of text. William R. Prince wrote in his catalog for 1835, "The strongest proofs the proprietors can give the public ... are the *precise descriptions* contained in the Treatises recently published ... any person, however ignorant on the subject, cannot fail to know if he has been deceived."[73] Americans were used to paying attention to such descriptions. Just as American orchards were crammed with dubious trees, American markets and pocketbooks were flooded with dubious banknotes. In the absence of a national paper currency or, for most of the period, a strong central bank, more than seven hundred small banks issued more than ten thousand kinds of banknotes.[74] This wilderness of notes created a comfortable habitat for counterfeiters, who faked existing bills, changed denominations on real bills, and issued "spurious" bills on entirely imaginary banks. By 1862, the *New York Times* claimed, more than four-fifths of the notes in circulation were false.[75]

In this mess of fakery, skilled observation became increasingly important. The same eyes searching for telltale mottling on fruit regularly searched for the wavery borders and imprecise shadings that would show that a particular note had emerged from the empire of counterfeiters sitting just across the border in British Canada. In his agricultural survey, Ebenezer Emmons raised this parallel directly. "To succeed as an observer, it is necessary that the observing powers should be highly cultivated; that they should be educated," he argued. "There is no fact probably more striking than the ability which is acquired of distinguishing counterfeit money. Clerks in mercantile establishments often acquire the power of detecting a spurious bill at a glance: they see a suspicious look without an effort; they have created as it were a new sense; and hence, what is nearly undistinguishable to other men is apparent to them."[76] In making a case for pomologists as observers, Emmons was attempting to invoke what David Henkin has called the "knowing clerk," a standard trope of urban fiction that would have been instantly recognizable to Emmons's readers.[77]

Parallels between counterfeit hunting and fruit identification became even stronger in the paper tools developed to deal with both. In their careful rendering of minute details as text, books of fruit description strongly resembled a tool developed to teach Americans to observe within a wider landscape of credibility: "counterfeit detectors." These were regular periodicals that packed thousands of bill descriptions into a few cramped pages. Even as periodicals like *Bicknell's Counterfeit Detector and Bank Note List* taught readers to scrutinize printed eagles, temples, and "females with vases of flowers" for minute discrepancies, they simultaneously encouraged them to expect falsehood and to see its detection as a necessary and a laudable skill.[78] Finding genuine varieties drew on related skills and values. Fixed by descriptions, nurserymen hoped, varieties would become transparent, true, and productive. This fixity, however, depended on the fixity of the organisms described, which, as Thomas pointed out, could not be counted upon.

"Limited and Local knowledge": Place and Shifting Form

Being desirable, relatively cheap, and easily shipped, fruit trees were in some ways a perfect early national consumer good. However, unlike the texts that accompanied them, they were not perfectly replicable. Variety forms differed, sometimes sharply, from place to place. European varieties frequently failed in the harsher winters and stronger light of the northeastern states. American pomologists knew this. Indeed, they had used this deterioration to justify the development of an American pomology in the first place. American conditions required new varieties, new books of descriptions, and new experts.

Nurserymen had also come to acknowledge the significance of changes of place in the movement of plants from the nursery. William R. Prince, for example, though scorning as a "prejudice" the idea that "trees, like cattle, when removed from a rich to a poorer soil cannot thrive" was careful to describe the intentional privations of his Flushing orchards. To be acknowledged as an appropriate place for variety testing, the Prince nursery had to display difficult but achievable conditions. Thus, Prince emphasized the cold winds of Flushing and declared in his catalog that "for many years he has not made use of as much manure on his grounds as is commonly put on the same quantity of ground by farmers in their usual course of agriculture." Instead he "substituted culture for manure," plowing his orchards deep to produce "thrifty" efficient trees that would do well in other places. By substituting labor for valuable material, he made the culture of good trees seem attainable.[79]

In "Hints on Describing Fruits," however, Thomas put his finger on a less easily solved problem. Unlike island-bound British pomologists, Americans faced a territory that stretched across many lines of latitude, connecting areas "almost as remote from each other as Norway and the Great Desert in the old world."[80] This fact threatened to undermine the national networks of plant distribution that American nurserymen were attempting to maintain. If an "Esopus" looked different in Albany than in Rochester, then the network of print and plant propagation could break down.

To combat this, Thomas outlined a program of research. Arguing that the methods of the London Horticultural Society would not answer in the varied climates of the United States, he suggested a network of pomological collections in different states, coordinated by the New York State Agricultural Society. This would help pomologists to determine the differing characters of varieties in different places and to see if some qualities were stable. This research could become the basis for a new standard of fruit description.

Thomas demonstrated such a system of description using flavor, the only category that seemed to him to be stable from place to place. He laid out a standard for the description of taste, describing, for example, the acidity of new varieties on a six-point scale from "sweet" to "very acid and austere," based on a hierarchy of older varieties that every pomologist might be expected to know.[81] This effort to promote flavor as a fixed character was not enormously successful. Flavor was a quality that depended on the experience of the taster. Moreover, the language of taste was limited to a remarkably short list of descriptive terms, "sweet," "acidic," "rich," and "sprightly" (though the list was longer than the one used by modern eaters of fruit). Finally, few nurserymen agreed that taste was in fact fixed. However, J. J. Thomas's search for a fixed character was a common one; his fellow nurserymen also proposed new systems based on purportedly fixed qualities of color and shape.[82]

Simple to make and to reproduce, the fruit outline or profile was one of the most successful of these. The concept was remarkably straightforward. To capture a profile, the pomologist had only to cut an apple or pear in half from stem to eye, put one half face down on a piece of paper and trace around it with a pencil, producing an unadorned but *accurate* line that could be reproduced in a woodcut at minimal expense. This method of representation, as well as a shadow box for the outlining of soft fruit (peaches, plums, and cherries), had been promoted in Loudon's *Gardener's Magazine* in Britain as early as 1828.[83] However, American authors of the 1840s refigured the concept to meet their need for a fixed character. In the volume of the *Agriculture of New York* devoted to fruit, Ebenezer Emmons postulated that while size might vary between specimens, the proportions of fruit, the ratio of height to the widths of the crown and base measured within a fruit profile was the true fixed character. Next to a bisected and labeled apple outline Emmons wrote, "If this is not true, we may despair of describing fruit so as to become useful to inquirers."[84]

With the fruit profile, other commercial modes of interpretation came again into play. Fruit profiles echoed the physiognomical systems with which nineteenth-century Americans struggled to tell the faces of honest strangers from those of confidence men.[85] The fruit profile's most successful advocate, Andrew Jackson Downing, justified it in terms that made clear its debt to the interest in character and facial structure. "The mere outline of a fruit," he wrote, "like a profile of the human face, will often be found more characteristic than a highly finished portrait in colour."[86] Profiles of fruit, like profiles of faces, could reveal character in a way that was both incontestably accurate (being taken from the fruit itself) and easily recognized, drawing on modes of interpretation

VARIETIES OF THE APPLE. 9

ORDER II. Breadth greater than the height.

B. Ends unequal.

BOROVITSKY. Fruit angular, medium size. Stripes faint.
BENONI. Fruit medium size. Skin deep red.
COLE. Fruit above the medium size, angular.

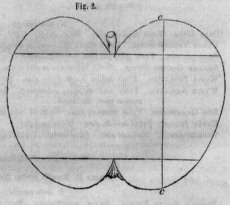

The line $a\ a$ (fig. 1) is drawn through the base, at the junction of the stem with the flesh at the bottom of the depression; and $b\ b$ is a line drawn through the crown, at the bottom of the calyx depression. When this depression is very shallow, the end is narrow, although it may speedily widen, and the two extremities may appear at first sight subequal : so when the stem depression is shallow, this end is often very narrow, especially when the depression is filled up, or obsolete. These proportions, I believe, will be found very important in determining the names of fruit; but as they have not as yet received much attention, it is impossible to make use of them except when an outline figure is given : these should be drawn with great care from the fruit itself.

In figure 2, the line $c\ c$ represents the height of the apple, and, like the lines representing the breadth at the extremities already referred to, will be found quite constant in its relations in the same varieties, although produced in different sections of this country. If this is not true, we may despair of describing fruit so as to become useful to inquirers.

[AGRICULTURAL REPORT — VOL. III.]

FIGURE 15. Ebenezer Emmons's 1851 plan to distinguish varieties by measuring the proportions of their profiles. "If this is not true, we may despair of describing fruit so as to become useful to inquirers." Author's collection.

with which Americans were already familiar. With the support of figures like Downing, such profiles appeared regularly in print.[87] However, despite their popularity, a system of fixed characters was never fully agreed upon.

Social agreement on quality and identity proved a more powerful tool. The networking of distant orchards that Thomas recommended was a promising project, not least because by the 1840s, agriculturists and horticulturists already had a strong social network. As we will see, the pomological convention, a social form of the 1840s and 1850s, embodied Thomas's imagined pomological networks, becoming a place where texts, fruits, and nurserymen were painfully reconciled.

"An Assemblage of Men and Fruits"

If Andrew Jackson Downing's orchard was a quiet court of varietal judgment, the first pomological convention of 1848 was a noisy one. The seventy delegates were a little overexcited, interrupting each other, and occasionally breaking into furious debate. Their phonographer, Oliver Dyer, sourly commented that his record was "as accurate and *faithful* as any *one* person could possibly have rendered it under the circumstances," since "we do not profess to be able to report more than *one* speech at a time."[88]

The delegates' agitation was perhaps understandable; where pomological gentlemen had previously had to make do with limited descriptions, drawings, and correspondence, now, at last, they were assembled in one place, carrying with them thousands of actual fruits for actual comparison. Together they would spend three days handing down definitive judgments on the new fruit varieties—determining identities and resolving synonyms, rejecting those varieties unworthy of cultivation, and commenting on the promise of new seedlings. Delegates were convinced that such a national meeting of minds and fruits would gradually sweep away all difficulties; as Lewis Falley Allen, noted author, president of the New York State Agricultural Society, and commercial apple grower, said with confidence in his opening address, "We must have uniformity, and we can obtain it."[89] The Buffalo delegates' sense of urgency was heightened by a sense of being part of a more general historical moment. Not one month later, a rival meeting, "the National Congress of American Fruit Growers," would be held in New York City.[90]

The rules of the convention confined the delegates to the serious business of varietal description and rating. During daylight, they were to examine varieties of fruit presented one at a time by a five-person fruit committee. Convention members were then to speak "brief statements of facts in each case" (though the limits of brevity can be seen in a rule limiting delegates to two ten-minute speeches). Other matters of "general pomological interest" were to be confined to "evenings and intervals."[91] The delegates attended, after all, not to share ideas, but to regulate novelty—to winnow down the number of new seedlings and circulating varieties to a small number of indisputable value.

The first of these was easier. According to the convention standards, five rules governed the production and assignation of new names. The first closed the ranks of the accepted varieties. To be given a name or a recommendation, a new seedling fruit had to be either better than all similar varieties or hardier and more productive than its competitors. The second laid out the boundaries of the new expertise. Names were not to be fixed until the fruit had been "accurately described in pomological terms" by an accepted authority: a fair's fruit committee, a "pomologist of reputation," or a journal or "some pomological work of some standard character."[92] The third and fourth rules laid out a checklist of identifying characteristics and defined the propriety of names, demanding simplicity and particularity. The final rule put J. J. Thomas's call for coordinated testing into action: "No new fruit can be safely recommended for general cultivation until the same has been tested and found valuable in more than one locality."[93] For a variety to achieve a name, and accepted qualities, it now had to be accepted, not just by its discoverer, but also by a larger social process. Trees had to be accepted over different times and in different places, and those that altered too easily were to be weeded out.

In the definition of expertise, the verdict of the meeting seems clear. Nurserymen dominated both the convention and the congress, and New York nurserymen dominated overall. Of the seven members of the fruit committee that gave the initial ratings and presented varieties to the group, four were New Yorkers.[94] Other New York nurserymen played active roles. In particular, William R. Prince spoke most frequently (and was the delegate most likely to break into impromptu pomological lectures contrary to the rules). In fact, despite efforts to give the Buffalo meeting a "national character," it was still clearly a New York production. Though delegates came from nine states and Upper and Lower Canada, fully half were New Yorkers. Others, including the convention's president, J. A. Kennicott, were former New Yorkers who had moved west. Not surprisingly, New York standards of fruit description and naming, generated by the New York State Agricultural Society, formed the basis for the convention's standards.

Both the convention and the congress were intended to produce a specific document—the varietal list. These were short lists of varieties judged as "first rate," for whose reputations nurserymen could vouch. The stakes of judgment were high, not only for the trees, but for the nurserymen and orchardists, whose stocks of thousands of saplings, if stigmatized as second rate, might have to be chopped off at the collar and regrafted with more accepted varieties. The standardization of fruit took two forms—first, fruit identities were to be fixed, and, second, varieties were to be graded as first, second, or third rate. (The New York City meeting, rather more optimistically, used "good," "very good," and "excellent.")

In creating varietal lists, nurserymen linked the language of natural characteristics to the language of credit, character, and reputation that saturated antebellum American market culture. Where books of description resembled counterfeit detectors and fruit profiles resembled phrenological manuals, varietal lists resembled a market tool developed only eight years before: the books of credit ratings produced by Lewis Tappan's Mercantile Agency in the wake of the Panic of 1837. Already by 1849, many of the nurserymen at the conference would have had entries in Tappan's ledgers in New York City, "ratings" of their character, drinking habits, and probable worth, produced by Tappan's network of almost two thousand secret correspondents, which could be accessed by their business contacts, for a fee. As the pomologists rated fruit, they themselves were being rated, first-, second-, or third-rate men, reputations that they could no longer escape by travel.[95] During the days of the meetings, and the annual meetings that followed, pomologists struggled to make reputations for organisms and, in doing so, to preserve their own.

Having laid out their rules, the delegates could now begin to judge the fruits themselves. This act of judgment, however, was not as simple as it had initially seemed. In the flesh, fruit specimens proved far from transparently meaningful. In part the problems were physical: many had been harvested before or after their peak period, and some were rotten, making their form difficult and their taste unpleasant to determine. In the candlelight of early evening, fuzzy peach skins proved too indistinct for careful observation, and many other fruits were bruised by hours spent in carriages taken over corduroy and plank roads.[96]

More consequential still were the differences caused by regional rivalries. Local cultures of judgment clashed at the national level. "There is such a diversity of experience, and consequently of opinion, respecting the merits of well-known varieties," wrote an observer at the New York meeting, "that many fruits which have long enjoyed the most irreproachable character in one part of the country, are found, on inquiry, to have the most indifferent reputation in another section."[97] The creation of fruit reputations, he argued, was as difficult as the creation of human reputations.

> Your committee have been reminded of the remark which an inexperienced politician once made to an eminent statesman in the political turmoil which

was going forward; "why," said he, "why make all this noise and trouble about a President, why not all agree on some good man and elect him at once."[98]

Certainly, some problems were resolved quickly. The Van Zandt's Superb presented to the Buffalo meeting was instantly pronounced an imposter. The Yellow Melocoton peach was unanimously "voted to be unworthy of a name or cultivation."[99] Many more, however, suffered assaults on their "reputation" and character.

The debates around the reputation of the Northern Spy, the most popular apple of western New York, were particularly revealing of both sectional rivalries and differences in place. Grown in the 1810s from seedlings on the farm of Oliver Chapin near Rochester, the Northern Spy quickly became a western New York staple. By the time of the Buffalo and New York meetings, counterfeit Northern Spys were, as we have seen, a common sight in western New York. Outside of western New York, however, its reputation was dubious.[100]

When the delegates examined the specimens of Northern Spy, they quickly fell into argument. William R. Prince (an easterner) declared "no apple calls for more ample investigation at the present time than the Northern Spy" and that "he was anxious to get at the truth in regard to it."[101] With thousands of specimens of Northern Spy in his nursery, Prince had reason to be nervous. Hodge, a western New Yorker, who had originally moved that the apple be passed as first rate, admitted that "he knew there were persons in the room who considered the apple a humbug." This was aimed at J. J. Thomas, who had commented publicly, that "out of ninety barrels of this apple only seventeen barrels were found fit to market." In the Northern Spy's defense, Hodge assured his peers that in his experience "the fruit was uniformly as fair as the Spitzenburg—five-sixths of them were fit for barreling." Thomas retreated, explaining that his initial experiences with the Northern Spy had been marred by poor cultivation. A small chorus of voices followed this conciliatory vein, but even as the conflict was smoothed over, Andrew Jackson Downing made a fresh attack: "Mr. Downing said Mr. Chapin, the originator, has forty acres of orchard, and he considers the Northern Spy so poor that he will not plant a single tree."

Instantly, pandemonium reigned at the pomological convention. The voices of the Spy's earlier defenders were swallowed up in the tumult; the stenographer complained that "owing to 'the noise and confusion' we could not catch the purport of them." Shouting down the crowd, Patrick Barry declared "that Mr. Downing's statement in relation to Mr. Chapin's orchard ought not to have any weight" and attempted to have it struck from the minutes. A Mr. Bissell added to Patrick's defense with an attack on both Chapin and his orchard. "A gentleman," he informed the convention, "visited [Chapin's] orchard and said it was the hardest looking orchard he ever saw. The trees had been neglected beyond all account." A Mr. Coit of Ohio, attempting to assist, then confused matters by praising the Spy's "fine color, which was a lightish green." Since the Spy was actually "a fine light red," the discussion ended in embarrassment, and the Spy remained in limbo.[102] This uncertainty continued at the New York meeting a few weeks later. Attempting to control the debate, the Rochester nurseryman James H. Watts sent a testimonial defending the Northern Spy, which was read into the minutes. Before listing its valuable qualities, he lamented, "Like every new thing now-a-days, to establish its character has been no small task."[103]

The Spy's character remained disputed—rejected by the first two meetings it appeared on varietal lists in an inferior character as a "variety good for particular places." However, the Spy's mixed reputation can be seen in two ways. First, it can be interpreted as an account of how instability over distance led to the adoption of more stable fruits. Certainly, this was the aim of the varietal lists. On the other hand, it might also be argued that the nurserymen's dispute emerged, not from changes in the bodies of the fruit itself, but from differences between regional cultures of fruit judgment—between the connoisseurs of Boston and the Hudson Valley and the more commercially focused nurserymen of the Great Lakes. In this reading, the account of the Northern Spy's changeableness, as well as its need for particular cultivation, was a way of resolving disputes that might lead to social rupture. Labeling the Northern Spy as first rate for western New York kept the stocks of western nurserymen valuable and kept the Spy on the market.[104]

The Ambiguities of Consumer Taste

Despite these internal wranglings, it is clear that by the time of these meetings an external factor had entered the judgment of varieties, one not entirely under nurserymen's control: the metropolitan market for fruit. Both the physical requirements of shipping and the seemingly fickle and uneducated demands of fruit consumers challenged nurserymen's concepts of quality. This appears most clearly in the frequent debates about the role of shipping and regular bearing versus taste in the identification of "first-rate" fruits. In the 1840s, marketability was not yet crucial to a "first-rate" ranking. For example, the first-rate Early Joe, which nurserymen agreed was "about the best eating apple they had ever known," had to be eaten in the first hours after it was picked and could not be moved.[105] Flavor, on its own, clearly qualified fruit for a reputation.

But what about marketable qualities like easy shipping and reliable bearing? In discussions of these qualities, fissures in pomology, between men who grew for the market and a dwindling group who were "curious in fruit" began to appear. An attack on the reliable Buffum pear at the New York meeting brought these tensions instantly to the surface. Nurseryman Patrick Barry sprang to the pear's defense. "Mr. BARRY . . . considered it hardly proper to insinuate any thing unworthy or knavish against gentlemen who spoke of fruits, and their qualities as 'market fruits.' . . . Every body knew that it was not always true, that a variety which stood highest in point of flavor, bore the same rank on the market list."[106] This question of commercial interest exposed embarrassing divides in the community of pomologists. Once the market for fruit was developed, questions of quality became inextricably bound to personal benefit, making the genteel applications to the "public good" that had justified the support of grafted fruit in the twenties and thirties increasingly difficult to sustain.

While nurserymen like Barry and Hovey might stalwartly defend the inclusion of marketable fruit on the lists, even they had to admit that the increasing role of consumers in determining the success of varieties made the assignation of value morally complex. By midcentury, nurserymen had come to expect that newly introduced varieties would become objects of public desire. In a report at the convention of 1852, James Watts declared that "since Horticulture has been made a study by the farmer . . . *the producers have found that consumers have become more particular about kinds*." The new orchards required "the choicest kinds of trees," specifically, the "Esopus, Spitzenberg, Baldwin, Roxbury Russet, Rhode Island Greening, Swaar, Talman Sweeting, Seek-no further, Pearmain, Twenty-Ounce Apple, and Vandevere." Since the counties in Watts's area produced two hundred thousand to two hundred and fifty thousand barrels of apples a year for urban markets, this was no small matter. Market demand indeed became an important part of the defense of dubious apples. Watts defended the Northern Spy by citing the astronomical "*nine dollars* per barrel" paid by "the good livers at the 'Astor House.'"[107] Here again, price could serve as evidence of quality.

Frequently, nurserymen embraced what they saw as increasingly educated consumer tastes, arguing that they heralded a consequent progress in varieties. In 1850, Charles Hovey cheerfully rejected the Juliene pear, though he had helped introduce it, and though it had been considered first rate for years. "Gentlemen," he declared, "were aware that their taste was progressing, and what was good ten years ago, was not so now—because they had other and better fruit."[108] The perpetual flow of novelties would lead to a perpetual progress in taste and a reshaping of demand.

TRUTH IN FRUIT 185

Simultaneously, however, the tastes of eaters made nurserymen uneasy. The Red Astrachan, a popular market apple, proved a particular sore spot. Though attractive, the Red Astrachan did not conform to many pomologists' standards. Patrick Barry admitted that it was "not a first rate eating apple" but pointed out that it "would bring more money in market than any other apple

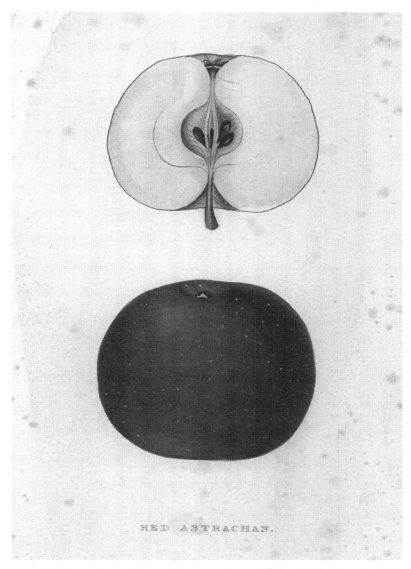

FIGURE 16. Ebenezer Emmons's image of the contested, potentially deceptive Red Astrachan. Author's collection.

of its season." J. J. Thomas was more dubious; "it had been remarked that the apple was good for market on account of its beautiful skin; when we get within its skin there is very little left."[109] Was the Red Astrachan then a sort of humbug? Were consumers lured away from their best interests by a deceptive skin? Should they be catered to, or educated out of their irrationality? Selling it, were nurserymen practicing on the ignorance of "the simple"?[110] Was the Red Astrachan, in short, a moral apple?

As they publicly defined quality, nurserymen were forced to address more general anxieties about the public good, the morality of the market, and the rationality of the consumer. Unsurprisingly perhaps, the Red Astrachan prevailed on variety lists for decades. However, it also is clear from these debates that varieties had achieved a commercial reality not just in pomological texts and pomologists' orchards, but also in market stalls and fruit seller's cries. That the Red Astrachan and other similarly "undistinguished" but marketable varieties remained on lists shows how market realities had begun to pull the definition of quality away from nurserymen's sole control, making the variety a subject of a wider social negotiation. It is this very loss of control that shows the broadening of the category and demonstrates the reality that the variety had achieved in the market.

★

Written in 1859, Henry David Thoreau's essay "Wild Apples" celebrated a vanishing landscape of seedling orchards, hedgerow trees, and feral trees growing in swamps and marshes and by the sides of roads.[111] His favorite trees, spread by apple-fed cattle, grew in the middle of meadows, clipped by cow teeth into dense, thorny pyramids that protected a fruited spike in the center. Again, improvement was his target. "We have all heard of the numerous varieties of fruit invented by Van Mons and Knight," he joked. "This is the system of Van Cow, and she has invented far more and more memorable varieties than both of them." Even as he memorialized this landscape, he mocked the one that had come to replace it. Pomological naming practices attracted his particular ire. He had "no faith in the selected lists of pomological gentlemen . . . their 'Favorites' and 'Non-Suches,' and 'Seek-no-farther's.'" These apples were tasteless and bland, chosen for size and bearing. Instead, he proposed his own list of apple names, some comic, some sentimental, for the hundred varieties of wild apple to be found in a single cider pile: "the Apple which grows in Dells in the Woods (sylvestrivallis), also in Hollows in Pastures (campestrivallis); the Apple that grows in an old Cellar-Holo (Malus cellaris)" and the "Beauty of the Air (Decus Aëris); December-Eating; the Frozen-Thawed (gelato-soluta)," as well as "the Slug-Apple (limacea)," not to mention "the Railroad-Apple, which perhaps came from a core thrown out of the cars," and finally "our Particular Apple, not to be found in any catalogue,—Pedestrium Solatium"[112]

This essay reveals not only the vivid details of the old wild apple landscape, but also, in its bitterness, the dominance of the new, market-driven, named, and cataloged one.[113] A letter to the *Genesee Farmer*, printed in 1860, confirms Thoreau's impression without his melancholy. Answering a query about which apples were best for market, D. A. A. Nichols, of Westfield, New York, had "no hesitation, as the market demands are uniform. The list is heard in every stall in every city market, at the season when the fruit is in market: *Rhode Island Greening, Esopus, Spitzenburgh, Baldwin, Newtown Pippin, Roxbury Russet,* and *Red Astrachan.*"[114] These repeated cries signified a radical shift in the farm landscape of New York. From the jumbled chaos of the wild fruit landscape, a small number of fruits had achieved not only fame, but also mass propagation. Through varietal lists and variety descriptions, commercial nurserymen had succeeded, at least in part. Diversity had been simplified, and the values assigned to fruit had changed.

Of course, this change was not as clean as the above paragraph might imply. As Cheryl Lyon-Jenness has shown, nurserymen faced intensified accusations of fraud during the midwestern horticultural boom of the 1850s and 1860s, a boom that New York nurserymen, particularly Barry and Ellwanger, supplied.[115] When traveling in northern Mississippi, the New York improver and landscape architect Frederick Law Olmsted interviewed a farmer who had been sold worthless seedling trees by an unscrupulous tree peddler. To establish his bona fides, the peddler had printed a fake catalog, listing a series of imaginary tree varieties with familiar sounding names, which Olmsted reproduced in his book.[116] While varieties might not be reliable, the appearance of reliability had come to inhere in the trappings of the nursery catalog; the categories of pomology were now needed to counterfeit fruit.

Thoreau focused his ire on names for the same reason that nurserymen obsessed over them: the names and descriptions attached to trees were vital to the markets that nurserymen built around them. By naming new seedling varieties and publishing works of varietal descriptions, American nurserymen bought credibility in and access to global botanical networks. Only such networks could provide the packed catalogs of fashionable varieties required to maintain the interest of refined buyers and market growers. Without names shared by a wide community, fruit varieties could be the object neither of meaningful taste and judgment nor of trade over distance. Both the free-for-all trade in grafted fruit and the seedling landscape shrank with the rise of printed tools for the regulation of names—particularly lists of rated varieties produced at pomological conventions and publicized through the nurseryman-dominated horticultural press. Their control was not total; pomological societies controlled neither the terms of value nor the continued movement of scions between neighbors nor the peddler trade. However, their effect was profound.

The story of the fruit tree variety does not provide a straightforward template for histories of the commercialization of plants. Other varietal stories played out differently, shaped by different biological possibilities and alternative cultural pathways. Threatened by the instabilities of reproduction, for example, seed-propagated varieties took decades longer to commercialize.[117] As apple and pear growers strove for fruit that could be identical across space, wine-grape growers invented the concept of *terroir*, embracing the variability created by different soils and seasons.[118] However, the story of fruit naming can help us see some of the ways that the marketing of living goods could be coordinated and accelerated. This in turn can help us understand both the rapid expansion of an antebellum agricultural economy and the creation of a landscape populated from a catalog.

The valuation of fruit varieties was highly shaped by their visibility and tangibility—fruit varieties were consumed with the eyes as well as the tongue. Pomological knowledge demanded the fine-tuning of these senses through words and images. Markets in turn depended on the slightness of differences between varieties as well as their diversity; as with wine, fruits' hard-to-perceive qualities stimulated the forms of connoisseurship, making sensory experience an expression of refinement. Machinery too had depended on direct experience, the theatrical, if highly debated public trials of efficacy. But as varieties and machines became standard elements of improvement, another form of value—fertility—by its nature invisible and perhaps unavailable to the senses, would come to preoccupy improvers, leaving quite different traces on the improving landscape.

8

The Balance-Sheet of Nature

On page 27, the American chemist George Waring's 1854 agricultural schoolbook abruptly flowers into narrative. The story's main character, an atom of nitrogen, sets off on a picaresque trip. Beginning in an animal, it falls into the sea and is eaten by a fish, from which it escapes as part of a "sea-insect," only to be captured by a whale. On the death and putrefaction of the whale, the atom rises into the atmosphere in the form of ammonia, falls from the sky in a raindrop, and is taken up by the roots of a food plant. "Could [the plant] speak as it lies on our table," Waring mused, "it could tell us a wonderful tale of travels, and assure us that, after wandering about in all sorts of places, it had returned to us the same little atom of nitrogen which we had owned twenty years before."[1] A year later, in 1855, Waring's fellow New Yorker Walt Whitman would famously echo this image of traveling atoms in the first lines of *Leaves of Grass*: "My tongue, every atom of my blood, form'd from this soil, this air." However, where for Whitman the atoms that formed the grass, his blood, and the air were pointedly ownerless—"every atom belonging to me as good belongs to you," he wrote—for Waring, matter was potential property. It passed like money from hand to hand, sometimes owned, but ultimately remaining "for ever as a part of the capital of nature."[2]

In the mid-nineteenth century, Waring's story of the atom felt new in ways that might not immediately strike us today. The connection between food and money is embedded in our language in the phrase "nutritional value." The idea of eating as the passage of specified quantities of matter from plants to animals to humans is foundational to our understanding of food, built into the tables inventorying the vitamins, minerals, fat, protein, and carbohydrates that many of us scrutinize on the backs of food packages. In the antebellum period, however, the cycle that Waring described represented two new ideas: a relatively recent understanding of organic life as composed of cycles of atoms that passed between living and nonliving matter and a new conception of nutrients as money equivalents, capital that could be owned or held in common. This was a different kind of value from value in fruit varieties, in land, in silk, or in machines. It did not emerge from the specific experience of flavor, the projection of consumer desire, or the publicly staged replacement of labor. Instead it would extend across the entire agricultural landscape, casting the secret processes of nourishment that bound farms together in monetary terms. Like fruit and machines, however, this form of value would eventually produce new improving goods. As Waring wrote, indeed, particular elements of nutrition were becoming commodities, justifying the prices of a new range of opaque and unsettling nutritional goods, a development that would have profound global implications for agricultural, ecological, and bodily knowledge.

This vision, these commodities, and Waring's story rested on a somewhat less poetic genre: the analytic table, the ancestor of the labels on our food packaging, which became central to American improvement in the 1840s. Its significance appears in the later volumes of Ebenezer Emmons's state-sponsored *Agriculture of New York*, in which Emmons shifted away from the god's-eye-view adaptation mapping of geology to the intimate specificities of the new analytic chemistry. Packed with hundreds of tables of chemical analysis, breaking different local soils, manures, varieties of potato, carrot, maize, peas and beans, hay and root crops, forest and fruit trees, and even flax, spearmint, yellow dock, and salt into lists of simpler substances—silica, lime, phosphates, magnesia, sulfuric acid, and carbonic acid—these later volumes comprise a seemingly exhaustive catalog of matter on antebellum farms.[3] This was Emmons's effort to trace the pathways of Waring's nitrogen atom, laboriously sketching them through columns of figures, which could, he hoped, also be translated into currency. Products of a new set of techniques that had recently arrived from continental Europe, analytic tables like these had become impossible to avoid in the agricultural press. They crowded agricultural journals, manuals, and advertisements and even bled into poetry and popular speech.[4] By midcentury, they had become one of the most publicly recognizable genres of American science, indeed organic chemistry itself was well on its way to monopolizing the title of "agricultural science." When William Seward, as governor of New York, championed the "principles" of the agricultural chemists "Liebig, Davy, Johnston and Dana" before an audience of thousands at the New York State Fair of 1842, he could assume that both names and principles would be recognizable.[5]

The rise and transformation of this kind of chemistry profoundly marked improvement. Organic analysis became foundational to improvers' vision of the farm as legible, measurable, and above all calculable. Through organic analysis, improvers hoped, the entire farm would be subjected to the accounting practices that were already central to their forms of experiment. By tracking the passage of nutrients, animals and plants and soils could be described in terms of their indebtedness to each other and to farmers. While this vision ultimately proved illusory, its strength both established new hierarchies of expertise within improvement and shifted terms of debate among improving constituencies. Ultimately, rather than following the transformations of the farm, the new analytic chemistry refocused on the world of improving goods, providing reasoning that underpinned new fertilizers and resetting nutritional debates to focus on fraud and industrial ingredients, creating a new, durable language of nutritional value. To examine these developments, however, we have to turn to the older theories that the new organic chemistry replaced and to the fundamental problem of eating on the mixed farm.

Vital Forces and Drunken Plants: Theories of Food in the 1820s

Farming is a system of production founded on living processes. Eating and growth are its central pillars. With their array of different crops and animals, most Northern farmers worked with complex interlocking economies of money and matter. Plants consumed fertile soil, manure, and swamp muck, air and water, transforming them into fodder, grain, or vegetables. These were sold or folded back into the soil as green manure or fed to cattle, sheep, horses, pigs, poultry, or people to become muscle, fat, and bone. Since most of the labor that farmers poured into their farms—plowing, harrowing, chopping feed, hauling manure, and cutting hay—coordinated these acts of eating, it is not surprising that an enormous proportion of improving experimentation focused on the control of food and fertility. New barn shapes channeled cow urine into troughs where it could be mixed with absorbent straw, keeping it from flowing away. Increasingly popular "green manures"—crops like clover or "lucerne" (what Americans now call "alfalfa")—were plowed under and rotted, transformed into new nourishing soils. Many of the most heralded new crops were not new human foods, but what we might call "producer foods"—foods needed for making other foods like meat and milk. When improvers planted rutabaga and brought in "English grasses," they were part of a movement to recapture British success in intensive foddering, which had, improvers believed, raised British cattle populations to unheard of heights.[6]

For improvers hoping to assess the profitability of their techniques, however, agricultural transmutations posed some complex problems. Every metamorphosis offered the farmer a bewildering array of options. Should cattle be set out to graze on harvested fields or eat grain? Would the cost of the additional flesh or the better milk outweigh the cost of the grain? Should animal dung be applied "long" (fresh) or "short" (rotted)? Farmers had only a limited amount of time, labor, and money to spend, but could speculate with them on a thousand products of fluctuating value. Oil cake, grain, and turnips produced more and better manure and meat than grass and hay did, and fields spread with slaughterhouse refuse produced better wheat flours, but how much of each did they make? Was there a conversion rate between manure and grass and grass and muscle? In short, what was "value in nourishment?" Even before the rise of the new agricultural chemistry in the 1840s, New Yorkers often turned to chemistry to answer these questions.

As improving journals pointed out, farming and chemistry were allied practices: both performed material transformations. Where farmers could make turnips into cows and pumpkins into pigs, chemists could turn sulfur, china clay, and sodium carbonate into lapis lazuli and, like New York chemist James Mapes, make "an ounce of sugar from a pound of linen rags."[7] Links between improvement and chemistry reached back to the Royal Society's "Georgical Committee" in the seventeenth century and had been strengthened by new texts produced by Scottish improvers in the eighteenth century. Moreover, though not yet as popular as geology, in the early decades of the nineteenth century, chemistry already had a powerful public presence in upstate New York.[8] Even as the Rensselaer Institute educated a generation of American chemists, in the provincial academies, the sons and daughters of wealthy farmers studied Jane Marcet's *Conversations in Chemistry*, and attended chemical demonstrations modeled on those fashionably performed in London by Sir Humphry Davy, then the most famous chemist in the English-speaking world.[9] By 1833, the provincial lecture circuit had become so lively that Amos Eaton complained about the surprising number of lecturers in Albany claiming to be descendants of the Scottish chemist William Cullen and railed against displays of exploding hydrogen and laughter-inducing "exhilarating gas" (nitrous oxide), what he called "the puppet show trinkets of chemistry."[10]

Until the 1840s, however, American agricultural chemistry did not much resemble Waring's description of circulating atoms or Emmons's tables of analysis. In the 1820s and 1830s, American improvers had generally followed a quite different model of soil fertility, popularized by Davy, a model of nutrition in which only organic matters—the bodies, excretions, and secretions of animals and plants—counted as true food. Nutriment in this model was described as separate from the soil itself—soil was a medium for conveying real food, that is, animal and vegetable matter, sometimes referred to as "humus," to the plant.[11] Humus could be measured, roughly, but it was hard to break down further into simpler components. Many early chemists argued in any case that humus was more than the sum of its components, that it had a particular quality or vital power that had come from its origins in living matter.

However, Davy's explanations were far from the only alternative available to American improvers. In 1834, for example, the *New York Farmer* reprinted a letter from a Southern planter, "R.R.H.," who stirred pints of his own blood into bushels of red clay in varying proportions and had them spread (presumably by enslaved people) on a series of corn hills for comparison. Alone, R.R.H. concluded, his blood was too powerful—a pint of it, undiluted, killed off a young cornstalk. But, excitingly, 1/64 of a pint, mingled with half a bushel of clay produced as much corn as a half-pint did. Further experiments with horse manure and clay led R.R.H. to a grand conclusion: that "animal excretions" were the "essence" of manure—providing fertility by breathing a manuring principle into the surrounding vegetable or mineral matter, irrespective of that matter's quantity.[12]

To R.R.H.'s readers, this might not have looked like fringe thought. Humoral medicine would have accustomed them to the drawing of pints of blood and to the notion of mysterious flows of forces transported by human bodily fluids.[13] Moreover, as R.R.H. pointed out, his invisible "principle" matched the behavior of other familiar materials in the nineteenth-century household. Yeast, given time, could "ferment all the grain on earth," and a piece of antimony, dropped clinking into a wineglass and then retrieved, could produce a powerful emetic without apparent loss to itself, over generations of family use.[14] The expandability and secret powers of R.R.H.'s principle could, he claimed, provide fertility "as inexhaustible as the water of the ocean . . . more valuable than the mines of Peru." In its reliance on a vital principle, at least, this model was not too far away from Davy's.[15] Such endlessly expandable power would mean that there was not a fixed conversion rate between manure, corn, and flesh.

New commodities raised further questions. Some of the most heated debates in the 1830s surrounded plaster of Paris, an odorless white powder made from gypsum. In the late eighteenth century, plaster of Paris had been one of American improvement's first visible successes, brought from France by Benjamin Franklin and adopted by well-off late eighteenth-century improvers. Scattered on soil, it seemed to have an "exhilarating" effect—fields grew visibly and dramatically greener and more luxuriant. Improvers treasured the story of Franklin's public experiment—sprinkling plaster to form the words "This field has been plastered" in a reluctant farmer's field, words that dissolved in the rain but then reappeared with the growth of the crop in a triumphant green.[16] By the 1830s, however, the scale of the plaster trade had hugely expanded, encompassing new quarries along the Erie Canal route: William Garbutt's plaster mill in western New York sold eight hundred tons in 1832 and 1,500 tons in 1833.[17] By the early 1840s, Mather estimated that Columbia and Dutchess Counties alone imported from fifteen thousand to thirty thousand tons of agricultural gypsum from Nova Scotia at five dollars per ton.[18] This expansion brought plaster's startling but sometimes spotty or transitory effects to the attention of a much broader group, spreading fundamental questions about the nature of plant nutrition.

Plaster's exuberance made a *Genesee Farmer* correspondent, "Plough Jogger," particularly suspicious in 1833, provoking a years-long debate. After an experimental disaster, Plough Jogger attacked "this stimulating, this intoxicating plaster of Paris" as a form of vegetal intemperance. "That larger crops may be grown with its aid for a year or two than can be without it, this I will not deny," he wrote, "I have no doubt but the drunkard can do more work with the aid of ardent spirits, for a few hours, than he could without it." This effect, however, was transitory: as "with the whiskey drinker . . . when the effect of this stimulating is over, the strength of the land is exhausted, and, like the drunkard, literally lays down in the furrow."[19] In revival-scorched western New York, such accusations were incendiary. Plaster's defenders, particularly the owner of the local mill and the nurseryman who had originally advocated plaster to the *Genesee Farmer*, united in their scorn for Plough Jogger. However, their enraged defenses did not coalesce into an orthodox explanation. In their letters, plaster featured variously and contradictorily as (1) "a true vegetable food"; (2) a solvent that made true vegetable food easier to absorb; (3) an absorbent substance attracting valuable moisture or an unnamed vegetating principle from the atmosphere; or (4) as a more wholesome "stimulant" or "condiment," like pepper or salt, piquing the appetites of wheat and corn plants so that they ate more enthusiastically.[20]

These explanations echoed arguments being made by European chemists and plant physiologists like Davy, Chaptal, and De Candolle, though they were often accompanied by accounts of experiments, expected to settle New Yorkers' questions in New York itself.[21] However, their answers also had fundamentally different implications for the question of plaster's value. If a food, it was, of course, valuable, but if it simply made real plant food more soluble or delicious to plants, then it might encourage plants to suck nourishment out of the soil more swiftly. To support this latter claim, plaster's opponents pointed to "plaster-sickness," fields rendered barren after the repeated use of plaster.

By the 1830s, other goods were beginning to complicate farm nutrition. Fertilizers were, of course, not new: manures had been crucial to British and American improving advice from its earliest period, and as fertilizer promoters sometimes pointed out, the classical Roman authors Varro, Pliny, Columella, and Cato had variously suggested pigeons' dung, lime, manure in covered pits, olive oil, and saltpeter. In the late eighteenth and early nineteenth centuries, Long Island farmers had adopted what they considered to be a Native American practice of spreading freshly caught fish over the fields and had begun to mine native villages more literally, crushing shells from middens for their soil-amending calcium.[22] In the 1820s and 1830s, this process accelerated, feeding on the waste goods of industrial and urban concentration. Long Island, within tantalizing reach of city markets and waste flows, became an incubator for experiments in fertilizing goods. Long Island farmers took advantage of New York's growing population of horses to replace their topsoil with manure shipped over from the city by specialized dealers.[23] In the early 1830s, Long Island's bone mills brought the enthusiasm for crushed bone fertilizer to North America—an enthusiasm that rapidly spread to Albany, Troy, and other major cities.[24] By 1837, farmers on Long Island and the lower Hudson could also buy powdered, dried, and deodorized human feces, which sounded better in catalogs under its French name, "Poudrette." The *New York Farmer* sold it for $1.50 a barrel from the Lodi Manufacturing Company on the Hackensack River.[25] As in Britain, foods for animals also appeared—industrial by-products, like crushed linseed "oil cake" or alcohol-sodden distillers' grains, joined the new varieties of grass seed, clover, vetch, turnips, swedes, and rutabagas.

Complicating matters, substances from farms themselves were also gaining market value. Ash, a fertilizer, had long been a part of international potash markets. But other markets, like the hay market for urban horses, were new. From an improving perspective, these markets were dangerous temptations. Manure, hay, and ash markets could encourage farmers to sell their fertility away—one commentator pointed out sourly "how very valuable (I mean money value) would it become were every farmer to rob his fields to supply the manure market?"[26]

The rising interest in new manures was partly a symptom of a larger fertility puzzle: the apparent crisis of soil exhaustion. Journals all over the North had printed cautionary tales of soil failure in long-colonized parts of New England and the southern Piedmont. In New York, old farmlands on Long Island and in the Hudson and Mohawk Valleys seemed to be losing their wheat-producing capacity.[27] In western New York, these disasters loomed more distantly—wheat production would continue there until the

1850s, but western farmers were well aware of the problem—many of them had grown up on farms that were suffering.[28] The cause of declining yields was not clear—was the soil "exhausted" in a bodily sense, requiring a rest? Was something being secretly extracted from it? Was wheat a fleeting phase of development rather than a permanent crop? More broadly, what was food, and what was fertility?

Amidst these debates, even individual improvers' concepts of fertility were not internally coherent. In the *Cultivator*, Jesse Buel drew on common bodily metaphors describing the area around Albany as "a worn down ox," now "exhausted" of its vital powers. Elsewhere, Buel extended these bodily metaphors, describing the appetites of different plants in class-inflected terms. Hoed crops, like corn, potatoes, beans, and turnips, were "strong hardy feeders, relishing, and thriving upon, the coarsest food." Grains, by contrast, "being more delicate feeders," needed manure that had been fermented and "parted with its grosser properties."[29] However, in almost the same breath, Buel employed what would soon become the dominant model for describing fertility. "Our ancestors," he wrote, "made annual drafts upon the riches of the soil, without making deposites [*sic*] to meet them—till neither wheat, nor hardly any thing else valuable, would requite them for their labor."[30] In this model, the land was not a body that labored and required care, it was a bank, and working it was akin to spending. The analytic vision that became popular in the 1840s would extend this vision of chemical value, one that seemed, for a time, more secure than the economy of money.

Organic Analysis Crosses the Atlantic

During the 1840s and 1850s, agricultural improvement carried chemistry to new heights. Agricultural journals carried chemistry columns and advertised self-proclaimed "popular treatises" on chemistry.[31] State legislatures sponsored agricultural chemists and distributed chemistry texts to schools, and lyceums and agricultural societies sponsored chemistry lectures. Historians have traditionally linked this burst of enthusiasm to the American publication of the works of the German chemist Justus von Liebig, particularly his *Organic Chemistry in Its Application to Agriculture and Physiology* (1840).[32] American improvers partly shared this perception. Though other European and American chemists became familiar names, it was Liebig's name that Americans attached to fertilizer companies ("the Liebig Poudrette Manufacturing Company") and analytic equipment ("Liebig's Apparatus for Organic Analysis").[33]

In fact, Liebig's dominance in the United States was an artifact of his negotiation of British improving networks and of American reliance on British print. Humphry Davy's preoccupation with electro-chemistry had led British scientists away from the refinements of analysis that occupied continental chemists. With Davy's death, Liebig briefly became the evangelist of European analytic chemistry.[34] British improvers read his book, and British manufacturers financed his (disastrous) commercial fertilizers.[35] Following British print, American agricultural journalists and authors freely abstracted, paraphrased, and digested Liebig's works, obscuring his contemporaries and muffling their debates.

In fact, Liebig's work was part of a wider movement in European chemistry during the 1820s and 1830s. Using a mixture of new analytic procedures and new "paper tools," chemists like Liebig, Friedrich Wöhler, and Jean-Baptiste Dumas extended the quantitative techniques used for finding the components of mineral substances to the more chaotic "jungle" of organic substances, which, like Davy's humus, had previously been considered both fundamentally different from inorganic substances and too complicated to analyze. In their hands a plant and animal chemistry deriving from the natural historical traditions of the eighteenth century shifted toward a new "carbon chemistry," which extended beyond substances extracted from living bodies to include artifacts created in the laboratory. Chemists came to describe organic substances in terms of "indivisible units of elements defined by their relative, invariable combining weights."[36]

Liebig and chemists like him denied the distinction between living and nonliving matter. This meant that they denied the existence of a vital principle, like that described in R.R.H.'s experiments with blood. It also meant that they extended the definition of food beyond dead organic matter to include all substances found in the bodies of organisms. Rather than inert media, condiments, or remedies for acidity, for example, lime and silica became the building blocks of plant and animal skeletons, joining the unchanging particles of nutriment that moved through the vast cycles of matter in Waring's textbook and Whitman's poetry.

Liebig contributed to the new quantitative mode mostly by making it easier to *be* quantitative, by developing and spreading changes in analytic technique. His practice centered on a piece of apparatus that he had invented in 1830: the *Kaliapparat*—five connected glass bulbs, partially filled with potassium hydroxide, which captured carbon dioxide as a solid rather than an unwieldy gas, allowing him to analyze much larger samples and describe their proportions with greater precision and to avoid the messy, tricky, and dangerous mercury-filled pneumatic trough used previously.[37] The *Kaliapparat* itself was part of a turn to cheaper blown-glass instruments, which opened the ranks of chemists to new practitioners.[38] Through such innovations, the practice of quantitative organic analysis accelerated and widened beyond a small coterie of virtuosi—under Liebig's tutelage, it seemed, even students could produce an analysis worthy of a Dumas or a Berzelius.[39] His cheaper, faster system of analysis and his laboratory full of industrious students made it possible to carry out the tricky task of organic analysis on a much larger scale.[40]

The Analytic Dream of Accounting

Since Liebig argued that organisms made themselves out of the same elements that minerals did, without recourse to a living force, historians sometimes describe Liebig as an "anti-vitalist" or a "mechanistic" thinker. To improvers, however, Liebig's work rendered the landscape in terms of a paper tool already central to their experimental culture: accounting. Agricultural experiment reporting already cast the fields and animals as debtors and creditors, owing the farmer for labor and paying him in marketable goods. But the complex flows of food in mixed farming meant that fields and animals were also indebted to each other. Cattle ate cornstalks but returned manure to the field. Characterizations of fertility as money were common enough by 1841 that a writer to the *Farmer's Visiter* fumed at the idea of indebtedness in food. "Why do farmers charge manure to their crops?" he demanded. "The soil is entitled to it—it belongs to his cornfield as a matter of right, and he has no right to fix a value on what though in his possession, is not his."[41] More numerous and orthodox articles focused on a different problem: How much should the manure pile be credited for the growth of the field?

Liebig's analytic methods, his supporters argued, helped answer this question definitively by casting eating itself as a measurable transaction. Plants subtracted a fixed weight of fertilizing atoms from the soil and the air and structured them into leaves and fruit; animals reconfigured these same materials into milk, meat, bone, and manure. These descriptions of the economy of nature drew equally from flows of currency and flows of matter. "Every element existing in nature," declared the Ohio chemist David Christy, "belongs to the treasures which the beneficent Creator has laid up in store for the use of man . . . ceaselessly moving in one great cycle."[42]

Central concepts of the new chemistry were easy to render in an accounting idiom. For example, in "The Chemical and Physiological Balance of Organic Nature," a table developed by the European chemists Dumas and Cahours, the relations between the animal and vegetable kingdoms appeared as balanced acts of "production" and "consumption." Vegetables produced "proteiniferous substances, fatty matters, sugar, starch and grain," while animals *consumed* them, producing in their turn "carbonic acid, water, ammonia" for the benefit of plants.[43] In American improver William D. Cochran's text *Agricultural Book-Keeping*, the Dumas-Cahours table appeared again, this time under a new title: "The Balance-Sheet of Nature."[44] In the prescriptive literature of agricultural accounting, the conservation of matter became evidence of a balanced economic relationship between living things. Plants and animals were good, mutually indebted neighbors.

Liebig had himself made clear that agricultural accounting should become a long-term research program. In his *Familiar Letters on Chemistry*, he instructed chemists to analyze the ash of every kind of plant as grown on every soil, to learn which elements each plant consumed in what quantities. "With this knowledge," he argued "the farmer will be able to keep an exact record of the produce of his fields in harvest, like the account-book of a well-regulated manufactory." Following a "simple calculation," the farmer would "be able to express, in pounds weight, how much of this or that element he must give to the soil in order to augment its fertility for any given kind of plant."[45]

During the 1840s, American chemists and improvers took up this project. By identifying the elements of nourishment, they argued, they could assign value to foods and soils, providing the data necessary to determine rates of conversion. Since there was seven times as much nitrogen in wheat as in straw, for example, farmers would need to manure with seven hundred pounds of rotten straw in order to make one hundred pounds of wheat. With the assistance of chemical analysis, even the gases that escaped rotting manure could receive a quantitative value. "For every pound of the strongly-pungent ammonia lost in the air," the improving publisher Joseph A. Smith contended (quoting Liebig without attribution), "a loss of at least sixty pounds of corn must correspondingly be sustained," just as "with every pound of urine a pound of wheat might be produced."[46]

Like Cochran, American chemists and improvers reconstructed chemical tables to fit them into accounting practices. The agricultural journalist Daniel Lee's ash analyses, for example, included an extra column listing the number of pounds of each substance needed to produce a ton of the plant in question. In an extended essay published with these analyses in the patent office's annual report, Lee made clear that these tables were intended to correct failures in accounting with crops.[47] "Let us suppose," he wrote, "a farmer produces crops worth one thousand dollars, and they cost him, including all expenses for labor, wear of implements, interest on capital, &c., eight hundred and fifty dollars." Ostensibly, his profits should equal $150. However, if he should take into account "as much of potash, soda, magnesia, phosphorus, soluble silica and other elements of crops, as both tillage and cropping had removed," then the balance sheet would read differently. "It is only by consuming the natural fertility of the land that he has realized any profit."[48] True profit, Lee declared, had to take into account the expenditure of farming's hidden capital base.

Promises of a transparent, accountable landscape fit well into antebellum economic narratives. At the individual level, the accounted-for farm looked a great deal like the accounted-for household. In the 1840s, texts on domestic economy increasingly encouraged middling New Yorkers to secure themselves against the economy's more dangerous gyrations by locating household "economies": places where money could be saved.[49] Agricultural chemistry likewise promised hoards of secret value—the disregarded fertility of swamp muck or the unexpected nutritional value of discarded cornstalks. It promised to insulate farm families from larger patterns of landscape destruction. Such arguments were particularly appealing to New York landlords and boosters concerned about the westward migration of local farmers—the seeming inevitability of population loss could be countered by an appeal to personal economic virtue, that is, by the development of regular habits of self-surveillance and self-control.[50]

As with the leaves of *Morus multicaulis*, which had failed only a few years before Liebig's works arrived, the elements of fertility gained metaphoric weight from references to gold: mines, treasures, and gold pieces. But stories of the trickling away of hidden stocks of value in soil resonated with more specific narratives of secret economic loss. The Bank War of 1834, had, on the whole, enriched New York, moving the locus of banking power from Chestnut Street in Philadelphia to Wall Street in New York. But it and the two successive panics that followed it closely in 1837 and 1839 had revealed underlying weaknesses in American banking and in the global financial system as a whole. During all three crises, banks around the nation had "stopped payment"—refusing to exchange specie for paper money. Even during booms, banks frequently failed, unable to come up with enough precious metal to cover all the paper notes that circulated in their name. Some, indeed, had nothing in their vaults at all, thus turning the performances of bank credibility (columns, elaborate images on the money, fine signatures) into functioning money.[51] Such a backdrop made stories about the secret disappearance of concrete value and sudden revelations of empty vaults compelling.

As in other forms of economic discourse, individual and business virtue could be scaled up to describe the wasteful habits of cities and nations. Writing for the patent office, Daniel Lee estimated that "fertilizing elements . . . equal to three hundred and fourteen million bushels of corn" were "in effect taken from American soils, of which next to none is ever returned in night-soil or liquid manure."[52] Just as individual farmers were unknowingly spending their capital, the nation as a whole was operating at a secret deficit, expending atoms into the sea. In Lee's ideal agriculture, the "husbanding of fertilizing atoms" would become the responsibility of the nation as well as of the farmer.[53]

Even as it offered explanations for failure, analytic chemistry also promised a cornucopian vision of redemption that seemed not only to promise an escape from recent soil damage but also to offer entire freedom from the constraints of adaptation.[54] At a state society meeting on wheat, for example, Lee suggested, not only that the degraded soils of the Hudson could be turned back to wheat, but that given enough "lime, sulphur, and phosphorus," the lands of the Southern Tier, poorly served by the Erie Canal and much less developed than northern parts of the state, could be turned to the reliable cash returns of wheat culture, though, Lee admitted, at the moment "the soil . . . [was] but poorly adapted to wheat culture." Even as he argued against adaptation language, Lee still found providential signs in the Southern Tier landscape. "I regard it as a fact of great practical importance," he observed, "that wood ashes, even leached ashes [ashes already used for making potash] contain all the earthy elements of this invaluable bread-bearing plant."[55] If such grand schemes of transformation seemed ridiculous, improvers had only to point to the terrible lands of Long Island, rendered fruitful by new flows of manure.

However, these appealing theories of value in nutriment also challenged other theories of economic value then contesting for primacy. Anti-Renters and Free Soilers had also been making claims about value in terms of precious metal, comparing it not to food but to labor. When the Spanish struck gold and silver in Peru and Mexico, the *Albany Freeholder* lamented in 1845, "Spain dropped her spade, her plow, and her hoe" and, as a consequence, had "almost ceased to be heard of except in her misfortunes." In happy contrast, it argued, by relying on agricultural labor, the Union had "grown so rich that the gold of the mines of Peru, gathered for twenty years, will not pay for the produce of our industry in *one year*." By setting the value of labor above even metallic wealth, the *Freeholder* was making a fairly standard assertion of the labor theory of value.[56]

The idea that human labor was responsible for the creation of economic value had become central to Anti-Renter and Workingmen's Party demands for citizenship and property rights and to standard improving addresses on the dignity of farming. Some earlier models of agricultural fertility had confirmed this vision of the power of labor. In his *New Horse-Hoeing Husbandry* of 1731 (still sometimes read in the nineteenth century), Jethro Tull had argued that plowing inherently increased fertility by grinding the soil into particles small enough for plants to feed on. According to this model, labor *created* value—working the land unlocked an inexhaustible store of materials for the construction of plants.[57] While Tull's theory had little currency by the 1840s, improving warehouses were full of subsoil plows, cultivators, soil rollers, stump haulers, and other machines that magnified human and animal labor and turned unbroken grass or unpromising fields into sources of profit. To tenants, these transformations were "improvements," lending legitimacy to their ownership claims.

On the other hand, tenants were well aware of the dangers of soil exhaustion; one of the first salvos of the Anti-Rent wars had been a 1839 letter from Rensselaer tenants centered on this very point: "owing to the sterility and roughness of the soil and country," the tenants complained, "it has become physically impossible to raise Wheat to pay our rents."[58] By recasting the act of tillage as an act of spending rather than an act of creation, however, the new models of nutrition raised potentially awkward questions for tenants. Rather than adding value through cultivation, perhaps tenants had spent their landlords' capital. Such concerns would legitimize leases requiring fertilizing practices—leases like those used by the Wadsworths of Geneseo in western New York from the late 1830s onward, which not only forbade the sale of manure, but insisted on its spreading and, by the late 1840s, required the purchase and spreading of plaster at one hundred pounds per acre of pasture.[59]

It is clear that the author of the *Freeholder* article had not only absorbed the implications of the new agricultural chemistry but that they troubled him. He not only admitted "the vast value of an accurate analysis of soils" but also directly echoed the banking language of improvers like Buel. He wrote, "Now, in our older fields we begin to see that in our excessive haste we have *overdrawn our bank*. Science and care must now be consulted to restore that vegetable power which has been *too profligately squeezed from the bosom of the earth!*"[60] However, he concluded with a reassertion of the significance of labor that recalled Tull's theories: "*the grand art remains where it was*. It is drilling the soil. . . . No art will ever render this constant stirring of the soil unnecessary.—*Man's labor* is bound to be forever mixed up in the products of agriculture." The "Farmers' Creed" reprinted in both the *Freeholder* and in the *Farmer and Mechanic* teetered between these two ideas. "We believe that the best fertility of the soil is the spirit of industry, enterprise and intelligence," it declared, "without this, lime and gypsum, bones and green manure, marl or plaster, will be of little use."[61]

Other radical agrarians took a more proactive stance, attempting to turn wasting agriculture into the same kind of economic crime as landlordism and to link the two together. "Our Creator has not commanded us to subdue and then exhaust the land, but to till, to improve it," the *Farmer and Mechanic* announced in 1846, "Is he not a robber who exhausts the land of its fertility? Is he not worse than a 'land monopolizer?' Can we not hitch the anti-exhausting principle onto the 'landsite' doctrine, and thus add strength to our friends who are striving to 'vote themselves a farm?'"[62] Ultimately such arguments would morph into a standard Anti-Renter narrative—that landlordism itself was responsible for soil deterioration. Accusations of soil wasting, however, would remain an uneasy current in the relationship between tenants and landlords.

Against the Senses: The Rise of the Consulting Chemist

If agricultural chemistry troubled relationships between landlords and tenants, it roiled hierarchies of improving knowledge even more. Even more than agricultural geologists or pomologists, the new analytic chemists would work to make farmers dependent on the services of experts. Unlike pomology, geology, or machinery judgment, the agricultural chemistry of the 1840s drew the valuation of soils, fodders, and foods away from fundamental sensory experience. Pomologists had worked to spread the forms of taste on which their markets depended. In geology, the canal cuttings and diggings that geologists used to observe strata were, as we have seen, literally open to the public; their processes required little in the way of specialized equipment. Survey geologists had characterized soils according to Davy's method; in fact, this method simply translated categories well known to farmers into learned language.[63] The major geological terms for soil—"argillaceous" and "siliceous"—were Latinate terms for "clayey" and "sandy." "Animal and vegetable matter," Davy's vegetable food, in practice meaning muck, mold, dung, and compost, made soil "loamy."

Though Davy proposed analytic methods for determining their quantities, these substances were all perceptible to informed senses. Where siliceous soils were harsh and gritty, the clay in argillaceous soils gave them "a smooth feel" and exhaled a peculiar damp odor when breathed upon. Where siliceous soils were dry and flowed, the clayey soils could be kneaded into little balls. Rotting animal and vegetable matter, of course, stank.[64]

It is not frivolous to acknowledge this stench—in an age of miasmatic medicine, it was crucial to early American theories of health. Rural Americans had long used their sense of smell to perceive not only the strength of their manure, but the healthiness of particular places and whole districts. In the 1840s and 1850s, the sense of smell was becoming increasingly formalized as sanitary reformers and physicians in American cities, what Melanie Kiechle has called "smell detectives," turned the pursuit of bad smells into an expert practice, hunting and mapping the odors and vitiated air that caused disease.[65] Agricultural improvers were well aware of this—indeed, many of them were deeply involved in these reforms: Waring, whose textbook opened this chapter, would gain much more fame as an urban sanitary reformer. So would the landscape designers Andrew Jackson Downing and Frederick Law Olmsted.[66]

Earlier agricultural chemists had used smell significantly in their own practice and encouraged farmers to do so too. The French chemist Jean-Antoine Chaptal had encouraged farmers to use their eyes and noses to analyze their soils—if deposits were brown, they contained "a mixture of animal or vegetable substances." If burned on a red-hot iron, the smoke arising from them would either have "the odor of burning, leather, hair, or feathers," indicating animal substance, or "of wood smoke," indicating the less valuable vegetable substance.[67]

For urban sanitarians, many of whom had chemical training, the accessibility of bad smells to the senses helped justify their prescriptions and gradually to form a platform on which they could build. But bad smells provided a very different set of problems for agricultural improvers—the very substances that were becoming the subjects of nuisance law in cities: manure, offal, rotted fish, swamp muck, and damp subsoils were the sources of fertility that improving authors asked their readers to reimagine as treasure. Walt Whitman expressed this sense of the peril of rot and corpses and of their transformation in "Poem of Wonder at the Resurrection of the Wheat" (1856), later retitled "This Compost."[68] "How can you furnish health, you blood of herbs, roots, orchards, grain?" he asked. "Are they not continually putting distempered corpses in the earth? Is not every continent worked over and over with sour dead? . . . Where have you drawn off all the foul liquid and meat?" Whitman, having reviewed Liebig enthusiastically in 1847, found an answer in chemistry.

> What chemistry!
> That the winds are really not infectious,
> . . .
> That when I recline on the grass I do not catch any disease,
> Though probably every spear of grass rises out of what was once a catching disease.[69]

However, new continental analytic chemists were drawing away from sensory experience.[70] In popular texts, chemists loved to show how chemical compounds concealed the properties of their constituent elements, observing, for example, that common table salt was composed of chlorine, a poisonous green gas, and sodium, an explosive metal, or pointing out that rose scent and lighting gas each had the same collection of atoms in the same ratio.[71] It is not surprising, therefore, that chemists would come to challenge the utility of smell as an aid in farming even as many of them supported the new sanitary reforms. In an 1850 letter to the *Cultivator*, Yale chemist John Pitkin Norton reminisced about "a somewhat celebrated Scotch Farmer," who had, from years of experience "come to like the odors of the most powerful manures." Given a bottle that even the laboratory-hardened Norton thought vile, the farmer's "countenance at once expanded in satisfaction, and he snuffed up the savoury fumes with undisguised delight; 'that'll be grand stuff,' said he at last, and at once inquired where it could be obtained."[72] Good farmers like bad smells, Norton explained, but that was not enough to perceive the difference between a good manure and a chemist's trick.

In their most confident moments, indeed, analytic chemists promised to render sensory experience of the landscape obsolete. Describing an imagined forest, Liebig claimed, "When we are acquainted with the nature of a single cubic inch of their soil, and know the composition of the air and rain-water, we are in possession of all the conditions necessary to their life."[73] In an 1850 lecture to the New York State Agricultural Society, Norton's teacher, the British chemist James F. W. Johnston, would explain that the laboratory surpassed simple human senses. "The chemist in his laboratory," he assured them "is better armed for the investigation of nature, than if his organs of sense had been many times multiplied."[74] For Liebig and Johnston the job of valuation had become a matter of glassware, controlled flames, and scales so sensitive that they had to be shielded from a stray breath. George Waring's textbook, bookended by advertisements for his services, maintained strongly that farmers could not perform analyses, since it took two years to learn the necessary skills. This was not an argument against farmer chemical knowledge, Waring maintained—to *use* analysis accurately would require his book and the evenings of one winter. Failing this, farmers could hire an interpreter, someone like Waring himself, to tell them how to turn their soil analyses into improving practice.[75] While farmers' minds might be capable of understanding chemistry, their bashed and thickened hands were "unfit for the most delicate manipulations."[76] Instead, chemists should post themselves in cities, protecting their fragile instruments and applying them to samples sent from around the state.

By 1852, one correspondent to the *Cultivator* complained that chemistry had become too powerful. "We protest against the use of the term scientific as applied solely to Agricultural Chemistry," "Cultor" wrote, "We claim that it has a wider and a more universal meaning, and that farmers are wronged by the exclusive and partial views so often made public on this subject."[77] To Cultor, agricultural science encompassed any rationalization of farming including "the principles of vegetation, the proper use and application of manures, the laws of farm husbandry and economy, the preparation and treatment of soils to adapt them to particular crops, and in short, the whole routine of farm labor . . . systematized and conducted upon rational principles."[78] Increasingly, however, chemists claimed the mantle of agricultural science.

In fact, as Waring and other chemists hoped, improvers' embrace of chemical valuation created a significant market for expert chemical labor, one of the first such markets for scientific work in the United States. Clamoring requests for analyses offered a vital haven, not only for scientists cast adrift by the end of the state geological surveys of the 1830s, but also for a newer generation.[79] Some, like Emmons, were hired by state legislatures or the patent office to conduct surveys or to receive specimens for analysis; a few, like John Pitkin Norton, entered high-profile professorships, particularly at Yale. However, many more operated in less elevated institutional contexts as "consulting chemists." By 1857, *Trow's Directory* would list forty-seven chemists in New York City alone, an enormous number compared to the few employed in academic posts.[80] This expansion was not limited to the metropolis: one New York chemist remarked, disapprovingly, "There are hundreds of analyzers scattered through the country, who will work very cheap, but their results cannot be confided in, and must necessarily lead the farmer into expensive and injurious operations."[81]

Even as the European organic chemists who had inspired them moved away from commercially significant materials and toward carbon chemistry and paper abstractions, American chemists moved the other way—collecting samples and occupying themselves with the assertion of value. As they did so, however, they undermined the promises of nutritional transparency that their kind of chemistry had made.

Cracks in the Analytic Dream

With his state salary, assistants, and laboratory space in Albany at the center of New York's improving networks, Ebenezer Emmons was among the luckiest of the new agricultural chemists. His hundreds of samples came from the state agricultural society's collections or were sent by its more prominent members. Their breadth allowed Emmons to perform several interlocking kinds of analyses: the ash analyses demanded by Liebig, which tracked the mineral substances that different plants removed from the earth; the soil analyses that described the proportions of these minerals in different places; and "organic analyses" of plants, that broke them into elements of animal nutriment, like starch, albumen, casein, and fiber. Emmons made sense of these lists of figures by comparing them to other published analyses of brains, teeth, cartilage, and dried muscle, with particular emphasis on blood and milk, substances that became valuable flesh. With the corruptly inflated printing budget granted him by the governor, Emmons could publish all these tables together to trace the tracks of nutriment on a grander scale.[82]

Emmons's resources also allowed him to drill down to a level of analytic detail that consulting chemists rarely had time for. Samples of the same plants grown on different soils received separate analyses, as did samples of corn at different stages of growth, and samples of milk from cows in different seasons with different feeds. Likewise, he produced individual tables for the center and each end of the of the merino potato. Such finely detailed analyses were increasingly common. As the crowds of chemists grew, those who had been formally trained in Europe, at Yale, or at the Rensselaer Institute differentiated themselves through virtuosic displays like Yale chemist John Pitkin Norton's "Analysis of the Oat," which won prizes in Scotland and was widely reprinted around New York.[83]

This obsessive detailing, however, revealed cracks in the accounting dream. Farming seemed too messy and bodies too variable to be simply accounted for. If corn changed its nutritional value depending on when it was cut, or valuable gases evaporated from piles of manure, how were the debts of the cow to the cornfield or the cornfield to the manure pile to be known? Given the variability suggested by these later analyses, academic and institutional chemists suggested, perhaps the consulting chemistry of their rivals was impossible to conduct. "No account is taken of $5 analyses," wrote Samuel Johnson from Yale, "A reliable analysis cannot by any means be made for so little money, *and support the analyst.*"[84] The hard realities of the market for labor, he claimed, revealed the fraudulence of cheap analysis.

Other problems also clouded the vision of clear columns of debit and credit with fields. Most fundamentally, the currencies of nutrition remained perhaps even more debatable than the currencies of banks—exactly which potential nutrient substances were analogous to gold? Liebig himself had stirred up controversy on this point by arguing that plants got two of their major elements, carbon and nitrogen, from the atmosphere itself. The first of these was generally accepted (and still is), but the second was more controversial. Nitrogen was a central element in meat, a "flesh-forming" element, the element that many manures were expected to supply. However, Liebig (and following him Waring) had argued in fact that the enormous stocks of nitrogen in the atmosphere became ammonia, then fell down as rain, and could be trapped by plaster and other soil amendments.

This theory held considerable appeal for both farmers and chemists. This becomes plain in a Liebig quotation reprinted by New York chemist, fertilizer manufacturer, and *Working Farmer* editor James Mapes: "One hundred and ten pounds of burned gypsum [an easy-to-handle odorless white powder] fixes as much ammonia in the soil, as six thousand, eight hundred and eighty seven pounds of horses' urine [distinctly not an easy-to-handle odorless white powder]."[85] For farmers it meant that only mineral manures like calcium and phosphorus had to be dragged to the fields. For chemists it meant that only relatively simple ash analyses would be relevant for plant growth, enormously simplifying the work of analysis. However, few European chemists agreed with Liebig, and new nitrogenous fertilizers would quickly raise doubts about his theories.

Questions remained. Were minerals true foods or stimulants? Did fertility come from the inexhaustible common stocks of the atmosphere or the increasingly fenced and divided landscape? Given these uncertainties, it's not surprising that some analysts hedged their bets; James Chilton's analysis of swamp muck in the *American Agriculturist* in 1846, for example, included a Liebig style ash analysis but also the quantities found of a substance from a conflicting model: "geine"—a vitalist category for vegetable and animal matter put forward by Liebig's American rival Samuel L. Dana.[86]

Other improving practices complicated nutritional transactions still further. A typical letter came to the Albany *Country Gentleman* in 1853. The author wanted to know if "fifty bushels of corn will be of equal value if fed to a horse, hog, or to neat cattle, for manure." If not, he asked, "which [animal] should have the preference?"[87] Even equipped with chemistry, the editor felt unable to give a clear answer. Such a response, he told Little, would require, first, "knowledge of the value of the manure," second, knowledge of "the value in nourishment imparted to the animal," and third, knowledge of the savings in other kinds of food. Worse, all of these would depend on the animal itself: "For example, a landpike [an inferior pig] may retain, assimilate and convert to flesh, one-half of the elements of corn"; by contrast, an imported pig, the Berkshire, would make flesh of three-fourths of the corn. "The landpike will of course yield the most manure from the fifty bushels of corn, yet be by far the most unprofitable animal on account of the little pork he manufactures."[88] As new "thrifty" breeds of pigs, cattle, and sheep arrived at auction houses, their capacity for transforming feed mostly into profitable meat or milk became the source of their appeal. Sold by agricultural warehouses like Emery and Sons, a new range of rollers, choppers, "masticators," and cookers also altered equations promising to render waste farm goods like cornstalks, wheat straw, and excess broom corn as more nourishing alternative foods for cattle. Digesting food outside the body, machines could make it more valuable inside the body.

Finally, by the early 1850s, some influential chemists had begun to argue that it was not possible to accurately assess the nutritional value of soil at all using their current methods. A particularly embarrassing moment came when David A. Wells analyzed the fertile soils of Ohio's Scioto Valley and found mineral compositions basically identical to those found in degraded and unproductive New England soils.[89] Such analyses, critics pointed out, failed to take into account the different ways in which basic elements of fertility actually manifested themselves—whether the plant would recognize them as food or not depended clearly on some other factor, perhaps their fineness, perhaps their origin.

Under such conditions, it was difficult to imagine how even the most eager student of Emmons might produce a coherent picture of nutritional transactions or develop a true picture of the tangled cycles of nutrition on a farm. Instead, the work of chemistry coalesced more and more around another function: the valuation of nutritional goods.

The Invention of Guanos

In 1846, laborers working for Robert Pell were making a sort of fertilizer cocktail. Laying Pell's wheat seed on his barn floor, they sifted a pharmacopeia over it: oyster shell lime, two kinds of charcoal, ashes, "Jersey blue sand," "Peruvian guano," "silicate of potash, nitrate of soda, and sulphate of ammonia," as well as brown sugar (perhaps to stick the mixture to the seeds). Outside, they sprinkled three dollars' worth of "30 different chemical substances over the field itself." Even on the degraded soils of eastern New York, Pell's wheat responded, he told the American Agricultural Association of New York later. It came up fast, green, and in enormous quantities—seventy bushels per acre at a time when farmers in the Genesee Valley "wheat country" were often content with twenty. Best of all, when analyzed by the American Institute, it contained extra gluten, which had recently been declared the institute's main criterion of flour quality, so it won that year's flour prize.[90] Pell's experiment further demonstrates the expanding dominance of organic chemical analysis. Gluten's rise itself had been a Liebigian story. Found, like muscle, to be full of nitrogen, it was deemed "meat-forming," superior in nutritive quality to carbon-based starch. (More prosaically, bakers found, flour with more gluten expanded into a greater volume and weight of bread.) Analytic values therefore meshed in Pell's experiment, as Pell had selected nitrogen-rich fertilizers for their theoretical capacity to make gluten.

However, the breadth of Pell's pharmacopeia also reveals the extent to which the market for fertilizing substances had expanded into a network of manufactories. During the 1840s and 1850s, such nutritional goods became more and more common. As they did so, markets would appear, trading the invisible substances that made up the capital of nature.

The substance that most powerfully brought these markets and debates to the fore was Peruvian guano. The dried dung of seabirds, deposited over centuries and preserved by the arid environments of western Peru, guano had been an element of Incan and pre-Incan agricultural practice, "discovered" on the global travels of Alexander von Humboldt. Davy had experimented with it in Britain in 1805, and so had the editor of the *American Farmer* in 1824, but it had made little public noise until the British began to trade it in bulk in 1840.[91] The first shipment had arrived in New York only two years before Pell's experiment, though news of rapturous British enthusiasm had been seeping in through the agricultural papers for five years.

Like other fertilizers, guano's main customers were in the exhausted Chesapeake, with a side market in Long Island. However, although fertilizers would mostly be consumed outside New York, New York mercantile interests would powerfully shape their development: the United States' trade in Peruvian guano was controlled by only two merchants, Samuel K. George in Baltimore and Edwin Bartlett in New York City. From these houses it was sold through dry goods merchants and through the network of agricultural journals, particularly the *American Agriculturist* in New York, which was by then reaching out for a national reputation for itself and for the goods from its warehouse.[92]

Like plaster, guano made a good, visible experiment—the effects of its application were immediately evident in shining, dark-green foliage. However, Peruvian guano was opaque to sensory judgment and could easily be adulterated. Smell in particular could be deceptive—water-damaged guano smelled more strongly of ammonia but had a weaker effect on plants. As a novelty, guano was also dependent on its print reputation and on the nonsensory knowledge of chemists. Bringing guano into New York, Edwin Bartlett set about spreading its fame in an eighty-page pamphlet crammed with analyses by consulting chemists and testimonials promising "pumpkins of enormous size."[93] These, however, mostly focused on a single component: ammonia, valuable, like gluten, for its flesh-forming nitrogen. To Dr. Gardener, for example, guano worked on the same principles as stable manure, but without the "vegetable rubbish": where stable manure contained 0.5 percent ammonia, guano contained 20 percent.[94] As Richard Wines has shown, Peruvian guano's experimental success solidified the reputation of nitrogen as a purchasable commodity, rather than, as Liebig and Waring would have argued, a freely available atmospheric "capital."[95]

Peruvian guano's popularity combined with the high prices set by the Peruvian government to stimulate political as well as theoretical shifts. American entrepreneurs rushed for alternative sources of dung on any uninhabited Pacific, Caribbean, and African islands where birds nested. In 1856, Franklin Pierce gave official support to this imperialist scramble by signing the Guano Islands Act, offering military protection to American citizens who discovered deposits of guano on uninhabited islands and thereby creating a framework for further Pacific expansion. Like the trade itself, the Guano Islands Act had a clear New York genealogy. It had been introduced by then-senator and crucial New York improver William Seward. It also strongly resembled an earlier resolution of the Farmers' Club of the American Institute of New York, of which Pell was a member. Behind this New York interest lay a secondary speculative market, not in guano itself but in guano rights to Pacific Islands, a market largely being established by the New York financier Alfred G. Benson, first through the American Guano Company, and then through its spin-off and rival the US Guano Company. Like Barrett, Benson had turned to the American Institute to promote annexed islands' potential. In attracting investors, these companies created anxious new audiences for chemical analysis.[96]

Supporters of Pacific guano also embarked on further debates about the relative value of nutrients. Initially, Benson's ships and their military escorts, seeing green vegetation and deducing rainwater, counted Pacific Island guano as worthless. The new island guanos, perpetually rained upon, had had their nitrogen leached away. However, analysis by Joseph Henry at the Smithsonian recovered Pacific guano's character, by raising an alternative source of value: phosphorus. An odorless mineral, phosphorus fit Liebig's ideas of mineral value comfortably, was already a minor nutritional good in the United States, and was an element in the "super phosphates" (bone dust treated with sulfuric acid) becoming popular in Britain and in the bone meal that improving farmers had begun to feed their dairy cattle to replace the minerals lost in milking.[97] To undermine Peruvian guano, Pacific guano manufacturers began to make a case that phosphorus was more valuable than nitrogen-bearing ammonia. The United States Guano Company, a phosphatic guano manufacturer, would flip old debates about plaster—now it was strong-scented and clearly organic ammonia that was called "stimulus" and compared to alcohol and phosphorus, the scentless white powder, that counted as "nutrition."[98]

Perhaps even more consequentially, phosphatic guano makers shifted markets in fertility to a new industrial mode. Peruvian guano could be used essentially as it was mined, but phosphatic guanos had to be processed. The factories that arose to do the processing began to repeat Pell's early experiments, formulating proprietary mixtures of phosphatic guano, ash, bone, offal, dried blood, plaster or lime, horseshoe crabs (called "Cancerine"), and the pressed leavings of fish oil factories. The term "prepared guano" began to refer to multiple products that contained no bird droppings but which hoped to mimic Peruvian guano's success.[99]

Such products made sense in terms of the accounting analyses of the mixed farm—Pell's gluten-producing fertilizer had been planned as a flesh-forming material. However, manufactured fertilizers also offered consumers another way of thinking about value: fertilizers were "real" only if they contained what the manufacturers promised, if they were not adulterated. Antebellum Americans were already used to worrying about adulterations. In the 1830s New Yorkers would famously debate adulterants and impurities in city milk, made by cattle kept in urban distilleries or, as some averred, manufactured wholesale from finely ground corn or magnesia, flavored with a dash of real milk.[100] Books cataloging chemical frauds warned of starch in the sugar, plaster of Paris in the honey, and oxide of lead in the wine.[101] Consulting analytic chemists got work attesting to the value of all kinds of goods: James Chilton analyzed soil but also tobacco and hair dye.[102] Fertilizers slipped easily into this broader culture of reasonable suspicion.

This created a simpler, more lucrative project for chemists—rather than analyzing messy soils, plants, and animal matter, they instead could value manufactured nutritional goods, which at least promised uniformity. As chemists refocused on commercial fertilizers, however, they began to value the substances of fertility in new ways. Back in the 1840s, improvers like Joseph Harris had hoped to value manures and animal foods in terms of the plants and animals they would become, that is, calculating the value of the ammonia in horse urine in terms of pounds of corn. However, actual monetary value for nutriment could not be calculated this way. Guano could be spread on any crop from wheat to potatoes to strawberries, which were worth different, fluctuating amounts.

The new fertilizer market offered an alternative way of calculating the value of nutriment. Manufactured fertilizers were made up of chemicals that were already commodities and thus already had an actual market value. An incident recorded in the *American Farmer* illustrates this. In 1849, the fertilizer manufacturers P. S. Chappell and Wm. H. Chappell attacked a rival firm that had opened up on the floor directly below them. To demolish their rivals' claims, the Chappells pointed out that the ingredients listed in the published analysis, if sold separately, would be worth more than their rivals charged for the fertilizer as a whole. "It is hardly to be supposed," they observed sourly, "that any person would be disposed to combine certain materials and sell them for $3 per barrel which could be disposed of readily, if sold *separately* for $8.20."[103] The cheapness of the fertilizer showed its analysis to be a lie.

This method would be championed by the chemist Samuel W. Johnson, whose postwar influence on the formation of the agricultural experiment stations has made him one of the best-known figures in nineteenth-century American science. The son of a merchant who had retired to a farm in Deer River, New York, in the 1830s, Johnson had made a name for himself writing for the *Cultivator*, continuing through much of his career, which would take him first to Yale, then to Germany, then back to Yale, where he became professor of analytic chemistry and chemist to the Connecticut State Agricultural Society in 1856.[104]

Johnson's career was propelled in part by his continual attacks on the fertilizer industry. By 1859, Johnson was asserting a "reasonable price" for phosphoric acid, ammonia, and potash by working backward from their cheapest commercial sources.[105] "If we divide the price per ton of Columbian guano, $35, by the number of pounds of phosphoric acid in a ton, which, at 40 per cent., amounts to 800 pounds," he wrote, "then we have the price of one pound as nearly 4 1/3 cents."[106] Invisible but measurable quantities of nitrogen, phosphorus and potassium could now determine dollar value. Six years after Waring's textbook story, ammonia, his "capital of nature," was publicly valued at fourteen cents a pound.

By 1878, the chemist Edmund Pendleton complained about a new kind of fraud. Manufacturers were spiking fertilizers with cheap, burned blood, which decomposed too slowly and was worth "nothing whatever to the farmer" but which was, under analysis, rich in nitrogen.[107] This signaled a worse problem: manufacturers and consumers now cared too much about chemical analysis. "It has now got to be a question with manufacturers, how shall my fertilizers analyze well, and not how shall they show the best results in the field," he wrote. This was to be blamed on ill-informed consumers, "a class who erroneously suppose that this is the key that unlocks the secret about the value of fertilizers."[108] In this branch of improvement, the expert chemical labor of chemists had come to dominate field experiments. Fraudulent fertilizer manufacturers no longer had to fool the senses; instead they merely created the false traces of an invisible good. Waring's nitrogen atom here was valued, but far from floating freely, it was embedded in new markets. However, while this form of valuation would become dominant, it would neither solve the questions raised about nutritional value nor immediately establish clear hierarchies of expertise.

A Fraud Crushed?

Working at the Yale Analytical Laboratory from 1856 onward, Samuel Johnson might look authoritative to modern eyes—he certainly claimed disinterestedness in a way that many chemists could not. In a letter to fertilizer agents, he refused to accept favors or conduct analyses for private fees "which would invalidate my claims to disinterestedness."[109] However, his assertions were far from law. Like machinery manufacturers, fertilizer manufacturers armed themselves with specially commissioned analyses and testimonials from trustworthy farmers.[110] The *Genesee Farmer* fretted that these analyses were too believable: "thousands will purchase any thing that the manufacturers may say is a manure similar to the one carefully examined by the chemist."[111]

Fractures within the culture of organic analysis became evident in the controversy surrounding the chemist James J. Mapes.[112] Mapes could make no claim to disinterestedness at all. Even as he had begun to edit the fertilizer-focused *Working Farmer* in New York, Mapes opened a factory for "Improved Superphosphate of Lime" and "Chilian Guano" in Newark. Mapes's soil analyses, his opponents later pointed out, often came with a prescription for Mapes's fertilizers. After Mapes responded angrily to an unfavorable fertilizer analysis commissioned by the *Genesee Farmer* in 1852, the editor enlisted Johnson for further analyses, beginning a decade of attacks from the journal intended to label Mapes as an intentional fraud.[113] Attacking Mapes, the *Genesee Farmer* raised specters from the world of confidence. "So long as the public supports quack doctors of every hue, Mormons, polygamy, spiritual rappers, the exhibitors of 'woolly horses,'" the *Farmer* wrote, "we have no right to complain of charlatans who sell twenty tons of their artificial manure a day, at $50 per ton."[114] More and more editors joined the attack, animated perhaps, as the *Prairie Farmer* observed, by Mapes's "attitude of superiority over the Agricultural Press, which is highly offensive to many of the most respectable members of the fraternity."[115] Throughout the 1850s, Mapes publicly exchanged fire—in the form of analyses, testimonials, and signed affidavits—with much of the eastern agricultural press.

But the Mapes story helps us question any simple story of the triumph of impartial scientific authority. Attacks from major journals and academic chemists failed to dislodge Mapes from his position or even to particularly damage his business. Both his warehouse and factory remained in operation until the Civil War, which cut fertilizer manufacturers off from their crucial Southern markets. After the war, his son restarted the business successfully. As with patent medicines, or Barnum exhibits, accusations of fraud in the antebellum fertilizer marketplace were clearly far from a market death sentence.

Indeed, Mapes's success can help us see powerful alternative structures of credibility within New York improvement. He survived in part because of his continued friendship with Horace Greeley, who published defenses of his fertilizer in the *Tribune* reaching an estimated readership of one hundred thousand people. And, though the members of the state society shunned Mapes, he remained respected in the manufacturer-dominated American Institute's Farmers' Club, the same body that wrote the first model of the Guano Island Act. The institute awarded Mapes the title "Professor" to match Johnson's, and the club's minutes, often containing extended quotations from Mapes's speeches, continued to circulate through many agricultural journals, not just Mapes's own *Working Farmer* or the American Institute–sponsored Mapes ally, the *Farmer and Mechanic*.[116] While state society members might snub him, advertising pamphlets for Mapes's superphosphate of lime proudly bore the Agricultural Institute's medal on the cover.[117]

In fact, Mapes's fertilizers were more challenging to fundamental questions about nutriment than his detractors would allow. By 1855, Mapes was arguing in favor of a new theory, "the progression of primaries," which built on narratives of living progress that had been a staple since Robert Chambers's evolutionary text *The Vestiges of the Natural History of Creation* had become a best seller in 1844.[118] Just as animals tended to higher and higher organization, progressing from worms, to fish, to reptiles, so too, Mapes argued, did elements change as they moved through the cycles of life. Having passed from a rock into lichen, a phosphorus atom had "progressed" enough to become food for a higher order of plant. Becoming grass, it could be invisibly reshaped to be suited for cattle. Following this logic, fertilizer from crushed bones or bullock's blood was more powerful as a fertilizer because it was more progressed and thus easier for bodies to absorb. This helped explain why, according to Mapes's supporter Judge Henry F. French, "the plants, with their instincts, sharper than man's reason, and more subtle than chemists' tests," seemed to prefer bone phosphates to mineral phosphates.[119] While chemists like Johnson claimed that New Jersey phosphatic rock was a good source of phosphorus, Mapes argued that mineral sources were useless, in part using arguments about solubility and absorbency that chemists like Johnson had raised in the first place.[120] If Mapes was correct, French declared, then phosphatic rock sales were the same as "the old illustration . . . of *asking for bread and receiving a stone!* Buying plant-food and receiving an indigestible rock."[121]

Continuing debates about the pathways of matter perpetuated disputes about the status of particular ingredients. An ad for "Prepared Guano" from S. F. Halsey's steam mills, for example, relied on Liebig's seemingly outdated atmospheric chemistry—like the fertilizers Liebig marketed himself, it promised absorbents to attract ammonia from the air, rendering "the 'Prepared Guano' durable and *permanently* nutritive for *years*."[122] According to alternative models of nitrogen movement, these absorbents were not saleable goods but adulterants.

Such fault lines also divided the anti-Mapes camp. While Johnson argued for phosphate rock, the *Genesee Farmer*, which the *Prairie Farmer* called a "downright manure paper," promoted conflicting theories about the proper origin of fertilizing goods. Rather than buying in fertilizers from distant islands, or from New Jersey, the *Farmer* argued, improvers should force municipal authorities to concentrate city wastes and distribute them to the places that their "grain, flour, meat, or cotton" came from. "In short," the *Genesee Farmer* declared, "the cultivators of the soil must make those that reside in cities and villages feed the land that both feeds and clothes the denizens of cities and villages. Not to do this is a violation of a law of God, and an offence which he will certainly punish."[123] While fertilizer markets might promote commodity-based valuation and chemical analysis, no table of analysis could eliminate the moral meanings and powers ascribed to particular ingredients.

★

The "SteakMaker" line of feed supplements currently sold by Purina is in some ways a set of commodities that antebellum Americans would have found profoundly unfamiliar. They are designed for a space that is nutritionally much simpler than the Northern mixed farm. The feedlot, where animals are fattened for slaughter, is nourished by a far-flung network of corn and soy fields, themselves fed by an even farther-flung network of fertilizer plants, which take nitrogen directly from the air using the Haber Bosch process and minerals from potash and phosphate mines. Metaphorical exchanges of nutritional "currency" have largely been replaced by actual market exchanges—purchased sacks of feed, alfalfa, corn, feed supplements, or fertilizer. By nineteenth-century standards, the cattle in the feedlot are unimaginably thrifty, remarkably good at turning feed into beef. However, their extraordinary numbers and their concentration in one spot mean that the manure that they do produce is not the precious substance antebellum improvers hoped to conserve but an industrial-scale problem. It flows into waterways, painting them thick with algal bloom and bursting into the Gulf of Mexico as deoxygenated dead zones. This is not the cycle of atoms that Waring hoped for.

Nineteenth-century improvers did not create this landscape nor could they have really imagined it. At the same time, the print of improving fingers appears all over SteakMaker. It is a nutritional good, of a sort that antebellum improvers were beginning to construct. Its claims of value come in the form of a "guaranteed analysis" that improvers would have found familiar in structure, though some of the entities listed there—"crude protein," "Vitamin A," and certainly drugs like "Bovatec" and "Cattlyst"—would not have matched any categories they knew. Its value is squeezed, too, from the kinds of sources that later improvers dreamed of—the wastes of industry: "Processed Grain Byproducts," "Dried Bakery Byproducts," blessedly unspecific "Animal Fat," "Animal Protein," pure mineral sources, and the products of industrial chemistry. It even contains urea, a nitrogen-rich substance, which in 1828 was the first "organic" substance to be synthesized in a laboratory, leading Liebig's contemporaries to question the existence of a boundary between living and nonliving substances.

Such guarantees are kept honest, buyers are expected to know, by the tests performed on SteakMaker and other supplements at land-grant universities and experiment stations, established by improvers after the Civil War. At the same time, Purina, like other fertilizer companies, continues to make claims about the particular qualities of their particular good, to establish communities of trust around SteakMaker, and to compete in the promises made for their particular collection of ingredients. Indeed, an appeal to the accounting dream has been incorporated in the SteakMaker product pitch. SteakMaker's line extension—its array of products for animals at different stages of growth and for farmers with different budgets—promises to answer the kinds of questions of variability that Emmons's analyses ran up against, particularly as they are buttressed by an online calculator allowing farmers to describe their animals and their own economic circumstances so that Purina can suggest the most fitting formulation.[124]

This kind of analysis is less visible to the average twenty-first-century American than the kind that comes on soft drink cans. In some respects, it is an inversion or reflection, promising fat rather than confessing to it, describing value through the presence of nutriment rather than its absence. At the same time, it is at least as present in our lives as the soft drink can. It is, in fact, the underpinning of the ingredients that make their way into the soft drink and thus into us. It is in the table on the back of the SteakMaker label, and in its antebellum roots, that we can find the foundation for the practices that shape the modern landscape, bringing cycles of nutrients and money together.

Epilogue

Agricultural improvers envisioned more than one kind of future and more than one has resulted, though none map clearly enough onto any improving vision to make a prophet of anyone. Perhaps the most traceable consequence of antebellum New York improvement was the post–Civil War rise of institutions that would reshape American science and American agriculture. We can see the joining point between old and new institutions quite clearly in February of 1882, when E. Lewis Sturtevant arrived alone on the farm that would soon be the New York State Experiment Station. The newly appointed director of the station, Sturtevant was clearly dismayed to find broken glass and dilapidated outbuildings. The land was sodden, "foul with weeds," and "in a reduced state of fertility."[1] "There has been," Sturtevant wrote, "a necessity for attending to trifling details to an extent detrimental to the thought work and the office work."[2]

Clearly part of the station's persuasive power was to come from its rapid transformation. After repairing the outbuildings, replastering the walls, and putting in shelves for hundreds of books, Sturtevant and his small staff began to dig pits: one for the lysimeter, an underground gauge that Sturtevant developed for measuring percolation of water through soil; the other a "gas pit" for the chemistry laboratory.[3] They bought soil thermometers for measuring temperature at different depths and built a tower for the silo, then still a new piece of equipment. They built a greenhouse and a laboratory with steam jets, ovens, a condenser, "handsome white marble mantles," and a "museum and reception room."[4] They were, in short, building the kind of space that antebellum improvers had asked for repeatedly, only to be rejected by state legislatures.[5]

Though the New York Experiment Station was not the first agricultural experiment station in the US, it was the first to break the mold of fertilizer-testing shop that Samuel W. Johnson had set up in the 1870s.[6] Certainly, it analyzed fertilizers and drew on the ideas of disinterestedness that Johnson had advocated. ("The Station," Sturtevant wrote, "must act impartially and justly between fertilizer manufacturers and the farmers.")[7] However, the station also addressed a broader range of priorities developed by antebellum improvers. For example, Stephen Moulton Babcock, the station chemist, immediately got to work on the sort of analyses that would have warmed the heart of nutritional accounting completists. As soon as the lab was functional, he analyzed pigweed in bloom, redroot, and cowpea; the day the cows arrived, he produced a table for "Mixed evening Milk, from three fatigued Jersey Cows."[8] The horticulturist began sorting varieties, not of fruit, but of beans—a foray into taming a seed industry still filled with wonderfully named novelties and deceptions.[9] That year they could not produce images, but the next year, Hattie Sturtevant, E. Lewis Sturtevant's wife, set to work fixing the identities of corn varieties with precise and beautiful profiles of their cross sections.[10]

State-funded experiment stations were a crowning feature of the new age of state funding for agricultural science that had begun shortly after the outbreak of the Civil War. In 1862, Lincoln had shifted the agricultural functions of government from the patent office to the newly created US Department of Agriculture. That same year, benefiting from the absence of the seceded Southern states, the Morrill Act passed at last, creating land-grant colleges, devoted to "such branches of learning as are related to agriculture and the mechanic arts."[11]

These new institutions made space for increasing numbers of people to make careers in agricultural science and to lay claim to the relatively new title "scientist." Credentialing of scientists was still not particularly rigid—Sturtevant himself had only an unused medical degree and several years working a model farm with his two brothers. But Babcock had done advanced work in chemistry at Cornell, a land grant, and had received a PhD from Göttingen (German universities remained at the center of agricultural chemistry, though Liebig had died in 1873). He brought the station a shelf of the latest German analytical instruments.[12]

This newly solidified institutional home also widened rifts between scientists and older communities of improvement. The weekly journal *Science*, founded only a few years before, would soon launch attacks on the agricultural journals and on scientists who published in them, becoming a voice for research scientists who wanted clearer professional boundaries.[13] In general, more identities existed after the war than before; in a world of "scientists" and "businessmen," the compendious identities of New York improvers were harder to maintain. Sturtevant was not one to launch attacks—his report was diplomatic but clearly drew boundaries between scientists and "practical farmers." "Special effort has also been made to instruct visitors," he wrote, "by showing them carefully the data upon which results were formulated." However, he clearly expected ideas to flow more than one way: "abundant hints of how to do, as well as what to do," he wrote, "have been received by us."[14] Marking this separation, experimental forms also differed: station experiments were measured in "merchantable yields" but not in explicit profit calculations.

It was perhaps Sturtevant's diplomacy that helped him gain approval from the agricultural societies that had lobbied for the station: the New York State Society, the State Grange, the American Institute Farmers' Club, the Central New York Farmers' Club, the Elmira Farmers' Club, and the Western New York Horticultural Society.[15] The gifts that soon came in to furnish the station made clear that producers of improving goods were also hoping to establish ties: Mahlon F. Smith sent samples of "Smith's Seed Corn preservers," Irving P. King sent a "North Western Corn Planter," and the Remington Agricultural Company sent two plows. The station also sent gifts—weekly bulletins went to the agricultural press, to other directors, and to "gentlemen who occupy no public position, but who are identified with agricultural progress."[16] Visitors came too, 626 in the first year.[17]

Touted by the agricultural journals, the New York experiment station would become a national model; even older stations would revise their missions to imitate it.[18] The station's influence spread in part through the movement of its personnel. When he moved to the University of Wisconsin, for example, Babcock would convert his Geneva experiments into the Babcock butterfat tester, permanently altering the dairy industry.[19] It also spread through a more general New York dominance of national agricultural scientific discourse, which again moved through old improving networks. In 1880, Sturtevant had established a national scientific organization: the Society for the Promotion of Agricultural Science. Not only were one-third of its members New Yorkers, three worked for the New York State Agricultural Society, and two were familiar from antebellum improvement: Patrick Barry, pomologist and nurseryman, still running Ellwanger and Barry in Rochester; and J. J. Thomas, no longer a frustrated describer of fruit but now the distinguished associate editor of the *Country Gentleman*.[20]

Improving connections are also easy to spot in the Hatch Act of 1887, the federal legislation that funded Agricultural Experiment Stations and linked them to the land grants. Most obviously, Norman Jay Colman, agricultural commissioner during the Hatch Act battle, had been an editor of the *Valley Farmer* before the war.[21] Before signing the act, President Cleveland had spent a stint working on the purebred cattle operation of Buffalo land speculator, commercial apple grower, agricultural author, and president of the New York State Agricultural Society Lewis Falley Allen, who was his uncle.[22] This didn't mean that the Hatch Act was the vision of improvers in institutional form. As Alan Marcus makes clear, experiment stations and the Hatch Act emerged from a multisided battle between different branches of improving farmers, station scientists, land-grant college presidents, and scientists hired by fertilizer companies. (Indeed, he shows, Sturtevant opposed the act in its final form.)[23] However, connections between each of these groups and improvement are not hard to draw.

Even now experiment research station agendas remain remarkably faithful to the concerns of antebellum improving networks. Cornell's experiment station at Geneva still has the largest apple breeding program in the country, and its new varieties are still registered with the American Pomological Society founded in Buffalo in 1848.[24] I have on my desk the *Soil Survey of Greene County, New York*, produced by the experiment station with the USDA in 1985, which breaks the county down into more than 150 soil types, classified by capability, and containing a foldout aerial photograph of Prattsville breaking it into dozens of bounded soil types.[25] The global reach of American agricultural science means that we can also follow genealogical networks of improvement out of the United States to the experimental farms established by the Meiji government in the 1870s through which American experts promoted fruit trees, purebred cattle, and the habit of eating dairy and beef. We can likewise trace them by following the "writer for *Wallace's Farmer*" who brought a new hog-breeding program to the Soviet Union.[26]

If we emphasize these lines of genealogy, we can tell a story of rationalization and disinterested scientific professionalization and show, as Ariel Ron has demonstrated, how agricultural societies worked to expand and alter the functions of the state.[27] This is an important and true story. But the boosters and the salespeople, the warehouses and the nurserymen also had descendants that intertwined with the better-known institutions described above.

We can start to see these connections by tracing the modern counterparts of agricultural prodigies. In 1845, William Ingalls of Oswego County won a prize for a monstrous, credibility-stretching yield of 142 bushels of corn on one acre.[28] This monster became a prophecy of sorts; in 2017, it was slightly less than the average New York yield of 161 bushels per acre.[29] Ingall's modern counterparts, the winners of the yield contest for the New York Corn and Soybean Growers Association, continue to stretch these boundaries in some of the same ways; the 2017 corn winner produced 319 bushels per acre. This yield contest certainly has some connection to state agricultural science—the Cornell Experiment Station sponsors and organizes it and demands a report resembling the accountings that prizewinners like Ingalls were required to provide in the 1840s.[30] But the yield contest reveals the continuation of commercial networks even more strongly. The winners of the corn contest are listed next to their town and county, the winning yield, and the hybrid brand and number of their seed variety. The winner of the 2017 contest used a DeKalb 64-87, from a seed company that is a subsidiary of Monsanto. By winning, he made claims about his own virtue, but also about their value.

Yield contest winners move into a more fully corporate world of knowledge making and communication. They receive a trip to the Commodity Classic, an annual convention and trade exhibition. Run by groups like the National Corn Growers Association, the American Soybean Association, and the Association of Equipment Manufacturers, the Commodity Classic promises winners access to "the answers, ideas, innovation, technology, equipment and expertise that can make a powerful difference on your farm." As improving journals did, the Commodity Classic organizers describe agricultural knowledge in monetary terms: "Investing a few days at Commodity Classic will pay dividends on your farm all year long!" They also promise a party. The 2017 Commodity Classic included a speech by Sonny Perdue, President Trump's secretary of agriculture, but also entertainment by the "World Classic Rockers" Santana, Lynyrd Skynyrd, Journey, Boston, and Steppenwolf and a set by Mark Mayfield, "the Corporate Comedian." Closeups of whiskey glasses and clinking ice punctuate the show's promotional video. These lures are worthwhile—having attracted farmers, the Commodity Classic sells their attention to its corporate exhibitors. "Last year's Commodity Classic in Anaheim, Calif.," the website boasts, "attracted thousands of America's best farmers with an average gross farm income of $1.51 million and average farm size of 2,850 total acres."[31] Testing facilities, analyses, and public contests remain as common features of agribusiness as they ever were of agricultural science.

If agricultural improvement's connection to industrial agriculture is clear, so too are its connections to critiques of industrial agriculture. We can see these connections in the writings of Jerome I. Rodale, who founded the core organic farming journal, *Organic Gardening and Farming*, in 1942. At first Rodale seems to come from a different line of descent. In the 1940s, for example, he explicitly identified Liebig's chemistry as his enemy. "The science of chemical fertilizing," he wrote, "dates back to the year 1840 when the famous German chemist, Liebig performed a defective experiment in which he burned a plant, thus destroying all its organic matter."[32] Attacking Liebig, however,

Rodale championed the vitalist tradition promoted by American improvers like James Mapes. He saw in microbes and organic matter the vital elements Liebig had attacked. In *The Organic Front* (1948), Rodale told a cautionary tale of an aquarium curator who tried to concoct artificial seawater from an analytic table and killed all the fish. But, Rodale explained, "when a tiny pinch of real sea-water was put in, the fish could live." Like the blood of the planter mixed with ever greater fractions of clay, Rodale's seawater had "a minuscule fraction in it, a spark, a gleam, but something that is absolutely vital to the life processes of an ocean fish." Even as he decried Liebig, moreover, Rodale borrowed the language of debt and property that American improvers had learned from Liebig's contemporaries. Writing about the "natives" in India, he wrote approvingly: "Every blade of grass that could be salvaged, all leaves that fell, all weeds that were cut down found their way back into the soil, there to decompose and take their proper place in Nature's balance sheet."[33]

Despite his stated disdain for agricultural scientists, Rodale's studies often came from institutions begun by nineteenth-century improvers, including the Rothamsted Agricultural Station, the Connecticut Agricultural Experiment Station, the *Country Gentleman* (an offshoot of the *Cultivator*), and from his idol, Sir Albert Howard, who began as a mycologist studying the effects of fungal infection on sugarcane in the British West Indies and rose to become imperial economic botanist to the government of India.[34] Though designed in opposition to "the NPK formula," which reduced fertility to nitrogen (N), phosphorus (P), and potassium (K), moreover, Rodale's composting language strongly resembles the improving advice common a century before, from the instructions for making compost, to the laments about wasted town materials—brewery waste, vegetable tops, and "fish cuttings, entrails, heads, etc."[35] This similarity has helped promote an organic fertilizer industry that depends on "traditional" fertilizers including guano, dried blood, and seaweed, all manures concentrated and marketed in the US by the same antebellum improvers who popularized Liebig.

Clearer still is the family resemblance between Rodale's multiple magazines and the agricultural journals themselves. Like the antebellum journals, Rodale's magazines acted as a clearinghouse for the experiments of readers and for those conducted at his own "organic gardening experimental farm." Like the antebellum journals, Rodale's magazines spread through subscription agents who were also readers, and they advertised and tested a rapidly expanding set of goods.[36] Like the antebellum journals, Rodale's magazines and advertisements, and those of other commercial organic gardening and health magazines, acted as an alternative site of credibility, sometimes independent of academic science, with their own adherents and theorists.

EPILOGUE

As this book has traced changing meanings of value, it is important to emphasize here that no single vision of value has triumphed. It is easy to think of capitalism as a rational grid of valuation placed on an unruly landscape and to expect one set of values to "win." But forms of value live as long as they have buyers and believers, and so too do the theories that undergird those forms of value.[37] Marketplaces full of competing goods can produce adherents loyal to particular goods. Antebellum improvement, like other marketplaces, developed a decentralized system of authority that "conventional" scientists could not control. Mapes, sometimes described as a fraud "exposed" by anti-vitalists, continued to support vitalist theories and to sell fertilizers to supporters who testified in favor of both goods and theories. (Just as in modern clickbait stories about one politician "crushing" another's argument, on a practical level no crushing had actually occurred.) Some of these forms of value fed on their opposites. At pomological meetings, admirers of flavor and marketability butted heads. However, the consequence was not a final victory for marketability over taste (the existence of the Red Delicious notwithstanding) but a continuing tension between both characteristics, supported by nurserymen who had to supply more than one market for trees and who saw themselves at once as businessmen and connoisseurs. In fact, markets in refined goods require the operation of alternative registers of value within which a smaller group scorns the less discerning judgments of a larger. (Buying a Pink Lady for three times the price of a Red Delicious is an act of identity construction.) If we can only imagine that a particular "rational" or "scientific" viewpoint has come to dominate a society by ignoring the great and swirling world of goods and the claims about nature embedded in those goods, then we are missing something.

As this book has traced visions of the future, it's important to note that its subjects have become identified with the past. Improving texts have increasingly become resources for modern alternative agriculture, which mines them for insights into preindustrial knowledge. Organisms and techniques that were once shattering in their novelty are recast as "traditional." Plows that signified modernity in antebellum museums are now garden ornaments in the "primitive" style. The apple varieties "discovered" by people like William Prince have shifted from novel property to "heirloom." "Apple detectives" comb old orchards to reaffix names to trees using the texts and paintings left by antebellum pomologists and their successors.[38] Like other old commodities, old varieties have become repositories of memory and singularity.[39]

It would be wrong to take these connections between improving knowledge and modern agriculture too far. Genealogy is also a good metaphor for the development of agricultural knowledge because, like systems of agricultural knowledge, children have more than one ancestor. Industrial agriculture and the land grants also built on the German research tradition and its agricultural schools, while variety testing came also from the French state botanical gardens.[40] Experiment stations worked with corn-planting dates and genetic material originally taken from the Haudenosaunee, with wheat and practices brought, as Courtney Fullilove has shown, from Russia by the Mennonites, an underrated but crucial set of agricultural modernizers.[41] Organic agriculture drew from the Steiner school of biodynamics, which based some of its reasoning on the astrological "moon-farming" that many improving texts actively attacked. Sir Albert Howard's "Indore Method of Composting" appropriated practices of South Asians under the Raj. Improvers themselves worked to obscure some of the lines of connection from their own period, concealing the names of women who developed new techniques of butter making, "discovering" organisms and techniques developed by free laborers and indigenous and enslaved people. Their methods of obscuring the labor and knowledge of others also forms part of their legacy.

Traces of New York improvement remain in the landscape itself. Upstate fences and yards are spiked with hacked-at stumps of white mulberry, though there is no sign of their more delicate cousin, *Morus multicaulis*. Lawns are infested with nitrogen-fixing white clover imported to improve soil and now cursed for its tendency to attract bees to what have become spaces for play. Outside the towns, cut forests and side-hill plows released soils into the rivers that will not be restored in our lifetime—bare rock protrudes from many former farms. As in much of the rain-fed North, farmers in New York maintain a mestizo landscape growing Native American corn, alternated, if the price of corn is not too high, with an East Asian crop—not silk, but soybeans, sometimes fed to dairy cattle, or to a few black Angus or part-bred Herefords sold for beef. The corn hills culture has made way for rows, curved rather than dead straight to prevent erosion, but placed by descendants of the mechanical seed planters that some improvers advocated.

Other parts of the landscape would be alien to nineteenth-century improvers. Corn and soy fields are mostly genetically modified to be "Roundup ready." Each year they are drenched in herbicide; in soy fields you can see cornstalks springing up, until recently the only weed that could survive the killing regimen. As a result, the fields do not have to be ploughed—"no-till" fields do not require the "great art of stirring the soil." The depopulation of the American countryside enabled in part by labor-saving machines, chemicals, and fertilizers birthed by the First and Second World Wars, might have horrified improvers. So too might the persistence of people they expected to disappear. Haudenosaunee and Algonquian peoples live in upstate New

York—some on the fifteen remaining reservations. Their agricultural knowledge, too, persists and changes. Through the Native American Seed Sanctuary, for example, the St. Regis Mohawk/Akwesasne Tribe of northern New York is preserving their seeds and their knowledge and developing new techniques.[42] Permaculturists and sustainable agriculture proponents now draw insight from indigenous planting techniques.[43]

What aspects of Northern improvement remain in agricultural capitalism more broadly? At first glance the links are not as clear as those between plantation slavery and modern agriculture: the rise of export-driven monoculture; long-running patterns of coerced labor; the rise of particular forms of market rationalization driven by massive, hungry markets for staples; and forms of labor coercion and rationalization that move from the plantation field to the factory. But antebellum improvement shows us features of capitalism with their own consequences. We have seen, for example, how ideas about luxury and quality have become connected to the specifics of place, in the stories of antebellum New York butter and dairy. If we look at modern supermarkets, we can see these developments taken to a profitable extreme. Based on a concept of imported luxury as well on ideas of "terroir" developed as a marketing form in the nineteenth century, it has long been possible to buy French wines in the United States. It still startles me, however, that I can buy butter from Denmark in my small Pennsylvania town. In recent decades ideas of luxury and regional taste are also becoming more local as part of the backlash against industrial foods that began, perhaps, in the 1960s and 1970s. Our local food movement relies on ways of determining value that were not a feature of a timeless traditional past, but a product of the new railroad and canal foodsheds of the nineteenth century. They are as fundamentally market driven as the futures markets and systems of commodity grading that developed in Chicago a few years later.[44]

While staple landscapes of wheat, corn, and cotton have become consistent pillars of modern consumption and modern agribusiness, they should not blind us to the more variable but equally capitalist landscapes driven by fashion and changing desire.[45] Kale is a punch line from the early 2010s *and* acres of land in the Central Valley and across the South (one distributor calls it the new "golden leaf," the successor to tobacco).[46] The North American quinoa boom fundamentally altered the agricultural economies of Bolivia and Peru. Enthusiasm for margaritas and Corona beer in the United States helped make limes into "green gold," valuable enough to be stolen by Mexican cartels.[47] The fashion for Greek yogurt injected life into western New York dairy for almost ten years, while creating a new form of pollution: thousands of gallons of leftover acidic whey.[48] As I write, however, New York dairy farms are shutting down, in response, they argue, to expanded consumer fears of lactose. Those same fears have helped to turn California almonds into an investment crop, sought by private equity. The holder of my retirement fund, TIAA-CREF, offers an almond investment guide, apparently unfazed by the threat to existing almond trees posed by droughts like those of 2006–10 and 2012–17. (Tree crops are particularly susceptible to drought, since they can't be just left to fallow. Trees need water in good and bad years).[49] These fluctuations are not aberrations from what are essentially steady markets; they are regular if not predictable features of capitalist agriculture.[50]

At the same time, intellectual property rights developed in the 20th century have allowed developers to patent new apple varieties like the SweeTango, bred at the University of Minnesota from their previous successes, the Honeycrisp (patent expired) and the Zestar, and licensed only to the growers of the "Next Big Thing cooperative."[51] Cornell hopes to catch up with its own varieties: Snapdragon and RubyFrost. The names are trademarked just as the fruits are patented, since the trademarks last longer than the patents. Even if other growers use the varieties, they won't be able to claim the name and so the reputation. The Honeycrisp, like other apples of this category, is finicky, so growers outside cooperative rules might not be able to claim its flavor either.

The illusion of improvement—the idea of the Northern family farm as a virtuous, sentimental, unchanging space—still shapes theories of American political legitimacy, warps farm policy, and thereby structures industry. Behind its appeal to an imagined past, the "family farm," like "the farmer," still lumps together a grossly unequal countryside. The USDA's definition of family farm refers to "any farm where the majority of the business is owned by the principal operator . . . and individuals related to the principal operator." In 2017, this definition covered 98.8 percent of all farms, including "Low Sales" small farms, with a gross cash farm income of under $150,000 and "Very Large Family Farms" with a gross cash farm income of $5 million or more. "Family farms" still include a large number of people with more than one identity: most farms in the US are either "retirement farms" (17.9 percent) where the owner is nominally retired or "off-farm occupation farms" (41.9 percent) where farmers work another job to support their farming, or, conversely, buy a farm as a tax shelter for their urban income.[52] Ideas of the household and home economies of labor also still hide people who make the countryside work, like the 94,327 nonfamily workers, 50 percent of whom were immigrants, who worked on US dairy farms in 2012. Appeals to family labor have allowed farmers to evade labor laws applied to other industries, in particular laws against child labor.[53] The romantic idea of the family farm also complicates the lives of small and alternative farmers struggling to operate in a context defined by markets, while performing an imagined past free of market logics.

Inheritances of improvement surround us in the present. Improvers, though, were interested in the future, and I find myself wondering what we might learn from that. I don't mean to take them as models—one reason to think about antebellum future stories is clearly to break away from them. But, like nineteenth-century New Yorkers, we are headed deeper into massive agricultural and environmental changes. Climate change is shifting the meanings of all our places in ways we are only beginning to see. Our regions and certainties, our commodity maps and transportation systems, our food processes, our weather, our systems of international trade, our patterns of landownership are all in flux. Antebellum New Yorkers, with their many agricultural systems, their deep experience of rural landscapes, and their sense of landscape politics had more ways to imagine what possible agricultural futures might look like than most Americans do now.

We are already consuming the stories of future agriculture, and there are more to come. Some are shining-surfaced technological fixes taken from a limited palette of futurism, like GMOs that promise higher yields and drought resistance, or skyscrapers full of microgreens, grown using red and blue LEDs (more efficient since plants can't consume green light). Others are restorations of the "traditional" landscape, going back in time to simpler values, often shaped by an imagined nineteenth century. Tracing the storytelling of agricultural improvement can help us see these projections as projections, not to disqualify them, for new landscapes are coming and plans to create them are needed, but to help us see their logics, to perceive their politics, to mark the ways that they are being sold as our inevitable future, and maybe to remind us to imagine agricultural story arcs of our own.

Acknowledgments

Without knowing what would come of them, I wrote the first words that would become this book back in 2005. It is a joy to think over the many people who have helped me since then—fifteen years do not fit easily into a few pages. In the History and Sociology of Science Department at the University of Pennsylvania, I'd like to thank Robert E. Kohler, Ruth Schwartz Cowan, Susan Lindee, Mark Adams, David Barnes, Steven Feierman, and Riki Kuklick, whose loss I feel very much. I am also grateful for the close-knit community at HSS. Jeremy Vetter, Andi Johnson, Corinna Schlombs, Hilary Smith, Babi Hammond, Ann N. Greene, Chris Jones, Paul Burnett, Samantha Muka, Matt Hersch, Dominique Tobbell, Agyeman Siriboe Boateng, Joanna Radin, Peter Collopy, Deanna Day, Eric Hintz, Damon Yarnell, Sejal Patel, Jessica Martucci, Christy Schuetze, Pat Johnson, Josh Berson, Kristoffer Whitney, Erica Dwyer, and Liyong Xing all challenged and sustained me.

The Program in Early American Economy and Society gave me an early research home at the Library Company of Philadelphia, where Wendy Woloson and Cathy Matson helped me think about the project. During a year's residence at the Chemical Heritage Foundation (now the Science History Institute) in Philadelphia, I explored my history of chemistry material. I would particularly like to thank Ronald Brashear, Anke Timmermann, Elsa Atson, Hilary Domush, and Augustin Cerveaux for creating the scholarly community there. Dan Richter graciously gave me a desk and an invigorating scholarly home at the McNeil Center for Early American Studies. I benefited enormously from the conversations and suggestions of Megan Walsh, Jo Cohen, Ken Owen, Zara Anishanslin, Gergely Baics, Shona Johnston, Laura Keenan Spero, Andrew Lipman, Jason Sharples, Jeffrey Edwards, Marie McDaniel, Vanessa Mongey, Brian Conolly, Christina Snyder, Hunt Howell, and Simon

Finger. I would also like to thank the audience of the McNeil Center's seminar for their many valuable thoughts on fruit.

Nine months at the American Antiquarian Society in Worcester expanded this project a great deal and started conversations that continue ten years later. Jessica Lepler, Honorée Fanonne Jeffers, Mary Beth Sievens, April Rose Haynes, Hex Hassinger, Ashley Cataldo, Megan Kate Nelson, Lauren Hewes, Elizabeth Pope, Diann Benti, Lloyd Pratt, Brigitte Fielder, Jonathan Senchyne, Caroline Sloat, Gigi Barnhill, Michael Winship, Meredith Neuman, and Danny Thompson helped me to rethink the history of bodies, the fluctuations of capitalism, and the principles and joys of scholarship. Paul Erickson created a beautiful community of fellows. He is forbidden to read any of the material on pages 67–68 or 186–87. Five months at the Smithsonian Institution's National Museum of American History allowed me to extend the story of silk. Among many generous scholars, I'd like to thank Pete Daniel and Deborah Warner in particular. At the University of Rochester, the Susan B. Anthony Institute for Gender, Sexuality, and Women's Studies offered a gracious scholarly community, the RUSH reading group sharpened my thinking on silk, and conversations with Dorinda Outram and Rachel Remmel materially shaped my thinking.

The research for this book would have been impossible without New York's remarkable network of archives and local history collections. I would like to thank the staffs of the New York State Historical Society, the New York State Archive, the New York State Library, the New York State Historical Association, the Cortland County Historical Society, the History Center in Tompkins County, the Jefferson County Historical Society, the St. Lawrence County Historical Society, the Oneida County Historical Society, the Onondaga Historical Association, the Skaneateles Historical Society, the Milne Library at SUNY Geneseo, the Folsom Library at Rensselaer Polytechnic Institute, the Dutchess County Historical Society, the Adriance Memorial Library, the Carl A. Kroch Library at Cornell, and the Rush Rhees Library at the University of Rochester. I am particularly grateful to Dr. Jim Folts at the New York State Archive; to W. Douglas McCombs and Jessica Lux, for help with images at the Albany Institute of History and Art; to Donna Eschenbrenner and Mary L. White at the History Center in Tompkins County, who went out of their way to help me with the McLallen Papers; and to Geoffrey Stein at the New York State Museum, who spent an afternoon taking me around the agricultural collections at the museum's storage warehouse.

I would also like to thank the thousands of amateur local historians and genealogists without whose interest in the past my sources would have ceased to exist and the dozens of amateur historians who kindly informed my re-

ACKNOWLEDGMENTS

search at every turn. Though I have seen images of their fingers often, I do not know the names of the many workers who digitized the resources that allowed me to continue my research at times when traveling was difficult. I acknowledge their labor here.

I am grateful to conference audiences at meetings of the History of Science Society, the American Society for Environmental History, the Society for Historians of the Early American Republic, the Agricultural History Society, the Workshop in the History of Environment Agriculture Technology and Society, the Joint Atlantic Seminar in the History of Biology, the "Nature's Accountability" conference sponsored by the German Historical Institute, the Cornell Conference on the Histories of American Capitalism, and the "The Culture of Print in Science, Technology, Engineering, and Medicine (STEM)" conference sponsored by the Center for the History of Print Culture in Modern America at the University of Wisconsin, for their comments and questions, which strengthened my work and helped me negotiate these disciplinary divides. Audiences at the Max Planck Institute (Lissa Roberts in particular) helped me think about my chemistry, those at the Harvard Business School Political Economy of Food Conference helped with butter and divinity, and those at Yale's workshop "Plants, Animals, and Ownership: Innovation and Intellectual Property Protection in Living Organisms since the Eighteenth Century" helped me think about fruit and silk. François Furstenberg also organized an extraordinary set of conversations around "Land" in Montreal in 2010, which shaped this book in many ways.

The group of scholars at these conferences also helped me along. In particular, conversations with Eric Stoykovich helped me think about the ramifications of improvement in the early days of this project, Benjamin Cohen offered consistent and gracious encouragement, and Margaret Rossiter kindly made time to help me think about sources. I am also grateful to Christopher Clark, Sarah T. Phillips, Tom Okie, Dael Norwood, Christine Keiner, Daniel Immerwahr, Gabriel Rosenberg, Kendra Smith-Howard, Jennifer Black, Walter Friedman, Emma Hart, Bart Elmore, Lisa Onaga, Thomas Finger, Taylor Spence, Ian Beamish, Joppe van Driel, Tamara Plakins Thornton, Sarah Greig, Helen Curry, Erica Hannickel Terry, Jamie Pietruska, Rebecca Woods, Katie Pryde, Jim Endersby, Courtney Fullilove, Jeremy Zallen, Sabine Höhler, Brett Mizelle, Will Mackintosh, Jeanette Marie-Therese Vaught, Camden Burd, Melanie Kiechle, Amrys Williams, Casey Lurtz, Bil Kerrigan, Nicole Welk-Joerger, Maura Capps, Jim Secord, Anna Foy, Marina Moskowitz, Paul Erickson, Hayley Goodchild, Jane Mt. Pleasant, and Philip Pauly. Willis and Cristina Wood, and the skilled interpreters at Old Sturbridge Village, also helped me think about farming's practicalities. Ariel Ron has long helped me sustain my sense that agricultural improvement is important and has shaped my thinking in many conversations, particularly in our planning for the "Grassroots Modernity" conference at Yale. Wendy Woloson, Brian Luskey, Joshua Greenberg, Hannah Farber, Jess Lepler, Eugenia Lean, Sarah Milov, Lee Vinsel, Lukas Rieppel, Lissa Roberts, and William Deringer helped me connect the history of capitalism literature to the history of science. A writing group with Vicky Albritton, Fredrik Albritton Jonsson, and Anya Zilberstein sharpened my thinking. Over recent years, Dan Kevles has generously given his time, his critical eye, and his counsel, for which I am very grateful. Conevery Bolton Valencius has inspired and encouraged me.

Colleagues and friends at Dickinson, particularly Hilary Smith, Chris Bilodeau, Crystal Moten, Regina Sweeney, Marcelo Borgés, Matt Pinsker, Elaine Mellen, Amy Farrell, Evan Young, Say Burgin, Greg Kaliss, Karl Qualls, Madeline Brown, Maria Bruno, Heather Bedi, Siobhan Phillips, Nikki Dragone, Amy Wlodarski, Antje Pfannkuchen, Julie Vastine, Sarah Niebler, Meg Winchester, Perin Gurel, Sarah Kersh, Sharon O'Brien, Susan Rose, Jerry Philogene, and Meta Bowman offered advice and read parts of the text. Christine Bombaro, Malinda Triller, James Gerencser, Don Sailer, and Deborah Ege helped with resources and images. As well as sharing their expertise, Jenn Halpin and Matt Steiman made space for me as a volunteer at the Dickinson College Farm, where I learned to think seriously about tilth. My students at Dickinson have asked the right questions, forced me to make connections, and taught me to tell stories.

I could not have done this research without the financial support I received: the American Antiquarian Society–National Endowment for the Humanities Long-Term Fellowship, a Smithsonian Postdoctoral Fellowship at the National Museum of American History, two Mellon-sponsored grants from the American Council of Learned Societies, a John C. Hass Fellowship and Roy G. Neville Fellowship from the Chemical Heritage Foundation, and short-term fellowships from the New York State Archives, the Program in Early American Economy and Society at the Library Company of Philadelphia, and the Hagley Center for the History of Business. My rural archives trips were sponsored by a grant from the National Science Foundation.

Much of the material on chemistry, fruit, and business networks appeared as articles in *Science as Culture* (December 2010), in *Business History Review* (Autumn 2016), and in a collaborative piece published with Conevery Bolton Valencius, David Spanagel, Sara Gronim, and Paul Lucier in the *Journal of the Early Republic* (Spring 2016). A research and development grant from Dickinson paid for indexing by June Sawyers. Any indexing errors are mine. At the University of Chicago Press, I was lucky to find a thoughtful, encouraging (and patient!) editor in Timothy Mennel. Many thanks also to Rachel Kelly, Susannah Marie Engstrom, Mark Reschke, Tyler McGaughey, and to the production team as a whole. Brian Chartier designed the cover using an image given by Paul Erickson.

When I cobbled a first draft together, Audra Wolfe of the Outside Reader read the whole thing and helped me think about what books are. In the final stages of writing, Denise Phillips, Chitra Ramalingam, Conevery Bolton Valencius, Jeremy Vetter, Catherine Kelly, Robert Suits, and Jessica Lepler read chapters. Jeff Wood reminded me that books are good things. Many other friends also supported me through the process, in particular Veronica Kitchen, Jenna Sindle, Charlotte Morrisson-Reed, Annushka Sonek-Wienert, Chitra Ramalingam, Kathy Nagel, David Schoppik, Michael Meeuwis, Kristen Fellowes, Sarah Neville, Kerry Dunn, and Llerena Searle.

I could not have done this work without the labor and care of other people. The teachers at the Dickinson College Children's Center, Little Lights Daycare, and Crestview Elementary helped raise my children. So did my friends on the Conodoguinet Creek. Lonna Malmsheimer, Kim Patten, Todd Alwine, Hannah and Halen, Susan Rose, Steve Brouwer, and Jan and Ari Brouwer created a community where we could all grow. I am also lucky in my family. My in-laws Mary Turner and Bill Turner have been supportive presences—Mary in particular has come to save us at many pinching moments. My parents, James and Christine Pawley, my sister, Alice Pawley, my brother-in-law Stephen Hoffmann, and my brother John Pawley all gave advice and critical support. My father taught me to pay attention to the act of sight and to notice the nonhuman world. While he did not live to see the book in print, his influence is here in it. My mother taught me to pay attention to the agency of readers, to structure my thoughts, and to say what I mean. My mother and sister read drafts, listened to rants, and dispensed sage counsel. My two children, Sam and Laura, born while I worked on the book, taught me many new kinds of joy and to multiply my attention rather than dividing it. As a friend of ours remarked once wonderingly, Roger Turner and I never run out of things to talk about, and this book has been enriched by our perpetual conversation. While learning to fly through stormy weather himself, Roger has carried me as well. He also is home at this moment keeping the children alive; I owe this bubble of time and space and the words I have made in it to him.

Abbreviations

Journals

AH	*Agricultural History*
AHR	*Agricultural History Review*
BJHS	*British Journal for the History of Science*
DSSNY	*Documents of the Senate of the State Of New York*
EAS	*Early American Studies*
EHR	*Economic History Review*
EH	*Environmental History*
GFGJ	*Genesee Farmer and Gardener's Journal*
JAC	*Journal of Agrarian Change*
JAH	*Journal of American History*
JAI	*Journal of the American Institute*
JER	*Journal of the Early Republic*
JEH	*Journal of Economic History*
JOC	*Journal of Commerce*
JPS	*Journal of Peasant Studies*
NYH	*New York History*
PMHB	*Pennsylvania Magazine of History and Biography*

Societies

NYSAS	New York State Agricultural Society
SPAAM	Society for the Promotion of Agriculture, Arts, and Manufactures

Archives

AML	Adriance Memorial Library, Poughkeepsie, NY
AAS	American Antiquarian Society, Worcester, MA
CAK	Carl A. Kroch Rare Books and Manuscript Library, Cornell University, Ithaca, NY
CCHS	Cortland County Historical Society, Cortland, NY

DCHS	Dutchess County Historical Society, Poughkeepsie, NY
HML	Hagley Museum and Library, Wilmington, DE
HCTC	History Center in Tompkins County, Ithaca, NY
JCHS	Jefferson County Historical Society, Watertown, NY
ML	Milne Library, SUNY Geneseo, Geneseo, NY
NAL	National Agricultural Library, College Park, MD
NYSA	New York State Archives, Albany, NY
NYSHA	New York State Historical Association, Cooperstown, NY
NYSL	New York State Library, Albany, NY
OHA	Onondaga Historical Association, Syracuse, NY
RPI	Rensselaer Polytechnic Institute Archives and Special Collections, Troy, NY
RRL	Rush Rhees Library, Rare Books, and Special Collections, University of Rochester, Rochester, NY
WC	Warshaw Collection, Smithsonian Institution Archives, Washington, DC

Notes

Introduction

1. "A Mammoth," *GFGJ* 5, no. 28 (1835): 218; "Fat Animals," *Cultivator* 2, no. 4 (1845): 112.
2. "Large Vegetables," *Cultivator* 8, no. 12 (1841): 188.
3. Neil Harris, *Humbug: The Art of P. T. Barnum* (Chicago: University of Chicago Press, 1981); Susan Nance, *Entertaining Elephants: Animal Agency and the Business of the American Circus* (Baltimore: Johns Hopkins University Press, 2013), 42–44.
4. See especially Lorraine Daston and Katharine Park, *Wonders and the Order of Nature, 1150–1750* (New York: Zone Books, 1998); Paula Findlen, "Inventing Nature: Commerce, Art, and Science in the Early Modern Cabinet of Curiosities," in *Merchants and Marvels: Commerce, Science, and Art in Early Modern Europe*, ed. P. H. Smith and P. Findlen (New York: Routledge, 2002); James W. Cook, *The Arts of Deception: Playing with Fraud in the Age of Barnum* (Cambridge, MA: Harvard University Press, 2001).
5. "Sixteenth Annual Show of the American Institute," *American Agriculturist* 2, no. 11 (1843): 261; "Yorkshire Gazette,—Extraordinary Pig," *American Agriculturist* 2, no. 12 (1843): 373; John B. Weeks, Diary, September 13, 1855, NYSHA, Special Collections, Cooperstown New York.
6. "Large Vegetables," 188.
7. Dexter Brewer, "Sheep Story," *GFGJ* 9, no. 9 (1839): 70.
8. For networks of commercial breeding, see Rebecca J. H. Woods, *The Herds Shot Round the World: Native Breeds and the British Empire, 1800–1900* (Chapel Hill: University of North Carolina Press, 2017). For the complex movements of seeds, see Courtney Fullilove, *The Profit of the Earth: The Global Seeds of American Agriculture* (Chicago: University of Chicago Press, 2017).
9. "The Newest and Greatest Humbug Yet Announced," *GFGJ* 9, no. 33 (1839): 261.
10. For this usage of "culture," see Philip J. Pauly, *Fruits and Plains: The Horticultural Transformation of America* (Cambridge, MA: Harvard University Press, 2007), 6.
11. "No End to Improvement," *New Genesee Farmer and Gardener's Journal* 2, no. 8 (1841): 119.
12. "No End to Improvement," 119.
13. Jane Mt. Pleasant and Robert F. Burt, "Estimating Productivity of Traditional Iroquoian Cropping Systems from Field Experiments and Historical Literature," *Journal of Ethnobiology* 30, no. 1 (2010): 52–79, 60.
14. "No End to Improvement," 119.
15. This is not a failing of Northern rural historiography, which addressed the complexities

of rural capitalism in the 1990s, but rather an inadvertent distortion created by the resurgence of interest in the relationship between slavery and capitalism and the new cultural history of Northern capitalism, which has mostly focused on cities. For the transition to capitalism debate, see Joyce Appleby, "Commercial Farming and the 'Agrarian Myth' in the Early Republic," *JAH* 68 (1982): 833–42; James A. Henretta, *The Origins of American Capitalism: Collected Essays* (Boston: Northeastern University Press, 1991); Michael Merrill, "Cash Is Good to Eat," *Radical History Review* 3 (1977): 42–71; Winifred B. Rothenberg, *From Market-Places to a Market Economy: The Transformation of Rural Massachusetts, 1750–1850* (Chicago: University of Chicago Press, 1992); Christopher Clark, *The Roots of Rural Capitalism: Western Massachusetts, 1780–1860* (Ithaca, NY: Cornell University Press, 1990); Alan Kulikoff, "The Transition to Capitalism in Rural America," *William and Mary Quarterly* 46 (1989): 120–44; Naomi Lamoreaux, "Rethinking the Transition to Capitalism in the Early American Northeast," *JAH* 90 (2003): 437–61. For New York's transition, see Donald H. Parkerson, *The Agricultural Transition in New York State: Markets and Migration in Mid-Nineteenth-Century America* (Ames: Iowa State University Press, 1995); Thomas S. Wermuth, *Rip Van Winkle's Neighbors: The Transformation of Rural Society in the Hudson River Valley, 1720–1850* (Albany: State University of New York Press, 2001); Martin Bruegel, *Farm, Shop, Landing: The Rise of a Market Society in the Hudson Valley, 1780–1860* (Durham, NC: Duke University Press, 2002). For the slavery and capitalism literature, see Sven Beckert, *Empire of Cotton: A Global History* (New York: Vintage, 2015); Walter Johnson, *River of Dark Dreams* (Cambridge, MA: Harvard University Press, 2013); Caitlin Rosenthal, "From Slavery to Scientific Management," in *Slavery's Capitalism: A New History of American Economic Development*, ed. Sven Beckert and Seth Rockman (Philadelphia: University of Pennsylvania Press, 2016); Caitlin Rosenthal, *Accounting for Slavery: Masters and Management* (Cambridge, MA: Harvard University Press, 2018). For Northern urban capitalism, see B. P. Luskey and W. A. Woloson, eds., *Capitalism by Gaslight: Illuminating the Economy of Nineteenth-Century America* (Philadelphia: University of Pennsylvania Press, 2015); Scott A. Sandage, *Born Losers* (Cambridge, MA: Harvard University Press, 2005); Jessica M. Lepler, *The Many Panics of 1837: People, Politics, and the Creation of a Transatlantic Financial Crisis* (Cambridge: Cambridge University Press, 2013); Stephen Mihm, *A Nation of Counterfeiters* (Cambridge, MA: Harvard University Press, 2009); Jane Kamensky, *The Exchange Artist: A Tale of High-Flying Speculation and America's First Banking Collapse* (New York: Penguin, 2008); Seth Rockman, *Scraping By: Wage Labor, Slavery, and Survival in Early Baltimore* (Baltimore: Johns Hopkins University Press, 2010); Seth Rockman, "What Makes the History of Capitalism Newsworthy?," *JER* 34, no. 3 (2014): 439–66.

16. Olga Elina, "Planting Seeds for the Revolution: The Rise of Russian Agricultural Science," *Science in Context* 15, no. 2 (2002): 209–37; David Arnold, "Agriculture and 'Improvement' in Early Colonial India: A Pre-History of Development," *JAC* 5, no. 4 (2005): 505–25; Stuart McCook, *States of Nature: Science, Agriculture, and Environment in the Spanish Caribbean, 1760–1940* (Austin: University of Texas Press, 2002); Peter M. Jones, *Agricultural Enlightenment: Knowledge, Technology, and Nature, 1750–1840* (Oxford: Oxford University Press, 2016); Denise Phillips and Sharon Kingsland, eds., *New Perspectives on the History of Life Sciences and Agriculture*, vol. 40. (New York City: Springer, 2015).

17. "Table of Contents," *Cultivator* 8, no. 12 (1841): 200.

18. For scientific institution building, see Margaret Rossiter, *The Emergence of Agricultural Science: Justus Liebig and the Americans, 1840–1880* (New Haven, CT: Yale University Press, 1975); Charles Rosenberg, *No Other Gods: On Science and American Social Thought* (Baltimore: Johns Hopkins University Press, 1997). For more on interrelations of science and agriculture, see

Carolyn Merchant, *Ecological Revolutions: Nature, Gender, and Science in New England* (Chapel Hill: University of North Carolina Press, 2010), 149–98.

19. My definition of science here is capacious, drawing on the literature on popular or vernacular science. Improving science emerged from multiple contesting sources, not only from European laboratories, British experiment stations, and enthusiastic American elites but also from the experience of a variety of farmers in dealing with new products, practices, and texts. See in particular Conevery Valencius's concept of "vernacular science" in Conevery Bolton Valencius, *The Lost History of the New Madrid Earthquakes* (Chicago: University of Chicago Press, 2013), 176–77, and the definitions in Conevery Bolton Valencius, David I. Spanagel, Emily Pawley, Sara Stidstone Gronim, and Paul Lucier, "Science in Early America: Print Culture and the Sciences of Territoriality," *JER* 36, no. 1 (2016): 73–123; but also Roger Cooter and Stephen Pumfrey, "Separate Spheres and Public Places: Reflections on the History of Science Popularization and Science in Popular Culture," *History of Science* 32, no. 3 (1994): 237–67; Anne Secord, "Science in the Pub: Artisan Botanists in Early Nineteenth-Century Natural History," *BJHS* 27, no. 3 (1994): 383–408.

20. For the scope of improvement in the US, see Donald Marti, *To Improve the Soil and the Mind* (Ann Arbor: University Microfilms, 1979); Albert Demaree, *The American Agricultural Press, 1819–1860* (New York: Columbia University Press, 1941); Margaret Rossiter, "The Organization of Agricultural Improvement in the United States, 1785–1865," in *The Pursuit of Knowledge in the Early American Republic*, ed. A. Oleson and S.C. Brown (Baltimore: Johns Hopkins University Press, 1976), 279–97; Steven Stoll, *Larding the Lean Earth: Soil and Society in Nineteenth-Century America* (New York: Hill and Wang, 2003); Benjamin R. Cohen, *Notes from the Ground: Science, Soil, and Society in the American Countryside* (New Haven, CT: Yale University Press, 2009); Ariel Ron, "Developing the Country: 'Scientific Agriculture' and the Roots of the Republican Party," PhD diss. (University of California, Berkeley, 2012); Ariel Ron, "Summoning the State: Northern Farmers and the Transformation of American Politics in the Mid-Nineteenth Century," *JAH* 103, no. 2 (2016): 347–74; Eric C. Stoykovich, "The Culture of Improvement in the Early Republic: Domestic Livestock, Animal Breeding, and Philadelphia's Urban Gentlemen, 1820–1860," *PMHB* 134, no. 1 (2010): 31–58. For Southern agricultural improvement, see John Majewski, *Modernizing a Slave Economy: The Economic Vision of the Confederate Nation* (Chapel Hill: University of North Carolina Press, 2011), 53–80; Joyce E. Chaplin, *An Anxious Pursuit: Agricultural Innovation and Modernity in the Lower South, 1730–1815* (Chapel Hill: University of North Carolina Press, 1993); William Thomas Okie, *The Georgia Peach: Culture, Agriculture, and Environment in the American South* (Cambridge: Cambridge University Press, 2016); Ian William Beamish, "Saving the South: Agricultural Reform in the Southern United States, 1819–1861" PhD diss. (Johns Hopkins University, 2013). For the tensions between Northern and Southern improvement, see Sarah T. Phillips, "Antebellum Agricultural Reform, Republican Ideology, and Sectional Tension," *AH* 74, no. 4 (2000): 799–822.

21. Most agricultural societies did not keep membership totals. However, sending out a questionnaire in 1858, the patent office received returns from thirty-five of New York's societies, some of which reported membership figures—totaling 13,213 people. However, this is certainly an underestimate since many societies did not list figures. The county treasurer's records for 1849 and 1850 report that fifty-one societies received state funds within those years, suggesting that sixteen societies did not bother to respond to the patent office. Furthermore, some societies that responded do not seem to have received state funding, and the report itself counts ninety-seven societies. "Condensed Reports," *Report of the Commissioner of Patents for the Year 1858*,

Part II: Agriculture (Washington, DC: William A. Harris, 1859), 91, 162–79; New York (State) Comptroller's Office, *Comptroller's Ledger of Accounts with County Agricultural Societies, 1841–1901, A1298*, NYSA.

22. Franklin B. Hough, *Census of the State of New York for 1855* (Albany, NY: Charles Van Benthuysen, 1857), 183.

23. "Report of the Executive Committee for 1851," *TNYSAS for 1851* 11 (1852): 17.

24. Herman Coons, Diary of Herman Coons, 1850–54, NYSHA NM-19.53, NYSHA.

25. Demaree, *American Agricultural Press*, 18–19.

26. *Subscriptions for the Cultivator and the Horticulturist*, Broadside, 1847, BRO1026+ NYSL.

27. Census figures retrieved September 15, 2006, from the University of Virginia's Historical Census Browser (2004). University of Virginia, Geospatial and Statistical Data Center: http://fisher.lib.virginia.edu/collections/stats/histcensus/index.html. As an eastern agricultural powerhouse, New York fits poorly into normal stories of antebellum agricultural capitalism. Historians of capitalism have usually faced south, detailing the appalling rationalization of cotton fields; west, examining territorial appropriation and grain production; or northeast, following the declining soils and rising factories of New England. Linking growing urban coastal markets to an expanding western hinterland, New York embodied the dynamics of Northern agriculture as a whole—its newly seized western lands rapidly became wheat boom land, a model for the development of states like Ohio and Indiana, before turning to more specialized crops. The weakening soils of the Hudson Valley resembled the trouble-plagued soils of New England.

28. Carol Sheriff, *The Artificial River: The Erie Canal and the Paradox of Progress, 1817–1862* (New York: Hill and Wang, 1996); Brian Murphy, *Building the Empire State: Political Economy in the Early Republic* (Philadelphia: University of Pennsylvania Press, 2015).

29. Nathaniel Hawthorne, quoted in Patricia Anderson, *The Course of Empire: The Erie Canal and the New York Landscape* (Rochester, NY: Memorial Art Gallery of the University of Rochester, 1984), 33.

30. George R. Taylor, *The Transportation Revolution, 1815–60* (New York: Routledge, 2015).

31. For urban horses, see Ann Norton Greene, *Horses at Work: Harnessing Power in Industrial America* (Cambridge, MA: Harvard University Press, 2004).

32. For dairy zone, see Sally McMurry, *Transforming Rural Life: Dairying Families and Agricultural Change, 1820–1885* (Baltimore: Johns Hopkins University Press, 1995), 13–15. For hops, see Thomas Summerhill, *Harvest of Dissent: Agrarianism in Nineteenth-Century New York* (Urbana: University of Illinois Press, 2005), 95. For hay, see Percy Wells Bidwell and John I. Falconer, *History of Agriculture in the Northern United States, 1620–1860* (New York: Peter Smith, 1941), and Ariel Ron, "King Hay" (forthcoming).

33. Parkerson, *Agricultural Transition*, 19.

34. Michael Williams, *Americans and Their Forests: A Historical Geography* (Cambridge: Cambridge University Press, 1992).

35. Jane Mt. Pleasant, "The Paradox of Plows and Productivity: An Agronomic Comparison of Cereal Grain Production under Iroquois Hoe Culture and European Plow Culture in the Seventeenth and Eighteenth Centuries," *AH* 85, no. 4 (2011): 460–92; Taylor Spence, "The Endless Commons: Indigenous and Immigrant in the British-American Borderland, 1835–1848," PhD diss. (Yale University, 2012); David Stradling, *The Nature of New York: An Environmental History of the Empire State*,(Ithaca, NY: Cornell University Press, 2010).

36. Pauly, *Fruits and Plains*, 33–50; Alan L. Olmstead and Paul W. Rhode, *Creating Abundance: Biological Innovation and American Agricultural Development* (Cambridge: Cambridge

University Press, 2008), 51–53; James E. McWilliams, *American Pests: The Losing War on Insects from Colonial Times to DDT* (New York: Columbia University Press, 2012), 26–55; Wendell Tripp, ed., *Coming and Becoming: Pluralism in New York State History* (Cooperstown, NY: NYSHA, 1991), 105–20.

37. Whitney R. Cross, *The Burned-Over District: The Social and Intellectual History of Enthusiastic Religion in Western New York, 1800–1850* (New York: Harper Torchbooks, 1950); Lawrence J. Friedman, "The Gerrit Smith Circle: Abolitionism in the Burned-Over District," *Civil War History* 26, no. 1 (1980): 18–38; Paul E. Johnson, *A Shopkeeper's Millennium: Society and Revivals in Rochester, New York, 1815–1837* (New York: Macmillan, 1978); Ann Braude, *Radical Spirits: Spiritualism and Women's Rights in Nineteenth-Century America* (Bloomington: Indiana University Press, 2001); Paul E. Johnson and Sean Wilentz, *The Kingdom of Matthias: A Story of Sex and Salvation in 19th-Century America* (Oxford: Oxford University Press, 1995); Carl J. Guarneri, "Reconstructing the Antebellum Communitarian Movement: Oneida and Fourierism," *JER* 16, no. 3 (1996): 463–88; April R. Haynes, *Riotous Flesh: Women, Physiology, and the Solitary Vice in Nineteenth-Century America* (Chicago: University of Chicago Press, 2015); Elizabeth Lowry, "Spiritual (R)evolution and the Turning of Tables: Abolition, Feminism, and the Rhetoric of Social Reform in the Antebellum Public Sphere," *Journal for the Study of Radicalism* 9, no. 2 (2015): 1–16.

38. My thinking about economic storytelling is influenced by Llerena Guiu Searle, *Landscapes of Accumulation: Real Estate and the Neoliberal Imagination in Contemporary India* (Chicago: University of Chicago Press, 2016).

39. Laurence M. Hauptman, *Conspiracy of Interests: Iroquois Dispossession and the Rise of New York State* (Syracuse, NY: Syracuse University Press, 2001), 36; Brian W. Dippie, *The Vanishing American: White Attitudes and US Indian Policy* (Lawrence: University Press of Kansas, 1982).

40. For staged modernization theory, see Timothy Mitchell, *Rule of Experts: Egypt, Techno-Politics, Modernity* (Berkeley: University of California Press, 2002), 232; for modernization in colonial improving thought, see Chaplin, *An Anxious Pursuit*, 48–50; for colonial New York, see Sara Gronim, *Everyday Nature: Knowledge of the Natural World in Colonial New York* (New Brunswick, NJ: Rutgers University Press, 2007), 81–87.

41. For recent work, see Sarah Wilmot, *"The Business of Improvement": Agriculture and Scientific Culture in Britain, c. 1770–c. 1870*, vol. 24, Historical Geography Research Series (Bristol: Institute of British Geographers, 1990); Mark Overton, *Agricultural Revolution in England: The Transformation of the Agrarian Economy* (Cambridge: Cambridge University Press, 1996), 148; Tom Williamson, *The Transformation of Rural England: Farming and the Landscape, 1700–1870* (Exeter: University of Exeter Press, 2002). For Scottish improvement, see T. C. Smout, "A New Look at the Scottish Improvers," *Scottish Historical Review* 91, no. 1 (2012): 125–49; T. M. Devine, *Clearance and Improvement: Land, Power, and People in Scotland, 1700–1900* (Edinburgh: John Donald, 2010); Fredrik Albritton Jonsson, *Enlightenment's Frontier: The Scottish Highlands and the Origins of Environmentalism* (New Haven, CT: Yale University Press, 2013); John Bohstedt, *The Politics of Provisions: Food Riots, Moral Economy, and Market Transition in England, c. 1550–1850* (London: Ashgate Publishing, 2013); J. M. Neeson, *Commoners: Common Right, Enclosure, and Social Change in England, 1700–1820* (Cambridge: Cambridge University Press, 1996). For nineteenth-century British improvement, see F. M. L. Thompson, "The Second Agricultural Revolution, 1815–1880," *EHR* 21, no. 1 (1968): 62–77; Nicholas Goddard, "The Development and Influence of Agricultural Periodicals and Newspapers, 1780–1880," *AHR* 31, no. 2 (1983): 116–31; Nicholas Goddard, *Harvests of Change: The Royal Agricultural Society of England* (London:

Quiller Press, 1989); Harriet Ritvo, *The Animal Estate: The English and Other Creatures in the Victorian Age* (Cambridge, MA: Harvard University Press, 1987). From an Americanist perspective, see James L. Huston, *The British Gentry, the Southern Planter, and the Northern Family Farmer: Agriculture and Sectional Antagonism in North America* (Baton Rouge: Louisiana State University Press, 2015).

42. Hugh C. Prince, "The Changing Rural Landscape," in *The Agrarian History of England and Wales*, vol. 4, *1750–1850, Part I*, ed. G. E. Mingay (Cambridge: Cambridge University Press, 1989); Robert Trow-Smith, *A History of British Livestock Husbandry, 1700–1900* (London: Routledge and Kegan Paul, 1959).

43. Maura Capps, "Fleets of Fodder: The Ecological Orchestration of Agrarian Improvement in New South Wales and the Cape of Good Hope, 1780–1830," *Journal of British Studies* 56, no. 3 (2017): 532–56; Jayeeta Sharma, *Empire's Garden: Assam and the Making of India* (Durham, NC: Duke University Press, 2011); Richard Drayton, *Nature's Government: Science, Imperial Britain, and the "Improvement" of the World* (New Haven, CT: Yale University Press, 2000); David Arnold, "Agriculture and 'Improvement' in Early Colonial India: A Pre-History of Development," *JAC* 5 (2005): 505–25.

44. Hezekiah Hull, Diary, June 14, 1837, NYSHA, Cooperstown, New York.

45. "British Agricultural Journals," *Cultivator* 9, no. 8 (1842): 121.

46. See, e.g., Frederick Law Olmsted, *Walks and Talks of an American Farmer in England*, vol. 1 (New York: G. P. Putnam, 1852); James F. W. Johnston, *Notes on North America: Agricultural, Economical, and Social*, 2 vols. (Boston: Charles C. Little and James Brown, 1851).

47. Tamara Plakins Thornton, *Cultivating Gentlemen: The Meaning of Country Life among the Boston Elite, 1785–1860* (New Haven, CT: Yale University Press, 1989).

48. "Measurements of the Great Oxen, Which Obtained the Two First Prizes at Brighton, Massachusetts, October 1817, Taken by Two of the Trustees, the 16th of December 1817, Compared with the Official Account of the Famous English Ox Called the Durham Ox," *Massachusetts Agricultural Repository & Journal* 5, no. 1 (1818): 75–76. I address this in Emily Pawley, "The Point of Perfection: Cattle Portraiture, Bloodlines, and the Meaning of Breeding, 1760–1860," *JER* 36, no. 1 (2016): 37–72.

49. Hauptman, *Conspiracy of Interests*; Barbara Graymont, *The Iroquois in the American Revolution* (Syracuse, NY: Syracuse University Press, 1975), 216–17.

50. Russell R. Menard, "Colonial America's Mestizo Agriculture," *The Economy of Early America: Historical Perspectives and New Directions*, ed. Cathy Matson (University Park: Pennsylvania State University, 2006), 107–23; Arturo Warman, *Corn and Capitalism: How a Botanical Bastard Grew to Global Dominance* (Chapel Hill: University of North Carolina Press, 2003); Thomas Paul Slaughter, *The Natures of John and William Bartram* (New York: Alfred A. Knopf, 1996).

51. Joanna Cohen, *Luxurious Citizens: The Politics of Consumption in Nineteenth-Century America* (Philadelphia, University of Pennsylvania Press, 2017), 124–25.

52. See David Maldwyn Ellis, *Landlords and Farmers in the Hudson-Mohawk Region, 1790–1850* (Ithaca, NY: Cornell University Press, 1946); Reeve Huston, *Land and Freedom: Rural Society, Popular Protest, and Party Politics in Antebellum New York* (Oxford: Oxford University Press, 2000); Charles W. McCurdy, *The Anti-Rent Era in New York Law and Politics, 1839–1865* (Chapel Hill: University of North Carolina Press, 2003); Summerhill, *Harvest of Dissent*, 95.

53. For a brief summary account, see J. H. French, *Gazetteer of the State of New York*, 8th ed. (Syracuse, NY: R. P. Smith, 1860): 101–2.

54. L. Ray Gunn, *The Decline of Authority: Public Economic Policy and Political Development in New York, 1800–1860* (Ithaca, NY: Cornell University Press, 1988); New York (State) Department of State, *Distribution Lists of State Publications, Subseries 7*, N-Ar/A0321, Box 5, folder 2, NYSA; Timothy E. Cook, *Governing with the News: The News Media as a Political Institution* (Chicago: University of Chicago Press, 1998), 28–29.

55. Sheila Jasanoff and Sang-Hyun Kim's concept of the "sociotechnical imaginary," that is, "collectively held, institutionally stabilized and publicly performed visions of desirable futures, animated by shared understandings of forms of social life and social order, attainable through, and supportive of advances in science and technology," sort of works here. However, since rural antebellum Americans didn't collectively hold an understanding of their future, agricultural improvement was not a single sociotechnical imaginary. Instead it was a stage on which multiple imaginaries overlapped and sometimes collided. See Sheila Jasanoff, "Future Imperfect: Science, Technology, and the Imaginations of Modernity," in *Dreamscapes of Modernity: Sociotechnical Imaginaries and the Fabrication of Power*, ed. Sheila Jasanoff and Sang-Hyun Kim (Chicago: University of Chicago Press, 2015), 1–33, 4.

56. J. H. French, *Gazetteer of the State of New York* (Syracuse, NY: R. P. Smith, 1860), 153; A. and S. D. Freer, *Cortland County Agricultural Warehouse* (Cortland Village, NY: Democrat Print, 1850); Dutchess County Agricultural Society, "Dutchess County Agricultural Society Minutes," Local History Collection, 1841–46, Doc. Box 630.074-D, AML.

57. Albany's *Cultivator* and Rochester's *Genesee Farmer* appeared in the 1830s. In the forties and fifties, *Moore's Rural New Yorker*, the *Country Gentleman*, the *Working Farmer*, the *Central New York Farmer*, the *Northern Farmer*, and the *American Agriculturist* followed.

58. For predictive science and meteorology, see Jamie L. Pietruska, *Looking Forward: Prediction and Uncertainty in Modern America* (Chicago: University of Chicago Press, 2017); for economic performativity, see Donald MacKenzie, Fabian Muniesa, and Lucia Siu, "Introduction," in *Do Economists Make Markets? On the Performativity of Economics*, ed. Donald MacKenzie, Fabian Muniesa, and Lucia Siu (Princeton, NJ: Princeton University Press, 2007), 1–18; Millennialist expectations that landscapes were intended for profit somewhat undermined the "conditions of uncertainty" that Beckert describes. Jens Beckert, *Imagined Futures: Fictional Expectations and Capitalist Dynamics* (Cambridge, MA: Harvard University Press, 2016). For business projection and the history of science, see Martin Giraudeau, "Proving Future Profit: Business Plans as Demonstration Devices," *Osiris* 33, no. 1 (2018): 130–48.

59. Here I am influenced by the concept of symmetry that underpins the Strong Programme in the sociology of scientific knowledge. See David Bloor, *Knowledge and Social Imagery* (Chicago: University of Chicago Press, 1991); See also antiteleological controversy studies like Martin J. S. Rudwick, *The Great Devonian Controversy: The Shaping of Scientific Knowledge among Gentlemanly Specialists* (Chicago: University of Chicago Press, 1988). For historiographical summary, see David J. Hess, *Science Studies: An Advanced Introduction* (New York: New York University Press, 1997), 86–88. However, symmetry applied to future objects works differently.

60. Studies of disinterestedness are often founded on Steven Shapin, *A Social History of Truth: Civility and Science in Seventeenth-Century England* (Chicago: University of Chicago Press, 1994).

61. Andrew Lewis shows that "interested science," in the sense of economic inventory of natural resources, was a feature of antebellum natural history more generally. See Andrew J. Lewis, *A Democracy of Facts: Natural History in the Early Republic* (Philadelphia: University of Pennsylvania Press, 2011), 46–71; Benjamin Cohen, "Surveying Nature: Environmental Dimen-

sions of Virginia's First Scientific Survey, 1835–1842," *EH* 11, no. 1 (2007): 37–69; Michele L. Aldrich, *New York State Natural History Survey, 1836–1845: A Chapter in the History of American Science* (Ithaca, NY: Paleontological Research Institution, 2000). By contrast, improving science was linked to individual personal benefit, resembling the dynamics described by Paul Lucier, *Scientists and Swindlers: Consulting on Coal and Oil in America, 1820–1890* (Baltimore: Johns Hopkins University Press, 2008). However, it also involved direct sales of goods, in which experts had an interest.

62. The state as disinterested observer can be found in James C. Scott, *Seeing Like a State: How Certain Schemes to Improve the Human Condition Have Failed* (New Haven, CT: Yale University Press, 1998). For the role of commercial nurserymen in botany in Britain and France, see Sarah Easterby-Smith, *Cultivating Commerce: Cultures of Botany in Britain and France, 1760–1815* (Cambridge: Cambridge University Press, 2017).

63. For the intersection of policy and the twentieth-century creation of particular notions of the family farm, see Gabriel N. Rosenberg, *The 4-H Harvest: Sexuality and the State in Rural America* (Philadelphia: University of Pennsylvania Press, 2015). See also Gabriel Rosenberg, "Fetishizing the Family Farm," *Boston Globe*, April 10, 2016, https://www.bostonglobe.com/ideas/2016/04/09/fetishizing-family-farms/NJszoKdCSQWaq2XBw7kvIL/story.html (accessed April 11, 2016).

Chapter One

1. For a summary of this debate, see Martin Bruegel, "Unrest: Manorial Society and the Market in the Hudson Valley, 1780–1850," *JAH* 82, no. 4 (1996): 1393–424.

2. See Ellis, *Landlords and Farmers*; Huston, *Land and Freedom*; McCurdy, *The Anti-Rent Era*. For role in Federalism and in creating the conditions for Martin van Buren's building of the Democratic Party, see John L. Brooke, *Columbia Rising: Civil Life on the Upper Hudson from the Revolution to the Age of Jackson* (Chapel Hill: University of North Carolina Press, 2010).

3. David I. Spanagel, *DeWitt Clinton and Amos Eaton: Geology and Power in Early New York* (Baltimore: Johns Hopkins University Press, 2014), 61.

4. The "military tract" in central New York was surveyed for grants to revolutionary veterans. Most sold their rights to speculators. William Wyckhoff, *The Developer's Frontier: The Making of the Western New York Landscape* (New Haven, CT: Yale University Press, 1988); Alan Taylor, "'Wasty Ways': Stories of American Settlement," *EH* 3 (1998): 291–310, 293. For elite land speculation, see Murphy, *Building the Empire State*, 70–71.

5. "Agrarianism," *Albany Freeholder*, January 21, 1846, AAS; "Equal Rights," *Albany Freeholder*, November 19, 1845, AAS; "Leasehold Tenures—Remarks of Mr. Harris of Albany in Committee of the Whole upon the Anti-Rent Question," *Albany Freeholder*, February 11, 1846, AAS.

6. "Senatorial Convention," *Albany Freeholder*, October 1, 1845, AAS.

7. Dixon Ryan Fox, *The Decline of Aristocracy in the Politics of New York* (New York: Harper and Row, [orig. 1919] 1965), 130.

8. Bruegel, "Unrest," 1402.

9. For the rise of British landed elites, see David Cannadine, *Aspects of Aristocracy: Grandeur and Decline in Modern Britain* (New Haven, CT: Yale University Press, 1994), 11–21. For eighteenth-century developments, see E. P. Thompson, *Whigs and Hunters* (New York: Pantheon Books, 1975). For timing debates, see Overton, *Agricultural Revolution in England*; R. C. Allen,

"Tracking the Agricultural Revolution in England," *EHR* 52, no. 2 (1999): 209–35. For development in Scotland, see Devine, *Clearance and Improvement*; Fredrik Albritton Jonsson, *Enlightenment's Frontier*; Brian Bonnyman, *The Third Duke of Buccleuch and Adam Smith: Estate Management and Improvement in Enlightenment Scotland* (Edinburgh: Edinburgh University Press, 2014). For an Americanist perspective, see chapters 1–2 of James L. Huston, *The British Gentry, the Southern Planter, and the Northern Family Farmer: Agriculture and Sectional Antagonism in North America* (Baton Rouge: Louisiana State University Press, 2015).

10. Cannadine, *Aspects of Aristocracy*, 11.

11. My definitions of agricultural capitalism are influenced by Immanuel Wallerstein, *The Modern World-System I: Capitalist Agriculture and the Origins of the European World-Economy in the Sixteenth Century* (Berkeley: University of California Press, 2011). For British capitalist agriculture, see Neil Davidson, "The Scottish Path to Capitalist Agriculture 3: The Enlightenment at the Theory and Practice of Improvement," *JAC* 5 (2005): 1–72; Pamela Horn, "An Eighteenth-Century Land Agent: The Career of Nathaniel Kent (1737–1810)," *AHR* 30 (1982): 1–16; G. E. Mingay, *Arthur Young and His Times* (London: Macmillan Press, 1975); H. Adams, "The Agents of Agricultural Change," in *The Making of the Scottish Countryside*, ed. M. L. Parry and T. R. Slater, 155–75 (London: Croom Helm, 1980).

12. For global plantations and capitalism, see Jason W. Moore, "The Capitalocene, Part I: On the Nature and Origins of our Ecological Crisis," *JPS* 44, no. 3 (2017): 594–630; for the coining of Plantationocene, see Donna Haraway, "Anthropocene, Capitalocene, Plantationocene, Chthulucene: Making Kin," *Environmental Humanities* 6, no. 1 (2015): 159–65.

13. For the original British agricultural revolution literature, see Baron Rowland Edmund Prothero Ernle, *English Farming, Past and Present* (London: Longmans, Green and Co., 1912), 176–89; J. D. Chambers and G. E. Mingay, *The Agricultural Revolution, 1750–1880* (London, 1966), 66–69. For responses, see Eric Kerridge, "The Agricultural Revolution Reconsidered," *AH* 43, no. 4 (1969): 463–76; Robert C. Allen, "Tracking the Agricultural Revolution in England," *EHR* 52, no. 2 (1999): 209–35; Mark Overton, "Re-establishing the English Agricultural Revolution," *AHR* 52, no. 2 (1996): 1–20; S. Todd Lowry, "The Agricultural Foundation of the Seventeenth-Century English Oeconomy," *History of Political Economy* 35, no. 5 (2004): 74–100; Nicholas Russell, *Like Engend'ring Like: Heredity and Animal Breeding in Early Modern England* (Cambridge: Cambridge University Press, 2007). For wider European roots, see Mauro Ambrosoli, *The Wild and the Sown: Botany and Agriculture in Western Europe, 1350–1850* (Cambridge: Cambridge University Press, 1997).

14. For these connections, see Andrea Wulf, *Founding Gardeners: How the Revolutionary Generation Created an American Eden* (New York: Random House, 2011).

15. George Washington, Arthur Young, and Sir John Sinclair, *Letters from his Excellency George Washington, to Arthur Young, Esq. F.R.S., and Sir John Sinclair, Bart.* (Alexandria, VA: Cotton and Steward, 1803); Rosalind Mitchison, "Sinclair, Sir John, First Baronet (1754–1835)," in *Oxford Dictionary of National Biography* (Oxford: Oxford University Press, September 2004; online ed., January 2008), http://www.oxforddnb.com/view/article/25627 (accessed March 18, 2008).

16. For publicizing of improvement, see, e.g., Lord Kames (Henry Home), *The Gentleman Farmer: Being an Attempt to Improve Agriculture, by Subjecting It to the Test of Rational Principles* (Edinburgh: W. Creech, 1776); Pamela Horn, *William Marshall (1745) and the Georgian Countryside* (Abingdon, Oxon: Beacon Publications, 1982); Mingay, *Arthur Young and His Times*.

17. Gronim, *Everyday Nature*, 92–102.

18. For American elite land speculation in the 1780s and 1790s, see John Lauritz Larson, *Internal Improvement: National Public Works and the Promise of Popular Government in the Early United States* (Chapel Hill: University of North Carolina Press, 2001), 9–38; Fox, *Decline of Aristocracy*, 123; Wyckhoff, *The Developers' Frontier*.

19. Alan Taylor, *William Cooper's Town: Power and Persuasion on the Frontier of the Early American Republic* (New York: Vintage, 1996); Alden Hatch, *The Wadsworths of the Genesee* (New York: Coward-McCann, 1959); Ellis, *Landlords and Farmers*, 70; Neil Adams McNall, *An Agricultural History of the Genesee Valley, 1790–1860* (Philadelphia: University of Pennsylvania Press, 1952), 58.

20. For improvement and coercion, see Drayton, *Nature's Government*, 54; Hauptman, *Conspiracy of Interests*, 82–87.

21. Marti, *To Improve the Soil and the Mind*, 5.

22. "Letters of the Agricultural Society to the Friends and Promoters of Rural Economy," *TSPAAM* 1 (1801): ix–xiv, xiv.

23. William Bacon, "History of the Agricultural Associations of New York from 1791 to 1862," *TNYSAS for 1863* 23 (1864): 143–68, 144. The British Library's copy of the SPAAM *Transactions* is inscribed as a gift from David Hosack to Joseph Banks. *Transactions of the Society, Instituted in the State of New York, for the Promotion of Agriculture, Arts and Manufactures, Part 1 and 2* [bound together] (New York: Chills and Swaine, 1792–94). Livingston also likely supported an independent agricultural society in Dutchess County starting in 1807, which consisted mostly of his political allies and landlords of the manor region. Brooke, *Columbia Rising*, 425–6.

24. For more on these efforts, see Victoria Johnson, *American Eden: David Hosack, Botany, and Medicine in the Garden of the Early Republic* (New York: W. W. Norton and Co., 2018).

25. Mark A. Mastromarino, "Elkanah Watson and the Early Agricultural Fair," *Historical Journal of Massachusetts* 17, no. 2 (1989): 105.

26. G. W. Featherstonehaugh to Sir John Sinclair, in *The Correspondence of the Right Honourable Sir John Sinclair, Bart*, vol. 1. (London: Henry Colburn and Richard Bentley, 1831), 278–79.

27. McCurdy, *The Anti-Rent Era*, 12; Daniel S. Dupre, "The Panic of 1819 and the Political Economy of Sectionalism," in Matson, *The Economy of Early America*.

28. Donald B. Marti, "Early Agricultural Societies in New York: The Foundations of Improvement," *NYH* 48, no. 4 (1967): 313–31.

29. Ariel Ron, "Summoning the State," 352.

30. Summerhill, *Harvest of Dissent*, 34.

31. Many scholarly texts draw on Arthur H. Cole, "Agricultural Crazes: A Neglected Chapter in American Economic History," *American Economic Review* 16, no. 4 (1926): 622–39; for contemporary use of "mania," see *Antidote to the Merino-Mania Now Progressing through the United States* (Philadelphia: J. and A. Y. Humphreys, 1810).

32. Woods, *Herds Shot Round the World*, 58–59.

33. Ellis, *Landlords*, 144; for the global context of the Merino mania, see Michael Lawson Ryder, *Sheep and Man* (London: Gerald Duckworth and Co., 1983), 611–12; Woods, *Herds Shot Round the World*, 60–72. For the American mania, see Ariel Ron, *Grassroots Leviathan: Agricultural Reform and the Rural North in the Slaveholding Republic* (forthcoming from Johns Hopkins University Press); Benjamin Hurwitz, "Grazing the Modern World: Merino Sheep in South Africa and the United States, 1775–1840," PhD diss. (George Mason University, 2017); Robert R. Livingston, "Letter from Robert R. Livingston to Arthur Young," *TSPAAM* 1, no. 1 (1792): 142–48; Robert R. Livingston, *Essay on Sheep* (Concord, NH: Daniel Cooledge, 1813).

NOTES TO CHAPTER ONE 249

For use of British texts, see, e.g., "Extract from Lord Somerville's Essay on Sheep," *Agricultural Museum* 1 (December 7, 1810): 178.

34. Woods, *Herds Shot Round the World*, 77; Drayton, *Nature's Government*, 87, 97.

35. Andrea Sutcliffe, *Steam: The Untold Story of America's First Great Invention* (New York: Palgrave Macmillan, 2004), 161–91.

36. Spanagel, *DeWitt Clinton and Amos Eaton*, 98–116.

37. Amos Eaton, "Circular: To Gentlemen Residing in the Vicinity of the Erie Canal" (Albany, NY: s.n., 1822), HML, 9; Amos Eaton and T. Romeyn Beck, *A Geological Survey of the County of Albany* (Albany, NY: n.p.); Amos Eaton, *A Geological and Agricultural Survey of Rensselaer County* (Albany, NY: E. and E. Horsford, 1822); Amos Eaton, *A Geological and Agricultural Survey of the District Adjoining the Erie Canal, in the State of New York*, vol. 1 (Albany, NY: Packard and Van Benthuysen, 1824); Daniel D. Barnard, *A Discourse on the Life, Services and Character of Stephen Van Rensselaer* (Albany, NY: Hoffman and White, 1839); Spanagel, *DeWitt Clinton and Amos Eaton*, 98–116.

38. Stephen Van Rensselaer, *Address of the General Committee of the Board of Agriculture of the State of New-York to the County Agricultural Societies, for 1820* (Albany, NY: S. Southwick, 1820), 16. See also account of Jay in Fox, *Decline of Aristocracy*, 8. For a comparative account showing the much lower rents in the US in the colonial period, see Huston, *British Gentry, Southern Planter, and the Northern Family Farmer*, 64–65.

39. "Obituary Notice of the Hon. Stephen Van Rensselaer," *American Journal of Science and Arts* 36 (1839): 156–64, 159.

40. Eunice Foote, "Circumstances Affecting the Heat of the Sun's Rays," *American Journal of Science and Arts* 22, no. 66 (1856): 382; R. P. Sorenson, "Eunice Foote's Pioneering Work on CO_2 and Climate Warming," *Search and Discovery Article #70092* (2011); Roland Jackson, *The Ascent of John Tyndall: Victorian Scientist, Mountaineer, and Public Intellectual* (Oxford: Oxford University Press, 2018), 483n7; Donald B. Marti, "The Purposes of Agricultural Education: Ideas and Projects in New York State, 1819–1865," *AH* 45, no. 4 (1971): 271–83; Spanagel, *DeWitt Clinton and Amos Eaton*, 74.

41. "Ralph and J. B. Fowler to Jas Wadsworth Lease, Home Dairy" (1834), Leases: Geneseo, Folder: Wadsworth Contracts, 1834, ML, Wadsworth Family Papers.

42. Summerhill, *Harvest of Dissent*, 34.

43. *Proceedings of the State Agricultural Convention Held at the Capitol in the City of Albany, on the 14th, 15th and 16th February, 1832: With the Constitution of a State Agricultural Society, Agreed to and Adopted, by the Said Convention* (Albany, NY: Webster and Skinners, 1832); Marti, "Early Agricultural Societies," 327.

44. Summerhill, *Harvest of Dissent*, 45.

45. Devine, *Transformation of Rural Scotland*, 119.

46. Stoll, *Larding the Lean Earth*, 143–50.

47. Huston, *Land and Freedom*, 21.

48. Standard accounts of this are Ellis, *Landlords and Farmers*, 225–312; Huston, *Land and Freedom*, 200; for older tensions, see Bruegel, "Unrest." For central New York, see Summerhill, *Harvest of Dissent*.

49. Huston, *Land and Freedom*, 28.

50. This argument was popular among informal occupants of land during the nineteenth century. Donald J. Pisani, "The Squatter and Natural Law in Nineteenth-Century America," *AH* 81, no. 4 (2007): 443–63; Huston, *Land and Freedom*, 83.

51. Front matter, *Albany Freeholder*, January 21, 1846, AAS.

52. For the agrarian roots of the labor theory of value, see Huston, *The British Gentry, the Southern Planter, and the Northern Family Farmer*; Jamie Bronstein traces the connections between American and British movements in Jamie L. Bronstein, *Land Reform and Working-Class Experience in Britain and the United States, 1800–1862* (Stanford, CA: Stanford University Press, 1999); Reeve Huston, "Multiple Crossings: Thomas Ainge Devyr and Transatlantic Land Reform," in *Transatlantic Rebels: Agrarian Radicalism in Comparative Context*, ed. Thomas Summerhill and James C. Scott (East Lansing: Michigan State University Press, 2004), 137–66. See also Seth Cotlar, *Tom Paine's America: The Rise and Fall of Transatlantic Radicalism in the Early Republic* (Richmond: University of Virginia Press, 2011), 123–27; "Works, Published, and for Sale Wholesale and Retail, by George M. Evans," *Radical*, December 31, 1841, AAS; "Anti-Rent Meeting," *Albany Freeholder*, March 11, 1846, AAS.

53. The classic examination of the rural wars is Eric Hobsbawm and Rudé, *Swing* (New York: Verso, 2014); see also Carl J. Griffin, *The Rural War: Captain Swing and the Politics of Protest* (Manchester: Manchester University Press, 2012).

54. See Hobsbawm and Rudé, *Swing*, Appendix II; Eric Richards, *Debating the Highland Clearances* (Edinburgh: Edinburgh University Press, 2007).

55. McCurdy, *The Anti-Rent Era*, 218.

56. "Senatorial Convention," *Albany Freeholder*, October 1, 1845, AAS.

57. "Senatorial Convention."

58. Daniel Lee, "Report on Agriculture," *Albany Freeholder*, December 3, 1845, and "Wheeler's Horse Power and Separating Threshing Machine," *Albany Freeholder*, October 20, 1847, AAS.

59. C. F. Bouton, "Agriculture," *Albany Freeholder*, June 23, 1848, AAS.

60. Jeremiah Allen, Allen Diaries, September 22, 1847, October 2, 1847, NYSHA.

61. "Anti-Rent Meeting," *Albany Freeholder*, March 11, 1846, AAS.

62. Huston, *Land and Freedom*, 27–33; Summerfield, *Harvest of Dissent*, 35.

63. Huston, *Land and Freedom*, 55.

64. "Harris's Address," *Albany Freeholder*, February 11, 1846, AAS.

65. "Harris's Address." See also "The Manorial Land Tenures," *Albany Freeholder*, October 15, 1845, AAS.

66. "Agrarianism," *Albany Freeholder*, January 21, 1846, AAS.

67. See Paul W. Gates, "The Role of the Speculator in Western Development," *PMHB* 66, no. 3 (1942): 314–33; Paul W. Gates, "Land Policy and Tenancy in the Prairie States," *JEH* 1, no. 1 (1941): 60–82; for the contrary view, see Jeremy Atack, "The Agricultural Ladder Revisited: A New Look at an Old Question with some Data for 1860," *AH* 63, no. 1 (Winter 1989): 1–25. See also Donald L. Winters, "Agricultural Tenancy in the Nineteenth-Century Middle West: The Historiographical Debate," *Indiana Magazine of History* 78, no. 2 (1982): 128–53; Frank Yoder complicates the ladder hypothesis, differentiating between tenants with local landed kin and those with none, and separating the tenancy patterns of different ethnic groups: Frank Yoder, "Rethinking Midwestern Farm Tenure: A Cultural Perspective," *AH* 71, no. 4 (1997): 457–78. For continued tenant resistance, see Summerhill, *Harvest of Dissent*, 175–78; McNall, *Genesee Valley*, 227.

68. William H. Seward, "Address of the Hon. William H. Seward (Governor of the State)," *TNYSAS for 1842* 2 (1843): 12.

69. Huston, *Land and Freedom*, 7.

70. Huston; Phillips, "Antebellum Agricultural Reform"; Adam Wesley Dean, *An Agrarian Republic: Farming, Antislavery Politics, and Nature Parks in the Civil War Era* (Chapel Hill: Uni-

versity of North Carolina Press, 2014); For the further political ramifications of this, see Ariel Ron, *Grassroots Leviathan*.

71. For estimate reasoning, see introduction, note 23; "Condensed Reports," *Report of the Commissioner of Patents for the Year 1858, Part II: Agriculture* (Washington, DC: William A. Harris, 1859). For sense of scale, see Rossiter, "The Organization of Agricultural Improvement," 279–97; New York (State) Comptroller's Office, *Comptroller's Ledger of Accounts with County Agricultural Societies, 1841–1901*, NYSA.

72. This was not a passive expansion of landlords' improving aims—new groups changed improvement's meaning. My thinking here influenced by Cooter and Pumfrey, "Separate Spheres and Public Places"; Andrew Cunningham and Perry Williams, "De-centering the 'Big Picture': The Origins of Modern Science and the Modern Origins of Science," *BJHS* 26, no. 4 (1993), 407–32.

Chapter Two

1. "Fig. 90: View of the New-York State Cattle Show," *Cultivator* 1, no. 10 (1844): 312.
2. "Great American State Fair and Cattle Show," *Illustrated London News* 5, no. 130 (1844): 260.
3. J. M. Sherwood, *J. M. Sherwood to Henry S. Randall, Dec. 2, 1841*, Letter, Folder, 1840–42, NIC 851 CCHS, Henry S. Randall Correspondence, 1828–1965.
4. George Bancroft, "Address, by Hon. George Bancroft of Massachusetts," *TNYSAS for 1844* 4 (1845): 12–19.
5. See Sarah Burns, *Pastoral Inventions: Rural Life in Nineteenth Century American Art and Culture* (Philadelphia: Temple University Press, 1989); Richard Hofstadter, "The Myth of the Happy Yeoman," *American Heritage* 7 (1956): 43–45; Martin J. Burke, *The Conundrum of Class: Public Discourse on the Social Order in America* (Chicago: University of Chicago Press, 1995); for the union of manufacturing and agricultural interests, see Lawrence Peskin, *Manufacturing Revolution: The Intellectual Origins of Early American Industry* (Baltimore: Johns Hopkins University Press, 2003).
6. Historians have inherited this snarl of ideas and overlaid it with historiographical binaries: following the transition-to-capitalism debates, we have adopted "market-oriented" and "not-market-oriented" farmer, though we now agree that there were no pure "subsistence farmers." While many historians have attacked Charles Sellers's contention that early American farmers were relatively untouched by the market, fewer have challenged his contention of an equality of condition founded in roughly equal distribution of land. Charles Sellers, *The Market Revolution: Jacksonian America, 1814–1846* (New York: Oxford University Press, 1994), 4. For debates about market participation, see Winifred B. Rothenberg, "The Market and Massachusetts Farmers, 1750–1855," *JEH* 41 (1981): 283–314; Rothenberg, *From Market-Places to a Market Economy*; Clark, *The Roots of Rural Capitalism*; Lamoreaux, "Rethinking the Transition to Capitalism in the Early American Northeast."
7. Catherine E. Kelly, *In the New England Fashion: Reshaping Women's Lives in the Nineteenth Century* (Ithaca, NY: Cornell University Press, 2002); David Jaffee, *A New Nation of Goods: The Material Culture of Early America* (Philadelphia: University of Pennsylvania Press, 2011).
8. Karen Halttunen, *Confidence Men and Painted Women: A Study of Middle-Class Culture in America, 1830–1870* (New Haven, CT: Yale University Press, 1982), 200.
9. My sense of strategic boundary work here comes from Thomas F. Gieryn, *Cultural Boundaries of Science: Credibility on the Line* (Chicago: University of Chicago Press, 1999).

10. Quoted in Bonnie Marranca, *A Hudson Valley Reader: Writings from the 17th Century to the Present* (Woodstock, NY: Overlook Press, 1995), 133; Thornton, *Cultivating Gentlemen*, 198; Pauly, *Fruits and Plains*, 51–79; John R. Stilgoe, *Borderland: Origins of the American Suburb, 1820–1939* (New Haven, CT: Yale University Press, 1988); William K. Wyckoff, "Landscapes of Private Power and Wealth," in *The Making of the American Landscape*, ed. Michael P. Conzen (Boston: Unwin Hyman, 1990), 344; Fox, *Decline of Aristocracy*, 24.

11. John Fowler, *Journal of a Tour in the State of New York in the Year 1830* (London: Whittaker, Treacher and Arnot, 1831), 33.

12. Sven Beckert, *The Monied Metropolis: New York City and the Consolidation of the American Bourgeoisie, 1850–1896* (Cambridge: Cambridge University Press, 2003), 25; on the new dominance of New York banking, see Daniel Walker Howe, *What Hath God Wrought: The Transformation of America, 1815–1848* (Oxford: Oxford University Press, 2007), 382.

13. Summerhill, *Harvest of Dissent*, 83.

14. Mark A. Mastromarino, "Cattle Aplenty and Other Things in Proportion: The Agricultural Society and Fair in Franklin County, Massachusetts, 1810–1860," *UCLA Historical Journal* 5 (1984), 50–75; Elkanah Watson, *History of Agricultural Societies on the Modern Berkshire System* (Albany, NY: D. Steele, 1820).

15. Harry J. Carman, "Jesse Buel, Early Nineteenth-Century Agricultural Reformer," *AH* 17, no. 1 (1943): 1–13.

16. Jacob R. Valk, "Country Seat and Farm," *Agriculturist* 5 (1846): 104.

17. Scott Sandage, *Born Losers*, 99–129; Beckert, *Monied Metropolis*, 44.

18. Raymond Williams, *The Country and the City* (New York: Oxford University Press, 1973), 18; Horace, "Beatus Ille," trans. Elizabeth Jones, *Arion: A Journal of Humanities and the Classics* 13, no. 2 (2005): 117–20.

19. Catherine Kelly, *Republic of Taste: Art, Politics, and Everyday Life in Early America* (Philadelphia: University of Pennsylvania Press, 2016); William Russell Birch, *The Country Seats of the United States of North America: With Some Scenes Connected with Them* (Springland, near Bristol, PA: s.n., 1808).

20. Thornton, *Cultivating Gentlemen*; Stilgoe, *Borderland*, 94.

21. Mrs. E. Wellnone, "From the Flag of Our Union Living in the Country," *Albany Freeholder*, November 28, 1849, AAS.

22. Frances F. Dunwell, *The Hudson River Highlands* (New York: Columbia University Press, 1991), 85; Melanie A. Kiechle, *Smell Detectives: An Olfactory History of Nineteenth-Century Urban America* (Seattle: University of Washington Press, 2017), 53.

23. William James MacNeven, *Introductory Discourse to a Few Lectures on the Application of Chemistry* (New York: G. and C. Carvill, 1825), 29.

24. Sandage, *Born Losers*, 8.

25. Samuel Willard Crompton, "Delafield, John," *American National Biography Online*, February 2000.

26. "Agriculturist," entry printed from *Oxford English Dictionary Online* (Oxford: Oxford University Press, 2009).

27. Sam Lawrence, *Sam Lawrence to H. S. Randall, December 21st 1850*, Letter, NIC 851 CCHS, Henry S. Randall Correspondence, 1828–1965.

28. "Who Would Not Be A Farmer?," *New-Yorker* 7, no. 14 (1839): 216.

29. "The Subscribers Offer for Sale the Following Valuable and Improving Property," *JOC*, February 13, 1839, AAS.

30. "Sale," *JOC*, January 5, 1839, AAS.

31. Courtney Fullilove, "The Price of Bread: The New York City Flour Riot and the Paradox of Capitalist Food Systems," *Radical History Review* 118 (2014): 15–41.

32. Beckert, *Monied Metropolis*, 64, 74, 90–92.

33. Gordon M. Winder, "A Trans-National Machine on the World Stage: Representing McCormick's Reaper through World's Fairs, 1851–1902," *Journal of Historical Geography* 33, no. 2 (2007), 352–76; D. Eldon Hall, *A Condensed History of the Origination, Rise, Progress, and Completion of the Great Exhibition of the Industry of All Nations Held in the Crystal Palace, London, during the Summer of the Year 1851* (Clinton Hall, NY: Redfield, 1852); William Carey Richards, *A Day in the New York Crystal Palace, and How to Make the Most of It* (New York: G. P. Putman and Co., 1853), 159–62.

34. On the expansion of British investment, see Beckert, *Monied Metropolis*, 25; Peter Temin, *The Jacksonian Economy* (New York: Norton, 1969), 64–90.

35. While Whigs and Republicans were more likely to support state funding for improvement, Democrats certainly participated. Henry S. Randall, the state society's corresponding secretary was also briefly New York secretary of state in a Democratic administration. Paul K. Randall, *Genealogy of a Branch of the Randall Family, 1666 to 1879* (Norwich, NY: Office of the Chenango Union, 1879), 2–3; Eliakim S. Weld, Eliakim S. Weld Journal, 1853–67, Diary, September 29, 1856, CCHS.

36. "From the Long Island Star," *Albany Evening Journal*, June 15, 1836, AAS.

37. "From the Vermont Patriot," *Albany Argus*, October 14, 1836, AAS.

38. "The Farmer of North Bend," *Albany Evening Journal*, August 24, 1836, AAS; Fox, *Aristocracy*, 412–13.

39. Elkanah Watson, Journal "E," Mixed Medley, GB13294, p. 60, NYSL, Elkanah Watson Papers, 1774–1885.

40. J. Halsey, "Report of the Committee on Agriculture," Doc. 110, March 26, 1834, *DSSNY* (Albany, NY: E. Croswell, 1834), 4.

41. "Book Farmers," *GFGJ* 2, no. 38 (1832): 297.

42. *Hoffman's Annual Advertiser and Albany Directory* (Albany, NY: n.p., 1844), 57–58, 132. F. Daniel Larkin, "Corning, Erastus," *American National Biography Online*, February 2000, http://www.anb.org/articles/10/10-00338.html (accessed March 1, 2016).

43. Sally McMurry, "Buel, Jesse (1778–1839), Agriculturist," *American National Biography Online*, March 2019.

44. Benjamin Rush, "Observations of the Duties of a Physician," in *Medical Inquiries and Observations* (Philadelphia: J. Conrad, 1805), 390; Paul Starr. *The Social Transformation of American Medicine* (New York: Basic Books, 1982), 65.

45. Gregory H. Nobles, "Commerce and Community: A Case Study of the Rural Broommaking Business in Antebellum Massachusetts," *JER* 4, no. 3 (1984): 287–93.

46. Salmon Bostwick, Diary of Salmon E. Bostwick, 1826–1831, Special Collection Diaries B657, NYSHA; Solomon Northup, *Twelve Years a Slave: Narrative of Solomon Northup* (New York: Miller, Orton, and Mulligan, 1855); Oliver J. Tillson, *Printed Maps and Correspondence, 1850–56*, Box 8, Archives 1458, CAK, Oliver J. Tillson Papers, 1803–98.

47. E.g., Hezekiah Hull, Diary of Hezekiah Hull, 1837, WPA Diary Transcript, FARM-34.55, June 14, 1837, NYSHA; Moses Eames, Diary, vol. 1, March 19, 1831–February 20, 1832, JCHS, Moses Eames Diaries.

48. Sally McMurry, "Who Read the Agricultural Journals? Evidence from Chenango

County, New York 1839–1865," *AH* 63 (1989): 1–18; Ronald J. Zboray, *A Fictive People: Antebellum Economic Development and the American Reading Public* (New York: Oxford University Press, 1993), 119. See also chapter 4, "Everyday Dissemination," in Ronald J. Zboray and Mary Saracino Zboray, *Everyday Ideas: Socioliterary Experience among Antebellum New Englanders* (Knoxville: University of Tennessee Press, 2006); Conevery Bolton Valencius, David I. Spanagel, Emily Pawley, Sara Stidstone Gronim, and Paul Lucier, "Science in Early America: Print Culture and the Sciences of Territoriality," *JER* 36, no. 1 (2016): 73–123; Richard R. John, *Spreading the News: The American Postal System from Franklin to Morse* (Cambridge, MA: Harvard University Press), 2009.

49. The 1850 census offers more actual detail, but also shows a more prosperous time following the recoveries from the Panics of 1837 and 1839. US Census of 1850, New York, Dutchess County, Roll: M432_497, P. 337A-369B RG 29 NARA, Ancestry.com.

50. *Commemorative Biographical Record of the Counties of Dutchess and Putnam, New York* (Chicago: J. H. Beers and Co., 1897), 391.

51. *Commemorative Biographical Record of the Counties of Dutchess and Putnam, New York*, 418.

52. US Census of 1850, New York, Dutchess County, Roll: M432_497, P. 337A-369B RG 29 NARA, Ancestry.com.

53. Johnson, *Shopkeeper's Millennium*, 31.

54. For the hidden scale of tenancy, see Jeremy Atack, "Tenants and Yeomen in the Nineteenth Century," *AH* 62, no. 3 (1988): 6–32.

55. Heman Chapin, Daybook, 1849–56, 198, RRL, Heman Chapin Family Papers.

56. Hough, *Census of the State of New York for 1855*, lii.

57. "Improved Agriculture of Delaware—Surprising Restoration of Worn-Out Land," *American Farmer*, 5, no. 32 (1843): 252.

58. US Census for 1850, New York, Dutchess County, Agricultural Schedule for Amenia 1850, NARA RG 29, Roll: M432_497, 337A-388A, Ancestry.com, 29; Federal Manuscript Census for Amenia 1840. Generational differences accounted for some of these distinctions; John Tanner, with $4,000, had perhaps received a portion of Briggs Tanner's land ($21,000); Clark, *Roots of Rural Capitalism*, 293.

59. For mortgages, see Jonathan Levy, "The Mortgage Worked the Hardest: The Fate of Landed Independence," in *Capitalism Takes Command: The Social Transformation of Nineteenth-Century America*, ed. Michael Zakim and Gary J. Kornblith (Chicago: University of Chicago Press, 2011), 42.

60. *A Gazetteer of the State of New-York* (Albany, NY: J. Disturnell, 1842), 61.

61. Graham Russell Hodges, *Root and Branch: African Americans in New York and East Jersey, 1613–1863* (Chapel Hill: University of North Carolina Press, 1999), 164.

62. William P. McDermott, "Slaves and Slaveowners in Dutchess County," *Afro-Americans in New York Life and History* 19, no. 1 (1995): 17; Hodges, *Root and Branch*, 165.

63. Bruegel, *Farm, Shop, Landing*, 93–84, 167.

64. Kelly, *In the New England Fashion*, 176–77.

65. While her early death denied us her published work, Sonya Marie Barclay's 2008 dissertation is one of the most sensitive readings of rural class in the nineteenth-century North, particularly in her attention to outdoor cultures of display. Sonya Marie Barclay, "Reading the Social Landscape: A Lexicon of Rural Class in Western Pennsylvania, 1790–1860," PhD diss. (Carnegie Mellon University, 2008), 218, 225, 246, 378, 392.

66. Herman Ten Eyck Foster, Diary, NYSHA; US Census of 1850, Fayette, Seneca County, New York, NARA, Roll M432_ 597, p. 163, Image: 325, household of Herman T. E. Foster, Ancestry .com.

67. A. B. Conger, *Memorial of Herman Ten Eyck Foster* (Albany, NY: NYSAS, 1870).

68. H. T. E. Foster, "Premium Farm," *Cultivator* 6, no. 9 (1849): 268.

69. Levi Weeks, Diary, vol. 1, August 30, 1855, Coll. No. 308, NYSHA, John B. and Levi Weeks Diaries, 1850–90.

70. Levi Weeks, Diary, vol. 2, June 2, 1855, Coll. No. 308, NYSHA, John B. and Levi Weeks Diaries, 1850–90.

71. J. S. Pettibone, "Agriculture of Vermont," *Cultivator* 5, no. 5 (1848): 143. Italics original.

72. Richard Lyman Bushman, *The Refinement of America: Persons, Houses, Cities* (New York: Vintage, 2011); David Jaffee, *A New Nation of Goods: The Material Culture of Early America* (Philadelphia: University of Pennsylvania Press, 2010).

73. Bushman, *Refinement of America*, 354–82; David Schuyler, *Apostle of Taste: Andrew Jackson Downing, 1815–1852* (Baltimore: Johns Hopkins University Press, 1996); Catherine E. Kelly, "'Well Bred Country People': Sociability, Social Networks, and the Creation of a Provincial Middle Class, 1820–1860," *JER* 19, no. 3 (1999): 451–71. On refinement, see Erica Hannickel, *Empire of Vines: Wine Culture in America* (Philadelphia: University of Pennsylvania Press), 57–58; Stilgoe, *Borderland*, 67–93; Pauly, *Fruits and Plains*, 53–54.

74. T. M. Niven, "Design for a Genteel Farm House," *Cultivator* 10, no. 1 (1843): 16.

75. See, e.g., Lewis Falley Allen, *Rural Architecture* (New York: C. M. Saxton, 1852); Sally McMurry, *Families and Farmhouses in Nineteenth Century America: Vernacular Design and Social Change* (Oxford: Oxford University Press, 1988); Bushman, *Refinement of America*, 245–47.

76. A. B. Allen, *Descriptive Catalogue of Horticultural and Agricultural Implements and Tools, and Field and Garden Seeds* (New York: A. B. Allen, 1846), 51.

77. Moses Eames, Diary, vol. 2, June 30, 1832, JCHS, Moses Eames Diaries; Levi Weeks, Diary, vol. 1, May 6, 1851, Coll. No. 308, NYSHA, John B. and Levi Weeks Diaries, 1850–90. On bookcases see, Bushman, *Refinement of America*, 283.

78. For this shift, see Catherine E. Kelly, "'The Consummation of Rural Prosperity and Happiness': New England Agricultural Fairs and the Construction of Class and Gender, 1810–1860," *American Quarterly* 49, no. 3 (1997): 574–602, 589; for Northern distinctiveness, Frederick Law Olmsted, *A Journey in the Seaboard Slave States, with Remarks on their Economy* (New York: Dix and Edwards, 1856), 79.

79. Sally McMurry, *Transforming Rural Life*; Joan M. Jensen, *Loosening the Bonds: Mid-Atlantic Farm Women, 1750–1850* (New Haven, CT: Yale University Press, 1986), 79–144; For the continuance of these patterns, see Nancy Grey Osterud, *Bonds of Community: The Lives of Farm Women in Nineteenth-Century New York* (Ithaca, NY: Cornell University Press, 1991).

80. A Mother, "What Can Mothers and Daughters Do to Make Farm Life Attractive to Their Sons and Brothers and Prevent Them from Leaving the Farm to Engage in Mercantile and Professional Pursuits," *GFGJ* 18, no. 3 (1857): 89.

81. For similar urban critiques, see John F. Kasson, *Rudeness and Civility: Manners in Nineteenth-Century Urban America* (New York: Hill and Wang, 1990), 176.

82. "Book Farmers," *GFGJ* 2, no. 32 (1832): 297.

83. "On the Proper Education of the Sons and Daughters of Farmers," *New York Farmer* 4, no. 10 (1831): 277.

84. T. C. Miner, untitled, *Northern Farmer* 1, no. 1 (1852): 1.

85. For evocations of new ideas of labor in this period beyond Anti-Rent, see, e.g., Sean Wilentz, *Chants Democratic: New York City and the Rise of the American Working Class, 1788–1850* (Oxford: Oxford University Press, [1984] 2004); Bronstein, *Land Reform and Working-Class Experience*.

86. Conger, *Memorial of Herman Ten Eyck Foster*, 11.

87. John A. Dix in "Address of Silas Wright," *TNYSAS for 1847* 7 (1848), 14.

88. Herman Coons, Diary of Herman Coons, 1850–54, NM-19.53, July 15, 1854, NYSHA.

89. J. R. Speed, *J. R. Speed to Henry S. Randall, August 16, 1846*, Letter, Folder, 1842–46, NIC 851 CCHS, Henry S. Randall Correspondence, 1828–1965.

90. J. S. Copeland, "Details of Operations in Farming," *Cultivator* 5, no. 5 (1848): 142.

91. "To the Farmers," *American Agriculturist* 2, no. 4 (1843): 104.

92. See "Analogy between Medical and Agricultural Education," *Cultivator* 5, no. 2 (1838): 32. For physicians' struggle for authority, see Starr, *The Social Transformation of American Medicine*; Albert Hazen Wright, *Cornell's Three Precursors: II New York State Agricultural College*, Studies in History (Ithaca: New York State College of Agriculture, 1958).

93. "List of Premiums Awarded," *TNYSAS 1844* 4 (1845), 28; US Census of 1850, Poughkeepsie, Dutchess County, New York, NARA Microfilm Publication, Roll M432_497, p. 116B, Image: 182, Household of Moses Humphrey, Ancestry.com.

94. Edythe Ann Quinn, "'The Hills' in the Mid-Nineteenth Century: The History of a Rural Afro-American Community in Westchester County New York," *Afro-Americans in New York Life and History* 14, no. 2 (1990): 35.

95. Bruegel, *Farm, Shop, Landing*, 111, 112; Clark, *Roots of Rural Capitalism*, 262.

96. Hodges, *Root and Branch*, 221, 236, 176.

97. In the late 1850s, the Northern states had 535,000 farm laborers but 1,523,000 farmers. By contrast, in England and Wales, 2.2 million agricultural laborers worked seasonally or permanently for 307,000 farmers. Huston, *Southern Planters*, 78; Edward Higgs, "Occupational Censuses and the Agricultural Workforce in Victorian England and Wales," *EHR* 48, no. 4 (1994), 700–716.

98. For strategies in the use of family labor, particularly local migration to allocate labor across multiple family farms, see Parkerson, *Agricultural Transition in New York State*; Della Roberts, "Woman Farming," *New England Farmer* 15, no. 7 (1863): 229. For women's field work, see Osterud, *Bonds of Community*, 139, 159; Bruegel, *Farm, Shop, Landing*, 116–22; Jensen, *Loosening the Bonds: Mid-Atlantic Farm Women*, 36–56.

99. Myra B. Young Armstead, *Freedom's Gardener: James F. Brown, Horticulture, and the Hudson Valley in Antebellum America* (New York: New York University Press, 2012); Peter Mickulas, "Cultivating the Big Apple: The New-York Horticultural Society, Nineteenth-Century New York Botany, and the New York Botanical Garden," *NYH* 83, no. 1 (2002): 34–54.

100. Autumn Stanley, *Mothers and Daughters of Invention: Notes for a Revised History of Technology* (New Brunswick, NJ: Rutgers University Press, 1993): 78–80.

101. James McLallen, Diary/Daybook and House Ledger, V 1-8-15, p. 154, HCTC.

102. Bancroft, "Address," 14.

103. Bancroft, 17.

104. "List of Premiums Awarded," *TNYSAS for 1844* 4 (1845): 288.

105. "List of Premiums Awarded," 283; Kelly, "The Consummation of Rural Prosperity and Happiness," 589.

106. "The State Fair at Poughkeepsie," *Cultivator* 1, no. 10 (1844): 312–16, 313; "Report on the Poughkeepsie Fair," *TNYSAS for 1844* 4 (1845): 29.

107. David Nelson, Diary, vol. 1, OHA 6730, September 29 and 30, 1859, OHA, David Nelson Diaries.

Chapter Three

1. Alson Ward, Alson Ward Diary, 1843–47, Diary Typed Transcript, DCHS, 23.
2. Ward, 68.
3. Ward, 47.
4. Ward, 19.
5. J.M., "Letter 1," *GFGJ* 3, no. 13 (1833): 99; Experimenter, "Communications for the Genesee Farmer: Experiments," *GFGJ* 1, no. 31 (1831): 244.
6. Peter Dear, "Narratives, Anecdotes, and Experiments: Turning Experience into Science in the Seventeenth Century," in *The Literary Structure of Scientific Argument: Historical Studies*, ed. Peter Dear (Philadelphia: University of Pennsylvania Press, 2015), 135–63; Steven Shapin, "Pump and Circumstance: Robert Boyle's Literary Technology," *Social Studies of Science* 14, no. 4 (1984): 481–520; Steven Shapin, "The Invisible Technician," *American Scientist* 77, no. 6 (November–December 1989): 554–63. In using Shapin to think about farm experiment, I am influenced by Christopher R. Henke, "Making a Place for Science: The Field Trial," *SSS* 30, no. 4 (2000): 483–511.
7. In thinking about textbooks and representations of science, I am influenced by Sharon Traweek, *Beamtimes and Lifetimes* (Cambridge, MA: Harvard University Press, 2009), 74–105.
8. Linda Nash addresses these differences in Linda Nash, *Inescapable Ecologies: A History of Environment, Disease, and Knowledge* (Berkeley: University of California Press, 2006), 142–43. For field-lab distinctions that develop later, see Robert E. Kohler, *Landscapes and Labscapes: Exploring the Lab-Field Border in Biology* (Chicago: University of Chicago Press, 2002); Jeremy Vetter, *Field Life: Science in the American West during the Railroad Era* (Pittsburgh: University of Pittsburgh Press, 2016).
9. As a result, Fussell concludes that they are not experiments. G. E. Fussell, "Agricultural Science and Experiment in the Eighteenth Century: An Attempt at a Definition," *AHR* 24, no. 1 (1976): 44–47.
10. Soraya De Chadarevian, "Laboratory Science versus Country-House Experiments: The Controversy between Julius Sachs and Charles Darwin," *BJHS* 29, no. 1 (1996): 17–41; Iwan Rhys Morus, "The Two Cultures of Electricity: Between Entertainment and Edification in Victorian Science," *Science & Education* 16, no. 6 (2007): 593–602.
11. Sydney Ross, "Scientist: The Story of a Word," *Annals of Science* 18, no. 2 (1962): 65–85, 71–73; Paul Lucier, "The Professional and the Scientist in Nineteenth-Century America," *Isis* 100, no. 4 (2009): 699–732.
12. Barbara J. Shapiro, *A Culture of Fact: England, 1550–1720* (Ithaca, NY: Cornell University Press, 2003).
13. Issac Carr to Caleb Carr, November 1, 1857, Carr Family Letters, Box 1, Folder 1, CAK.
14. See, e.g., Patricia Cline Cohen, *The Murder of Helen Jewett* (New York: Vintage, 2010).
15. Levi Weeks, Diary, February 27–28, 1856, Weeks Family Papers, NYSHA.
16. Nathaniel Tallmadge, *Remarks of Mr. Tallmadge in Defence of the People of New-York against the Charge of Bank Influence in the Result of Their Election* (Washington, DC: Madisonian Office, 1838), 6.
17. *First Annual Report of the Oneida Association* (Oneida Reserve: Leonard and Co., Print-

ers, 1849), 20; Charles Petit McIlvaine, *The Evidences of Christianity, in Their External Division* (London: H. Fisher, R. Fisher, and P. Jackson, 1833), 325.

18. Society of Friends, *The Case of the Seneca Indians in the State of New York* (Philadelphia: Merrihew and Thompson, 1840), 38; Robert Wiltse, Document No. 92, March 14, 1834, "Report, &c. A Detailed Statement of the Government, Discipline, &c. of the New-York State Prison at Mount-Pleasant," in *DSSNY, 57th Session 1834, Vol. 2 from No. 51-126 Inclusive* (Albany, NY: E. Croswell, 1834), 1–45, 6.

19. "Report of the Select Committee on the petition of Jasper P. Allaire and Others," in *DSSNY, Fifty-Seventh Session, 1834* 2, no. 96 (March 13, 1834): 1–2.

20. Lepler, *Many Panics*, 76–83; Hannah Farnham Sawyer Lee, *Three Experiments of Living: Living within the Means, Living Up to the Means, Living beyond the Means*, 21st ed. (Philadelphia: George S. Appleton, [1836] 1846).

21. Hannah Farnham Sawyer Lee, *Living on Other People's Means, or, The History of Simon Silver* (Boston: Weeks, Jordan and Co., 1837).

22. Mihm, *A Nation of Counterfeiters*, 192–93.

23. Edward Johnson, Diary, Vol 2., 1851–56, June 18, 1852, June 3, 1852, NYSL.

24. Jarvis W. Brewster, "Corn—Potatoes—Rutabaga," *Cultivator* 2 (December 2, 1835): 149–50.

25. Henry David Thoreau, *Walden, or, Life in the Woods* (Boston: Houghton Mifflin Co., [1854] 1893), 87.

26. Thoreau, 87–90.

27. Jesse Buel, "Art. 59: Experiments on the Potato, by Arthur Young, Communicated for the American Farmer by J. Buel, Esq," *New York Farmer* 1, no. 3 (March 1828): 71; Arthur Young, *A Course of Experimental Agriculture*, 4 vols. (Dublin: J. Exshaw, [1770] 1771).

28. Young, *Course of Experimental Agriculture*, vol. 1, vii.

29. E.g., S. W. Johnson, "Foreign Correspondence," *Cultivator* (Third Series) 2, no. 6 (June 1854): 169; E. John Russel, "Rothamsted and Its Experiment Station," *AH* 16, no. 4 (1942): 161–83; David F. Lindenfeld, *The Practical Imagination: The German Sciences of State in the Nineteenth Century* (Chicago: University of Chicago Press, 2008), 77; Mark R. Finlay, "The German Agricultural Experiment Stations and the Beginnings of American Agricultural Research," *AH* 62, no. 2 (1988): 41–50. For improving efforts to create new state functions, see Ron, "Summoning the State"; Timothy K. Minella, "A Pattern for Improvement: Pattern Farms and Scientific Authority in Early Nineteenth-Century America," *AH* 90, no. 4 (2016): 434–58.

30. Shapin, "Pump and Circumstance," 491.

31. Sophia Rosenfeld, *Common Sense: A Political History* (Cambridge, MA: Harvard University Press, 2011), 9; George H. Daniels, *American Science in the Age of Jackson* (New York: Columbia University Press, 1968), 65; Lewis, *Democracy of Facts*, 34. Experiential knowledge is central to Cohen's "Georgic Science"; see *Notes from the Ground*, 41–51.

32. Shapin, "Pump and Circumstance," 497.

33. J.L., "With Effect of Stirring the Surface of the . . . Relief against Drouth," *GFGJ* 2, no. 36 (September 8, 1832): 283.

34. C., "Guano vs. Worn-Out Land," *American Agriculturist* 9, no. 4 (Apr 1850): 117.

35. Young, *Course of Experiment*, vol. 1, vi.

36. "A Decided Specimen of Book Farming," *Cultivator* (Third Series) 1, no. 1 (1853): 18.

37. H. M. Collins, *Changing Order: Replication and Induction in Scientific Practice* (Chicago: University of Chicago Press, 1985); Steven Shapin and Simon Schaffer, *Leviathan and the Air-Pump: Hobbes, Boyle, and the Experimental Life* (Princeton, NJ: Princeton University Press, 1985), 225–83.

38. Christopher Henke addresses this lab-field distinction (and makes the connection to Shapin) in terms of modern field trials in *Cultivating Science, Harvesting Power: Science and Industrial Agriculture in California* (Cambridge, MA: MIT Press, 2008), 122.

39. Halsey, "Report of the Committee on Agriculture," 10.

40. "Experiments Proposed," *New York Farmer* 7, no. 10 (1834): 318.

41. Valencius, *The Lost History*, 16, 178, 197; Valencius et al., "Science in Early America," 108.

42. Shapiro, *Culture of Fact*, 63–85; Vladimir Janković, *Reading the Skies: A Cultural History of English Weather, 1650–1820* (Manchester: Manchester University Press, 2000), 78–81.

43. Thoreau, *Walden*, 87.

44. Brewster, "Corn—Potatoes—Rutabaga," 149–50.

45. Lorraine Daston, "Attention and the Values of Nature in the Enlightenment," in *The Moral Authority of Nature*, ed. Lorraine Daston (Chicago: University of Chicago Press, 2004), 100–26; Bruce Hevly, "The Heroic Science of Glacier Motion," *Osiris* 11 (1996): 66–86.

46. US Census of 1850, Trenton, Oneida County, New York, Roll 314, p. 306, Household of J. W. Brewster NARA Microfilm Publication, Ancestry.com.

47. "Experiments," *Cultivator* 2, no. 12 (1835): 132.

48. Steven Shapin, "The Invisible Technician," *American Scientist* 77 (1989); Henke, *Cultivating Power*, 127.

49. D.T. [possibly David Thomas], "Experiments: On the Culture of Squashes and Melons," *GFGJ* 3, no. 14 (1833): 106. Similar erasures and appropriations have been examined in colonial natural history by Susan Scott Parrish, *American Curiosity: Cultures of Natural History in the Colonial British Atlantic World* (Chapel Hill: University of North Carolina Press Books, 2012).

50. John Lorain, *Nature and Reason Harmonized in the Practice of Husbandry* (Philadelphia: H. C. Carey and Lea, 1825), 439; Jane Mt. Pleasant, "Food Yields and Nutrient Analyses of the Three Sisters: A Haudenosaunee Cropping System," *Ethnobiology Letters* 7, no. 1 (2016): 87–98.

51. C.B., "Pumpkins" *GFGJ* 4, no. 17 (1834): 130.

52. Paul Lucier, "The Origins of Pure and Applied Science in Gilded Age America," *Isis* 103 (2012): 527–36; Daniel S. Greenberg, *The Politics of Pure Science* (Chicago: University of Chicago Press, 1999 [1967]); Audra J. Wolfe, *Freedom's Laboratory: The Cold War Struggle for the Soul of Science* (Baltimore: Johns Hopkins University Press, 2018).

53. Young, *Course of Experimental Agriculture*, vol. 1, vii.

54. Young, xiii.

55. A Friend and Well Wisher, "For the Farmer's Monthly Visitor," *Farmer's Monthly Visitor* 3, no. 2 (1841): 23.

56. "Farm Accounts," *Cultivator* 3 (1837): 165. This article also appeared in the *Maine Farmer*, the *Boston Cultivator*, the *Southern Agriculturist*, and the *Register of Rural Affairs*.

57. "Inaccuracy in Farming," *Cultivator* 9, no. 4 (1852): 129.

58. See, e.g., Niagara, "Last Leaf of a Farmer's Leger [sic]," *GFGJ* 6, no. 7 (1836); S. Porter Rhoades, "Farm Account—Balance Sheet," *Cultivator* 7, no. 2 (1840), 33; Anonymous, "A Profitable Cow," *GFGJ* 7, no. 7 (1837): 56.

59. Brian P. Luskey, *On the Make: Clerks and the Quest for Capital in Nineteenth-Century America* (New York: New York University Press, 2010), 60; Lepler, *Many Panics*, 74.

60. "Farm Accounts," 165.

61. For the promotion of self-control through accounting, see Lepler, *Many Panics*, 70, and Sandage, *Born Losers*, 45–47.

62. In rural New England and the Mid-Atlantic states, competence at reading, writing, and basic arithmetic are estimated to have been greater than 85 percent between 1812 and 1830.

William J. Gilmore, *Reading Becomes a Necessity of Life: Material and Cultural Life in Rural New England, 1780–1835* (Knoxville: University of Tennessee Press, 1992), 120–23.

63. For slaveholder accounting, see Caitlin Rosenthal, *Accounting for Slavery*. For reading farm accounts, see Rothenberg, *From Market-Places to a Market Economy*, 56–79. For a more recent treatment, see chapter 1, Richard Bushman, *The American Farmer in the Eighteenth Century: A Social and Cultural History* (New Haven, CT: Yale University Press, 2018).

64. J. H. Hanson (1844–58) Account Book, Manuscript Collections, HML.

65. Salmon Bostwick, Diary of Salmon E. Bostwick, 1826–31, Diary Typescript, Diaries B657, NYSHA.

66. For use of accounts in litigation, see Daniel Adams, *Adams's Book-Keeping* (Keene, NH: J. H. Spalter and Co., 1849), 9–10.

67. Harvey P. Badger, Badger Family Record Book, 1843–83, Account Book, Archives 2025, CAK; Levi Weeks, Diary, vol. 1, Coll. No. 308, NYSHA, John B. and Levi Weeks Diaries, 1850–90.

68. Levi S. Fulton and George Washington Eastman, *A Practical System of Book-Keeping by Single Entry; Containing Three Different Forms of Books; Designed Respectively for the Farmer, Mechanic, and Merchant*, 2nd ed. (New York: A. S. Barnes and Co., 1848), 16.

69. George T. Sprague, George T. Sprague Diary, 1854–55, April 9, 1854, Diary, Diary Sp72, NYSHA.

70. McLallen, Diary/Daybook and House Ledger, HCTC.

71. "Farm Accounts," *Cultivator* 4, no. 2 (1837): 31.

72. Young, *Course of Experimental Agriculture*, xxi.

73. J.B. [Jesse Buel], "Letter 2," *Cultivator* 2 (1834): 65.

74. "The Science of Agriculture and Book Farming," *GFGJ* 4 (1834): 345.

75. Brewster, "Corn—Potatoes—Rutabaga," 149–50.

76. Stoll, *Larding the Lean Earth*, 146–48.

77. Peter Tone, *Peter Tone to William Cushing, Oct 30th 1848*, Letter, Box 5 Folder 9, ML, Wadsworth Family Papers.

78. "Report of a Majority of Committee on Memorials for a Limitation of the Quantity of Land That Any Individual May Hereafter Acquire in This State; Also for the Occupation of the Homestead of Each Family from Alienation for Any Future Debt or Liability," *Albany Freeholder*, October 20, 1847, AAS.

79. Bronstein, *Land Reform and Working-Class Experience*, 62; Feargus O'Connor, *A Practical Work on the Management of Small Farms*, 5th ed. (Manchester: Abel Heywood, 1847), 120, 151, 156–58, 164; Ray Boston, *British Chartists in America* (Manchester: Manchester University Press, 1971), 54; Thomas Ainge Devyr, *The Odd Book of the Nineteenth Century, or "Chivalry" in Modern Days, A Personal Record of Reform—Chiefly* LAND REFORM *for the Last Fifty Years* (Greenpoint, NY: Published by the Author, 1882), 75.

80. Henke, "Making a Place for Science," 487, 490.

81. Rosenberg, *No Other Gods*, 147, 161.

Chapter Four

1. J. S. Wright, *Atkins' Automaton, or, Self-Raking Reaper and Mower, Invented by Mr. Jearum Atkins' of Chicago Ill.* (London: Warrington and Sons, 1853), 12.

2. Wright.

3. On global networks, see Gordon M. Winder, *The American Reaper: Harvesting Networks*

and Technology, 1830–1910 (Farnham, UK: Ashgate Publishing, Ltd., 2013); for reversal, see C. Andrew Jewell, "The Impact of America on English Agriculture," *AH* 50, no. 1 (1976): 125–36.

4. Jonathan Harwood suggests that "product testing" also became a common form of agricultural experiment in Europe in this period. Jonathan Harwood, "Comments on Experimentation in Twentieth-Century Agricultural Science," *History and Philosophy of the Life Sciences* 37, no. 3 (2015): 326–30, 326.

5. Benjamin P. Johnson, *Report of Benj. P. Johnson: Agent of the State of New York, Appointed to Attend the Exhibition of the Industry of All Nations, Held in London, 1851* (Albany, NY: C. Van Benthuysen, 1852), 77–87.

6. The thing on the Soviet flag is a sickle, used for reaping; the thing the Grim Reaper carries is a scythe, making him likely to be a Grim Mower.

7. W. A. Armstrong, "Labor I: Rural Population Growth, Systems of Employment, and Incomes," in *The Agrarian History of England and Wales*, vol. 6, *1750–1850, Part II*, ed. G. E. Mingay (Cambridge: Cambridge University Press, 1989), 641–727, 682.

8. Jonathan Brown and H. A. Beecham, "Arable Farming Practices," in *The Agrarian History of England and Wales*, vol. 6, *1750–1850, Part II*, 467–68.

9. Many farmers had attempted to manage this pressure by shifting from the sickle to the broader strokes of the scythe. B. A. Holdernesse, "Prices, Productivity, and Output," in *The Agrarian History of England and Wales*, vol. 6, *1750–1850, Part I*, 84–189, 102; Edward J. T. Collins, "Harvest Technology and Labour Supply in Britain, 1790–1870," *EHR* 22, no. 3 (1969): 453–73.

10. G. E. Fussell, *The Farmer's Tools: The History of British Farm Implements, Tools, and Machinery before the Tractor Came* (London: Andrew Melrose, 1952), 115–27.

11. Johnson, *Report of Benj. P. Johnson*, 77–87.

12. *The New York State Agricultural Society's Trial of Implements at Geneva, July 1852* (Albany, NY: Charles Van Benthuysen, 1852), 4–5.

13. NYSAS, *Second National Trial of Mowers, Reapers, Horse Powers, Etc.* (Albany, NY: Charles Van Benthuysen and Sons, 1866).

14. NYSAS, *Trial of Implements at Geneva*, 16.

15. Wright, *Atkins' Automaton*, 12–16.

16. Such claims perhaps appeared as a way of justifying high prices. Manny and Co., "Great American Triumph at the Paris World's Fair: Manny's Celebrated Reaper and Mower Victorious!!" (Springfield, MA: Samuel Bowles and Company Printers, 1855), 36.

17. Thomas Morton, *Speed the Plough: A Comedy in Five Acts* (London: Longman and O. Rees, [1798] 1800), 12, 40. For the other British trials, see Bennet Woodcroft, *Appendix to the Specifications of English Patents for Reaping Machines* (London: Eyre and Spottiswoode, 1853), 6–50. Ploughing competitions, part race, part test of skill, part education, had a longer history in both Britain and North America. Jones, *Agricultural Enlightenment*, 87; Catharine Anne Wilson, "A Manly Art: Plowing, Plowing Matches, and Rural Masculinity in Ontario, 1800–1930," *Canadian Historical Review* 95, no. 2 (2014), 157–86.

18. Woodcroft, *Appendix to the Specifications*, 17.

19. Daniel Ott, "Producing a Past: McCormick Harvester and Producer Populists in the 1890s," *AH* 88, no. 1 (2014): 87–119.

20. Leo Rogin, *The Introduction of Farm Machinery in Its Relation to the Productivity of Labor in the Agriculture of the United States during the Nineteenth Century* (Berkeley: University of California Press, 1931), 30–31. For American farm machinery and implements until 1830, see Peter D. McClelland, *Sowing Modernity: America's First Agricultural Revolution* (Ithaca, NY: Cornell University Press, 1997).

21. Alan Olmstead, "The Mechanization of Reaping and Mowing in American Agriculture, 1833–1870," *JEH* 35, no. 2 (1975): 327–52, 334–40; Sprague, Diary, 1854–55, July 30 and September 11, 1856.

22. H. B. Inches, *H. B. Inches, to W. W. Wadsworth, September 13th 1831*, Folder 2, letter, ML, Wadsworth Family Papers.

23. Hough, *Census of the State of New-York, for 1855*, 181–95.

24. Levi Weeks, Diary, vol 1., September 26, 1851, Coll. No. 308., NYSHA, John B. and Levi Weeks Diaries, 1850–90.

25. Rogin, *Introduction of Farm Machinery*, 30–31.

26. Rogin, 76–82.

27. B. Zorina Khan, "Property Rights and Patent Litigation in Early Nineteenth-Century America," *JEH* 55, no. 1 (1995): 58–97, table 1, 63.

28. X, "Ploughing Exhibition: Steel 'Himself Again!,'" *Spirit of the Times* 9, no. 11 (1839): 294.

29. Here I use the standard definition of robot as a machine that mimics human labor. R. Harmon and E. Comstock, "Report of the Committee on the Trial of Plows at the State Fair at Poughkeepsie," *TNYSAS* 4 (1844) 147–49; for mechanical objectivity, see Lorraine Daston and Peter Galison, "The Image of Objectivity," *Representations* 40, Special Issue (1992): 82–128.

30. X, "Ploughing Exhibition," 294.

31. Manny and Co., "Great American Triumph," 1–15, 37.

32. This was also true of other major journals: Yankee Farmer, "Agricultural Papers and Warehouses," *New Genesee Farmer and Gardeners' Journal* 1, no. 7 (1840): 99.

33. Luther Tucker, *Annual Descriptive Catalogue of the Agricultural Implements, Horticultural Tools, and Field Grain and Garden Seeds, for Sale at the Albany Agricultural Warehouse and Seed-Store* (Albany, NY: C. Van Benthuysen, 1847–48). The *Cultivator* stayed in place even after Tucker sold the warehouse to the manufacturer Horace Emery: Horace L. Emery, advertisement, "Albany Agricultural Warehouse," *American Journal of Agriculture and Science* 7, no. 5 (1848), back matter.

34. Yankee Farmer, "Agricultural Papers and Warehouses," 99.

35. Yankee Farmer, 99.

36. Luther Tucker, *Annual Descriptive Catalogue 1847–48*, front matter.

37. Horace Emery, advertisement, "Hay and Harvest Tools and Machinery," *Cultivator* 7 (1848): 232.

38. "State Agricultural Society: Executive Committee," *Central New York Farmer* 3, no. 5 (1844): 147–48; "Report of the Executive Committee for 1858," *TNYSAS for 1858* 18 (1859): 3–31, 17; "Geological Rooms," in *The New-York-State Register for 1845*, ed. O. L. Holley (New York: J. Disturnell, 1845), 333–35.

39. Lyman H. Low, "Hard Times Tokens," *American Journal of Numismatics* 33, no. 4 (1899): 118–22, 121.

40. Richard Peters, "A Plan for Establishing a Manufactory of Agricultural Instruments," *American Register* (January 1, 1810), 6.

41. "Proposed Agricultural Museum, at Rochester," *New Genesee Farmer and Gardeners' Journal* 1, no. 1 (1840): 2; "Agricultural Museums," *Cultivator* 3 (1836): 89; Peter Lawson and Son, *The Agriculturists' Manual* (Edinburgh: William Blackwell and Sons, 1836).

42. Willis Gaylord, "The Agriculture of Onondaga County," *TNYSAS for 1842* 2 (1843): 174–86, 185.

43. Yankee Farmer, "Agricultural Papers and Warehouses," 99.

44. "Annual Report—for the Year 1844," *TNYSAS for 1844* 4 (1845): 4.

45. J. L. Tappan, "The New York Agricultural Museum," *Michigan Farmer* 1, no. 45 (1859): 370.

46. Ariel Ron, "King Hay: Nationalism, Energy, and Economic Development in the Nineteenth-Century Northern United States" (paper delivered at the History of Economics Society Meeting, Toronto, 2017); Greene, *Horses at Work*, 196.

47. For timing and threshers, see Hobsbawm and Rudé, *Captain Swing*, appendix IV.

48. NYSAS, *Trial of Implements at Geneva*, 12.

49. NYSAS, 36.

50. Manny and Co., "Great American Triumph," 14–16.

51. Manny and Co., 14–16.

52. NYSAS, *Trial of Implements at Geneva*, 12.

53. NYSAS, 28.

54. NYSAS, 18.

55. Manny and Co., "Great American Triumph," 8–9.

56. Winder, *The American Reaper*, 48.

57. Winder, 38.

58. William H. Seward, *The Reaper: Argument of William H. Seward in the Circuit Court of the United States Oct. 24, 1854* (Auburn, NY: William H. Finn, 1854), 11.

59. Mayher and Co., "Catalogue with Full Description of Agricultural and Horticultural Implements at Mayher and Co.'s United States Agricultural Warehouse and Seed Store, Eleventh Edition" (New York: Baker, Godwin, and Co., 1854), 78.

60. Winder, *The American Reaper*, 17.

61. Manny and Co., "Great American Triumph," 5–6.

62. Christopher Beauchamp, "The First Patent Litigation Explosion," *Yale Law Journal* 125, no. 4 (2016): 796–1149.

63. Khan, "Property Rights," 58–97, 65.

64. Khan, 68.

65. Chaim M. Rosenberg, *The International Harvester Company: A History of the Founding Families and Their Machines* (Jefferson, NC: McFarland and Co., 2019), 17; Seward, *The Reaper*.

66. Seward, 8.

67. Seward, 7.

68. Seward, 11.

69. For this meaning of automata, see John Tresch, "The Machine Awakens: The Science and Politics of the Fantastic Automaton," *French Historical Studies* 34, no. 1 (2011): 87–123.

70. Benjamin C. Howard, "McCormick v. Talcott et al.," *Reports of Cases Argued and Adjudged in the Supreme Court of the United States, December Term, 1857*, vol. 20 (Washington, DC: W. H. and O. H. Morrisson, 1858), 408–12.

71. Levi Weeks, Diary, vol. 1, Weeks Family Papers, July 17, 1855, Coll. No. 308, NYSHA, John B. and Levi Weeks Diaries, 1850–90.

72. Weeks, 57–58.

73. Stanley, *Mothers and Daughters of Invention*, 79.

74. Levi Weeks, Diary, vol 2., August 28, 1855, Coll. No. 308, NYSHA, John B. and Levi Weeks Diaries, 1850–90.

Chapter Five

1. "Nicholas Biddle, Esq.," *New-Yorker* 9, no. 20 (1840): 313.

2. Untitled, *Silk Grower and Farmer's Manual* 1, no. 6 (1838): 132; John Quincy Adams,

Memoirs of John Quincy Adams, Comprising Portions of His Diary from 1795 to 1848, vol. 10, ed. Charles Francis Adams (Philadelphia: J. B. Lippincott and Co., 1876).

3. Numbers of trees were roughly estimated by tabulating advertisements in the *Journal of Commerce* (henceforth *JOC*) 1838–39. Occupations of speculators were found by cross-referencing advertisements with *Longworth's American Almanac, New York Register and City Directory* (New York: Thomas Longworth, 1838). For the tragedian, see "Good Fashions," *JOC*, August 23, 1839, AAS.

4. Scott Sandage, *Born Losers*, 24–27; Charles P. Kindleberger and Robert Z. Aliber, *Manias, Panics and Crashes: A History of Financial Crises* (New York: Palgrave Macmillan, 2011); Edward J. Balleisen, *Navigating Failure: Bankruptcy and Commercial Society in Antebellum America* (Chapel Hill: University of North Carolina Press, 2001); Lepler, *Many Panics*; Mihm, *A Nation of Counterfeiters*. Though I would argue that believers in silk had reasons for their belief, the timing of the bubble supports William Deringer's argument that bubbles emerge particularly at "*epistemic crises*, moments in which the complexity of financial invention outstrips the capacity of the financial community to understand what is going on." William Deringer, "For What It's Worth: Historical Financial Bubbles and the Boundaries of Economic Rationality," *Isis* 106, no. 3 (2015): 646–56, 655.

5. One of the Humbugged, "Hobbies and Humbugs," *GFGJ* 9, no. 87 (1839): 293.

6. "The Silkworm and Its Productions," *Family Magazine* 3 (1835–36): 460–62. For early European efforts at silk culture, see chapters 2 and 4 of John Clarke, *A Treatise on the Mulberry Tree and Silkworm and on the Production and Manufacture of Silk, Embellished with Appropriate Engravings* (Philadelphia: Thomas, Cowperthwait and Co., 1839); many of these drew on Dionysius Lardner, *A Treatise on the Origin, Progressive Improvement and Present State of the Silk Manufacture* (London: Longman, Rees, Orme, Brown, and Green, 1831).

7. For the distribution of silk notions in backcountry stores in the 1770s and early 1800s, see Ann Smart Martin, *Buying into the World of Goods: Early Consumers in Backcountry Virginia* (Baltimore: Johns Hopkins University Press, 2008), 150–66.

8. For eighteenth-century Russian silk culture, see William Tooke, *View of the Russian Empire during the Reign of Catharine the Second and to the Close of the Eighteenth Century*, vol. 3 (Dublin: P. Wogan, 1801), 210–17; for Britain, see Linda Levy Peck, *Consuming Splendor: Society and Culture in Seventeenth-Century England* (Cambridge: Cambridge University Press, 2005), 91–99. See also, Shichiro Matsui, *History of the Silk Industry in the United States* (New York: Howes Publishing Co., 1930), 5.

9. Lisbet Koerner, *Linnaeus: Nature and Nation* (Cambridge, MA: Harvard University Press, 1999), 3, 136. For German cameralism, see Henry E. Lowood, "The Calculating Forester: Quantification, Cameral Science, and the Emergence of Scientific Forestry Management," in *The Quantifying Spirit in the 18th Century*, ed. Tore Frangsmyr, J. L. Heibron, and Robin E. Rider (Berkeley: University of California Press, 1991), 315–42; Marcus Popplow and Torsten Meyer, "'To Employ Each of Nature's Products in the Most Favorable Way Possible'—Nature as a Commodity in Eighteenth-Century German Economic Discourse," *Historical Social Research* 29, no. 4 (2004): 4–40. For cameralism's predecessor, localism, see Alix Cooper, "'The Possibilities of the Land': The Inventory Of 'Natural Riches' in the Early Modern German Territories," *History of Political Economy* 35, no. 5 (2003): 129–53. For more on economic and botanical networks, see Drayton, *Nature's Government*; Emma Spary, *Utopia's Garden: French Natural History from Old Regime to Revolution* (Chicago: University of Chicago Press, 2000).

10. Koerner, *Linnaeus*, 112–39, 117, 133–34.

11. Drayton, *Nature's Government*, 85–124; Spary, *Utopia's Garden*, 49–98; Jim Endersby, *Imperial Nature: Joseph Hooker and the Practices of Victorian Science* (Chicago: University of Chicago Press, 2008).

12. Stuart McCook, *States of Nature: Science, Agriculture, and Environment in the Spanish Caribbean, 1760–1940* (Austin: University of Texas Press, 2002), 77–85.

13. John Smith, quoted in Charles E. Hatch Jr., "Mulberry Trees and Silkworms: Sericulture in Early Virginia," *Virginia Magazine of History and Biography* 65, no. 1 (1957): 3–61, 4; cocoon reference in Thomas Hariot, *A Briefe and True Report of the New Found Land of Virginia* (London: R. Robinson, 1588); Joseph Ewan, "Silk Culture in the Colonies, with Particular Reference to the Ebenezer Colony and the First Local Flora of Georgia," *AH* 43, no. 1 (1969): 129–42.

14. Hatch, "Mulberry Trees and Silkworms," 10.

15. Jacqueline Field, Marjorie Senechal, and Madelyn Shaw, *American Silk: Entrepreneurs and Artifacts, 1830–1930* (Lubbock: Texas Tech University Press, 2007), 9–10; Hatch, "Mulberry Trees and Silkworms," 12; David John Rossell, "The Culture of Silk: Markets, Households, and the Meaning of an Antebellum Agricultural Movement," PhD diss. (SUNY Buffalo, 2001); Samuel Hartlib, *The Reformed Virginian Silk-Worm* (London: Giles Calvert, 1655).

16. Hatch, "Mulberry Trees,"18.

17. Hatch, 43; Rossell, "Culture of Silk," 21–22.

18. Hatch, "Mulberry Trees," 6; Peck, *Consuming Splendor*, 91–109.

19. Chaplin, *An Anxious Pursuit*, 159–61.

20. Megan Kate Nelson, *Trembling Earth: A Cultural History of the Okefenokee Swamp* (Athens: University of Georgia Press, 2005), 12–13; Mart A. Stewart, *"What Nature Suffers to Groe": Life, Labor, and Landscape on the Georgia Coast, 1680–1920* (Athens: University of Georgia Press, 1996), 60–69. For transatlantic silk culture, see Zara Anishanslin, *Portrait of a Woman in Silk: Hidden Histories of the British Atlantic World* (New Haven, CT: Yale University Press, 2016).

21. Nelson, *Trembling Earth*, 17; Rossell, "Culture of Silk," 23; Field, *American Silk*, 13; Nelson Klose, "Sericulture in the United States," *AH* 37, no. 4 (1963): 225–34, 25–26.

22. Field, *American Silk*, 13; Klose, "Sericulture," 26.

23. Felix Pascalis, *Practical Instructions and Directions for Silkworm Nurseries, and for the Culture of the Mulberry Tree, Dedicated to the American Institute of New York*, vol. 1 (New York: J. Seymour, Printer, 1829), 93–95.

24. Anishanslin, *Portrait of a Woman in Silk*, 70.

25. Klose, "Sericulture," 24.

26. John D'Homergue, *The Silk Culturist's Manual* (Philadelphia: Hogan and Thompson, 1839), xiv; Field, *American Silk*, 14–15.

27. A good history of the early republic silk craze can be found in Ben Marsh, "The Republic's New Clothes: Making Silk in the Antebellum United States," *AH* 86, no. 4 (2012): 206–34. For manuals, see, e.g., J. H. Cobb, *A Manual Containing Information Respecting the Growth of the Mulberry Tree, with Suitable Directions for the Cultivation of Silk in Three Parts* (Boston: Carter, Hendee and Babcock, 1831); Samuel Whitmarsh, *Eight Years' Experience and Observations in the Culture of the Mulberry Tree and in the Case of the Silk Worm* (Northampton: J. Butler, for the author, 1839).

28. Lardner, *A Treatise on the Origin*, 41–43.

29. Stephen van Rensselaer III, "Report: Mulberry—Silk Worm," in *Letter from the Secretary of the Treasury Transmitting the Information Required by a Resolution of the House of Representa-*

tives, of May 11th 1826, in Relation to the Growth and Manufacture of Silk (Washington, DC: Duff Green, 1828), 6.

30. "Rural Economy," *American Farmer* 7, no. 29 (1825): 229.

31. "Announcement of Eighth Annual Fair of the American Institute," *JAI* 1, no. 1 (1836): 48.

32. "Rural Economy," 229.

33. David Danbom, *Born in the Country: A History of Rural America* (Baltimore: Johns Hopkins University Press, 1995), 74. See also Beckert, *Empire of Cotton*.

34. "Silk," *JAI* 1, no. 1 (1836): 15.

35. "Review," *American Monthly Review* 1, no. 3 (1832): 230.

36. Burke, *The Conundrum of Class*, 34–39; Peskin, *Manufacturing Revolution*, 119–20.

37. Livingston, *Essay on Sheep*, 6.

38. Laurel Thatcher Ulrich, *The Age of Homespun: Objects and Stories in the Creation of an American Myth* (New York: Alfred A. Knopf, Inc., 2001), 11–40.

39. Pascalis, *Practical Instructions*, 30.

40. "Seed of Morus Multicaulis," *JAI* 2, no. 4 (1837): 181; American Silk Grower, "The Morus Multicaulis," *GFGJ* 8, no. 26 (1838): 207.

41. Senate, New York (State) Legislature, "An Act in Relation to the State Prisons, Chapter 302, Laws of 1835," in *DSSNY*, 1835, 339.

42. Clarke, *Treatise on the Mulberry Tree*, 180.

43. Spellings of Perrottet's name in American print have variable numbers of *r*'s and *t*'s, and occasionally unexpected *i*'s. Searches for him should be conducted with this in mind.

44. Perrottet's report was translated and reprinted for American audiences in H. A. S. Dearborn, "Silk," *New England Farmer* 9, no. 4 (1830): 28.

45. Other attendees included Benjamin Silliman of Yale, editor of the *American Journal of Science*; David Hosack, physician, botanist and horticulturist; John Torrey, soon to be botanical surveyor of New York; and physician and naturalist Samuel Mitchill. *Celebration at Flushing of the Birthday of Linnaeus, by the New-York Branch of the Linnaean Society of Paris* (New York: Office of the Statesman, 1824), 3–5.

46. William Prince, *Catalogue of Fruit and Ornamental Trees and Plants, Bulbous Flower Roots, Green-House Plants, &C. &C.: Cultivated at the Linnaean Botanic Garden* (New York: T. and J. Swords, 1823), 6.

47. Jesse Buel, "On the Horticulture of the United States of America," *Gardener's Magazine and Register of Rural and Domestic Improvement* 5, no. 15 (1828): 193–97, 195; Drayton, *Nature's Government*, 50; Margaret M. Quinn, "Bayer, Adèle Parmentier," http://www.anb.org/articles/15/15-00876.html; *American National Biography Online*, February 2000 (accessed January 20, 2011).

48. Clarke, *Treatise on the Mulberry Tree*, 121.

49. Unsigned review, "Practical Instructions on the Culture of Silk, and of the Mulberry, Felix Pascalis," *American Journal of Science and Arts* 18, no. 2 (1830): 278.

50. Felix Pascalis and D. C. Wallace, "To the Editor of the New-York Farmer—Morus Multicaulis," *New York Farmer* 4, no. 7 (1831): 163; W. M. Ireland, "A List of Premiums Awarded by the N.Y. Horticultural Society, for 1831," *New York Farmer* 5, no. 5 (1832): 40.

51. Advertisement, "Morus Multicaulis," *New England Farmer* 10, no. 39 (1832): 311.

52. Letter reprinted in Clarke, *Treatise on the Mulberry Tree*, 129–32.

53. For the panic, see Temin, *The Jacksonian Economy*; Jessica Lepler, *Many Panics*.

54. F. K. Boughton, *F. K. Boughton to W. R. Prince, September 27th 1838*, Letter, NAL, Prince Family Collection.

55. This catalog was disguised as a newspaper, presumably to decrease postage charges. "Mulberries, &c.," *Flushing Farmer and Silk Culturist* 1, no. 1 (1839): 3.

56. Wm. Prince and Sons, "Messrs Princes' Letter on the Mulberry," *New England Farmer* 16, no. 38 (1838): 300.

57. C. S. Rafinesque, *American Manual of the Mulberry Trees: 25 Species and 30 Varieties, Their History, Cultivation, Properties, with Hints on Procuring Silk out of the Bark &c.* (Philadelphia: Published by the Author for the Eleutherium of Knowledge, 1839), 93; James Endersby, "'The Vagaries of a Rafinesque': Imagining and Classifying American Nature," *Studies in History and Philosophy of Science Part C: Studies in History and Philosophy of Biological and Biomedical Sciences* 40, no. 3 (2009): 168–78.

58. C. J. Wolbert, *Morus Multicaulis Trees, at Public Sale*, Broadside, BDSDS 1839 (Philadelphia?: s.n., 1839), AAS.

59. Untitled, *JOC*, October 5, 1839, AAS.

60. William Kenrick, *William Kenrick to L. Bradford Prince*, Box 5, Letter 307, William R. Prince Papers, NAL.

61. *Flushing Farmer and Silk Culturist* 1, no. 1 (1839): 2.

62. "New Chinese Mulberry.—Morus Multicaulis," *New England Farmer*, 4, no. 8 (1831): 217.

63. P., "Superiority of the Morus Multicaulis," *New England Farmer* 11, no. 11 (1832): 81.

64. Advertisement, "Silk," *Fessenden's Silk Manual and Practical Farmer* 1, no. 5 (1835): 79.

65. E.g. (from the *Florence Mirror*), "How the Mulberry Grows in Georgia," *JOC*, June 1, 1839, AAS.

66. E.g., "Specimen of a Leaf of the Morus Multicaulis Tree for the Silk Grower" (Philadelphia: Charles Alexander, n.d.), Pocumtuck Valley Memorial Association, Deerfield, MA.

67. E.g., Rensselaer, "Report: Mulberry—Silk Worm," 10; Cobb, *A Manual Containing Information Respecting the Growth of the Mulberry Tree*, 23–41; Whitmarsh, *Eight Years' Experience*, 99.

68. F. G. Comstock, *A Practical Treatise on the Culture of Silk, Adapted to the Soil and Climate of the United States* (Hartford, CT: P. B. Gleason and Co., 1839); "Silk Culture," *Southern Agriculturist* 12, no. 2 (1839): 71.

69. Cobb, *Manual*, 41; Clarke, *Treatise*, 123. Martin Giraudeau describes this form of fictional futuristic accounting as virtual witnessing. Giraudeau, "Proving Future Profit."

70. Mihm, *Nation of Counterfeiters*.

71. "Multicaulis," "To My Durham," *Spirit of the Times*, October 19, 1839, 394.

72. Dennis, *Dennis's Silk Manual: Containing Complete Directions for Cultivating the Different Kinds of Mulberry Trees, Feeding Silk Worms, and Manufacturing Silk to Profit* (New York: Mahlon Day and Co., 1839), 52.

73. Clarke, *Treatise on the Mulberry Tree*, 119.

74. Clarke, 5.

75. "Silk Worm Mortality," *JAI* 1, no. 1 (1836): 128.

76. Clarke, *Treatise on the Mulberry Tree*, 128.

77. Untitled, *JOC*, August 9, 1839, AAS.

78. "New Chinese Mulberry—Morus Multicaulis," *New York Farmer* 4, no. 8 (1831): 217; see also Gideon B. Smith, "The Morus Multicaulis, and Mr. Whitman's Mulberry Seed," *Monthly Genesee Farmer* 2, no. 3 (1837): 40–41.

79. "200,000 White Mulberry Trees," *GFGJ* 3, no. 24 (1833): 192.

80. "Mulberries, &c.," *Flushing Farmer and Silk Culturist* 1 (1839): 3.

81. Advertisement, "Large Sale of MORUS MULTICAULIS Trees," *JOC*, March 22, 1839, AAS.

82. Wolbert, *Morus Multicaulis Trees, at Public Sale*.

83. Dennis, *Silk Manual*, 21.

84. Wm. G. Harrison, "On Saturday Morning," *JOC*, October 5, 1838, AAS.

85. Harrison, 6.

86. Silk Culturist, "One Bud," *GFGJ* 7, no. 28 (1837): 221.

87. "Multiplication of the Multicaulis," *Cultivator* 6, no. 4 (June 1839): 66; William Prince, untitled, *Flushing Silk Journal* 1, no. 4 (1839): 29.

88. William Kenrick, "Successful Modes of Rearing the Morus Multicaulis: Causes of the Great Failure," *American Farmer* 1, no. 19 (1839): 150; for failures, see, e.g., T., "Morus Multicaulis," *GFGJ* 9, no. 25 (1839): 193.

89. Anti-Puff, "Princeana," *Farmer's Register* 7, no. 1 (1839): 55.

90. Since "Sharp," the name of the buyer, is another word for confidence man, it is possible that this story is a joke about humbugs. Northampton Gazette, "A Great Story," *JOC*, September 14, 1839, AAS.

91. "Mulberry Sales," *JOC*, August 10, 1839, AAS.

92. "Multiplication of the Multicaulis," *Cultivator* 6, no. 4 (1839): 66.

93. "Scraps of Information with Regard to the Culture of the Chinese Mulberry," *Family Visiter and Silk Culturist*, 1, no. 12 (1839): 147.

94. "Address to Those Engaged in the Silk Culture," *Flushing Silk Journal* 1, no. 4 (1839): 25–26. Attention to statistics of global trade were already common among US planters. Brian Schoen, "The Burdens and Opportunities of Interdependence: The Political Economies of the Planter Class," in *The Old South's Modern Worlds: Slavery, Region, and Nation in the Age of Progress*, ed. L. Diane Barnes, Brian Schoen, and Frank Towers (Oxford: Oxford University Press, 2011), 66–86, 76. This meshes well with Joanna Cohen's description of manufacturers' tariff-influenced shift to an idea of "luxury goods for universal consumption." *Luxurious Citizens*, 133–34.

95. *JOC*, October 5, 1839, AAS.

96. "Ed. Far. Reg." [Edmund Ruffin], "The Second Physick Humbug," *Farmers Register* 8, no. 6 (1840): 380.

97. *JOC*, October 17, 1839, AAS.

98. *JOC*, October 2, 1839, AAS.

99. John J. Wallis, "What Caused the Panic of 1839?," *NBER Working Paper Series on Historical Factors in Long Term Growth*, Historical Paper 133, April 2001.

100. H.C., "Morus Multicaulis," *New England Farmer* 18, no. 19 (1839): 170.

101. Kenrick, *William Kenrick to L. Bradford Prince*.

102. Anne Goldgar, *Tulipmania: Money, Honor, and Knowledge in the Dutch Golden Age* (Chicago: University of Chicago Press, 2007).

Chapter Six

1. For dairy and New York regionalization generally, see McMurry, *Transforming Rural Life*, 11–15; for hops, see Summerhill, *Harvest of Dissent*; for theories of regionalization, see John C. Hudson, *Making the Corn Belt: A Geographical History of Middle-Western Agriculture* (Bloomington: Indiana University Press, 1994); and, of course, William Cronon, *Nature's Metropolis: Chicago and the Great West* (New York: Norton, 1991).

2. Bidwell and Falconer, *History of Agriculture in the Northern United States*, 238.

3. Here again I am influenced by Searle, *Landscapes of Accumulation*, 49–72.

4. For boosterism, see Carl Abbott, *Boosters and Businessmen: Popular Economic Thought and Urban Growth in the Antebellum Middle West* (Westport, CT: Greenwood Press, 1981); Cronon, *Nature's Metropolis*, 36–38; Sheriff, *The Artificial River*, 52–78.

5. [Zadock Pratt], *Chronological Biography of Hon. Zadock Pratt, of Prattsville, N.Y.* (Kingston, NY: William H. Romeyn and Sons, 1866), 36. For Democratic articles, see, e.g., "Hon. Zadock Pratt," *Mechanic's Advocate*, 1, no. 16 (1847): 121. Distressingly, Pratt's papers at the Zadock Pratt Museum in Prattsville were damaged by Hurricane Irene in 2011 and at the time of writing remain in frozen storage while the curators gather funds to redeem them.

6. "Abbott Lawrence and Zadok [sic] Pratt: Or, City Success and Country Success," *American Phrenological Journal* 48, no. 1 (1868), 8.

7. On Pratt and tanning, see David Stradling, *Making Mountains: New York City and the Catskills* (Seattle: University of Washington Press, 2007), 30–31; P. E. Millen, *Bare Trees: Zadock Pratt, Master Tanner and the Story of What Happened to the Catskill Mountain Forests* (Hendersonville, NY: Black Dome Press Corp., 1995).

8. Henry Colman, *First Report on the Agriculture of Massachusetts* (Boston: Dutton and Wentworth, 1838), 4; Francis Lieber, a German visitor claimed that Goshen butter was shipped as far away as Malta. Francis Lieber, *Letters to a Gentleman in Germany: Comprising Sketches of the Manners, Society, and National Peculiarities of the United States* (Philadelphia: Carey, Lea and Blanchard, 1835), 273.

9. By an Amateur [James Kirke Paulding], *The New Mirror for Travelers: And Guide to the Springs* (New York: G. and C. Carvill, 1828), 33; see also John M. Salviani, "Butter Money," *Numismatist* 118, no. 1 (2005): 46–48.

10. Antediluvian, "The Story of the Old Blue Butter Pail," *Creamery Journal* 24, no. 1 (1913): 4, 25.

11. "Fraudulent Stamping Butter?," *Merchant's Magazine* 40, no. 5 (1859); Col. K. P. McGlincy, "The Private Dairyman and Farmer," *Transactions of the Wisconsin State Agricultural Society* 22 (1883–84): 20–26, 26.

12. Zadock Pratt, "The Dairy Farming Region of Greene and Orange Counties, New York, with Some Account of the Farm of the Writer," *Report of the Commissioner of Patents for the Year 1861: Agriculture* (1862): 411–27, 411.

13. William Paley, *Natural Theology, or Evidences of the Existence and Attributes of the Deity* (London: R. Faulder, 1803). As in Britain, *Natural Theology* became a basic text for North American science education. See Lily A. Santoro, "The Science of God's Creation: Popular Science and Christianity in the Early Republic," PhD diss. (University of Delaware, 2011). For postmillennialism more generally, see James Moorehead, *World without End: Mainstream American Protestant Visions of the Last Things, 1880–1825* (Bloomington: Indiana University Press, 1999); for millennialism, see Johnson, *A Shopkeeper's Millennium*; Cross, *The Burned-Over District*. For Whiggish postmillennialism, see Daniel Walker Howe, *The Political Culture of the American Whigs* (Chicago: University of Chicago Press, 1984), 152.

14. Charles Glidden Haines, *Considerations on the Great Western Canal, from the Hudson to Lake Erie: With a View of Its Expence, Advantages, and Progress* (Brooklyn: Spooner and Worthington, 1818), 35; Sheriff, *Artificial River*, 10.

15. Quoted in Elkanah Watson, *History and Progress of the Rise, Progress, and Existing Condition of the Western Canals in the State of New-York, from September 1788, to the Completion of*

the Middle Section of the Grand Canal in 1819 Together with the Rise, Progress, and Existing State of Modern Agricultural Societies, on the Berkshire System, from 1807, to the Establishment of the Board of Agriculture in the State of New York, January 10, 1820 (Albany, NY: D. Steele, 1820), 67.

16. "Message of the Governor to the Legislature of the State of New-York," Albany Argus, January 1, 1839, AAS.

17. "Herkimer," "The Three Railroads," Albany Argus, January 8, 1839, AAS.

18. Janet Browne, The Secular Ark: Studies in the History of Biogeography (New Haven, CT: Yale University Press, 1983), 27.

19. William W. Townsend, The Dairyman's Manual (Vergennes, VT: Rufus W. Griswold, 1839), 5.

20. "Farms for Exchange," New-York Spectator, May 11, 1837, AAS.

21. Wyckhoff, Developer's Frontier, 34; Conevery Bolton Valencius, The Health of the Country: How American Settlers Understood Themselves and Their Land (New York: Basic Books, 2002), 34.

22. Johnston, Notes on North America, 430. Watching the growth of commercial agriculture in Scotland, David Hume declared in 1777 that the world had been made with various "geniuses, climates and soils," to promote "mutual intercourse and commerce." Margaret Schabas, The Natural Origins of Economics (Chicago: University of Chicago Press, 2005), 59; for Adam Smith, see Fredrik Albritton Jonsson, "Rival Ecologies of Global Commerce: Adam Smith and the Natural Historians," AHR 115, no. 5 (2010): 1352. For Antebellum economic nationalism, see Ariel Ron, "Henry Carey's Rural Roots, 'Scientific Agriculture,' and Economic Development in the Antebellum North," Journal of the History of Economic Thought 37, no. 2 (2015): 263–75, 268.

23. Townsend, The Dairyman's Manual, 6–7.

24. Zadock Pratt, "Address of Hon. Z. Pratt, President, &c. to the Members of the Greene County Agricultural Association," Albany Freeholder, October 22, 1845, AAS.

25. Watson, History and Progress, 37; Murphy, Building the Empire State; Hauptman, Conspiracy of Interests, 64–68.

26. Ebenezer Emmons, Geology of New York, Part II, Comprising the Survey of the Second Geological District (Albany, NY: W. and A. White, and J. Visscher, 1842), 5; Watson, History and Progress, 37.

27. C. A. Bruce, "Report on the Manufacture of Tea, and on the Extent and Produce of the Tea Plantations in Assam," Document 1. Transactions of the Agricultural and Horticultural Society of India 7 (1840): 36; See Sharma, Empire's Garden, 31–34; Sven Beckert's concept of "War Capitalism" dependent on agriculture and the imperial expropriation of land provides a useful framing tool. Beckert, Empire of Cotton, xvi.

28. Sharma, Empire's Garden, 31–41.

29. Horatio Gates Spafford, A Gazetteer of the State of New-York (Albany, NY: H. C. Southwick, 1813), 328; William M. Beauchamp, "Aboriginal Place Names of New York," New York State Museum Bulletin 108 Archeology 12 (Albany, NY: New York State Education Department, 1907), 140.

30. Hauptman, Conspiracy of Interests.

31. Richard Drayton makes this argument with respect to British improvement in Nature's Government, 87; Valencius, Health of the Country, 237. For colonial roots see, Joyce E. Chaplin, Subject Matter: Technology, The Body, and Science on the Anglo-American Frontier, 1500–1676 (Cambridge, MA: Harvard University Press, 2009), 117, 176.

32. Wyckhoff, Developers' Frontier, 23.

33. Wyckhoff, 33; Beauchamp, "Aboriginal Place Names," 22.

34. Beauchamp, "Aboriginal Place Names," 35–37, 76; Hauptman, *Conspiracy of Interests*, 77.

35. Amos Eaton, *To Gentlemen Residing in the Vicinity of the Erie Canal*, (Troy, NY: n.p., 1822), 8.

36. Gillen D'Arcy Wood, *Tambora: The Eruption That Changed the World* (Princeton, NJ: Princeton University Press, 2015), 219–21.

37. For accounts of historical ideas of changing climate, see Jan Golinski, "American Climate and the Civilization of Nature," in *Science and Empire in the Atlantic World*, ed. Dew and Delbourgo (New York: Routledge, 2008), 153–74; James Rodger Fleming, *Historical Perspectives on Climate Change* (Oxford: Oxford University Press, 2005). For a critical summary of ideas of increasing warmth, see N. Webster Jun, "A Dissertation on the Supposed Change in the Temperature of Winter," *Connecticut Academy of Arts and Sciences* (January 1, 1810): 1–68. For transatlantic climate theories, see Anya Zilberstein, *A Temperate Empire: Making Climate Change in Early America* (Oxford: Oxford University Press, 2016); Anya Zilberstein, "Inured to Empire: Wild Rice and Climate Change," *WMQ* 72, no. 1 (2015): 127–58; Fredrik Albritton Jonsson, "Climate Change and the Retreat of the Atlantic: The Cameralist Context of Pehr Kalm's Voyage to North America, 1748–51," *WMQ* 72, no. 1 (2015).

38. William Maclure, *Observations on the Geology of the United States of America: With Some Remarks on the Effect Produced on the Nature and Fertility of Soils, by the Decomposition of the Different Classes of Rocks; and an Application to the Fertility of Every State in the Union, in Reference to the Accompanying Geological Map* (Philadelphia: Abraham Small, 1817), 65.

39. Ebenezer Emmons, *Agriculture of New York: Comprising the Classification, Composition and Distribution of the Soils and Rocks, and the Natural Waters of the Different Geological Formations; Together with a Condensed View of the Climate and the Agricultural Productions of the State*, Natural History of New York, 5 vols., vol. 2, (Albany, NY: C. Van Benthuysen & Co., 1846), 199.

40. McMurry, *Transforming Rural Life*, 10.

41. Isaac Carr to Children, June 21, 1857, Carr Letters, CAK.

42. Pratt, "Address" (1845).

43. Joan M. Jensen, "Butter Making and Economic Development in Mid-Atlantic America from 1750 to 1850," *Signs* 13, no. 4 (1988): 813–29, 820; Osterud, *Bonds of Community*, 283–85.

44. McMurry, *Transforming Rural Life*, 113–14, 103.

45. Jensen, "Butter Making," 822.

46. Bruegel, *Farm, Shop, Landing*, 118–19, 122.

47. Paying attention to masculine butter-making work, I am influenced by Ruth Schwartz Cowan's idea of the "work process"; *More Work for Mother: The Ironies of Household Technology from the Open Hearth to the Microwave* (New York: Basic Books, 1983), 11–14.

48. Thomas F. Gordon, *Gazetteer of the State of New York* (Philadelphia: Printed for the author, 1836), 59.

49. B. P. Johnson, "Manufacture of Navy Butter for Foreign Stations," *TNYSAS for 1847* 7 (1848): 43–45; Charles Mason, *An Elementary Treatise on the Structure and Operations of the National and State Governments of the United States* (Boston: D. H. Williams, 1842), 163; "The Dairy," *TNYSAS for 1847* 7 (Albany, NY: C. Van Benthuysen 1848): 570–80; B. P. Johnson, "Navy Butter," *TNYSAS for 1848* 8 (1849): 10–13.

50. Johnson, "Manufacture of Navy Butter," 43–44.

51. Johnson, 43–44.

52. B. P. Johnson, "Visit to County Societies," *TNYSAS for 1847* 7 (1848): 51–54, 51.

53. Joseph E. Bloomfield, "Agriculture of Oswego County," *TNYSAS for 1849* 9 (1850): 424–28, 428.

54. B. P. Johnson in untitled section, *Report of the Commissioner of Patents for the Year 1848* (Washington, DC: Wendell and Van Benthuysen, 1849), 386–89.

55. J. J. Hawley in B. P. Johnson, "Butter for the Navy," *American Agriculturist* 7, no. 9 (1848) 283–84, 284.

56. McMurry, *Transforming Rural Life*, 46.

57. McMurry, 387.

58. Even Orange County dairywomen themselves, connected to western dairies by networks of kin, may have had little interest in attacking the reputation of western butter. Hawley in Johnson, "Butter for the Navy," 389.

59. Legget and Lapham et al., "To the Dairy Women of Our Country," *Cultivator* 5, no. 4 (1838): 79.

60. Hawley in Johnson, "Butter for the Navy," 11.

61. For a discussion of these dynamics in the "Dairy Zone," see McMurry, *Transforming Rural Life*, 12–15.

62. Townsend, *The Dairyman's Manual*, 6–7.

63. Bloomfield, "Agriculture of Oswego County," 427.

64. X. A. Willard, "The Prices of Fifty-Four Years," *Cultivator and Country Gentleman* 44, no. 1367 (1879): 235–36.

65. E. C. Frost and A. J. Wynkoop, "Agriculture of Chemung County," *TNYSAS for 1842* 2 (1843): 143–19, 144.

66. Edward Harold Mott, *Between the Ocean and the Lakes: The Story of Erie* (New York: John S. Collins, 1901), 406–9.

67. Zadock Pratt, "The Dairy Farming Region," 415.

68. Bruegel, *Farm, Shop, Landing*, 77.

69. For agricultural deforestation, see Williams, *Americans and Their Forests*, 112–14.

70. Coons, Diary, NYSHA.

71. Harry Miller, "Potash from Wood Ashes: Frontier Technology in Canada and the United States," *Technology and Culture* 21 (1980): 187–208; Philip L. White, *Beekmantown, New York: Forest Frontier to Farm Community* (Austin: University of Texas Press, 2011), 56.

72. Francis Hall, *Travels in Canada and the United States in 1816 and 1817* (London: Longham, Hurst, Rees, Orme, and Brown, 1818), 47.

73. Pratt, "The Dairy Farming Regions," 414.

74. John R. Stilgoe, "Fair Fields and Blasted Rock: American Land Classification Systems and Landscape Aesthetics," *American Studies* 22 (1981): 21–33; Wyckhoff, *Developer's Frontier*, 37–8.

75. Wm. O. Connor [Advertisement], "To Farmers and Others!," *Albany Freeholder*, May 20, 1846, AAS.

76. Alson Ward, Alson Ward Diary, August 26, 1845, DCHS, 53.

77. Lyman Chandler, *Lyman Chandler to Moses Chandler September 1825*, Letter, Correspondence, 1825–49, 1:1 RRL, Chandler (Lyman) Family Papers 1815–1905.

78. John Dryden, translation of Virgil in Joseph William Jenks, *The Rural Poetry of the English Language* (Boston: John P. Jewett and Co., 1856), 216.

79. Wyckhoff, *Developer's Frontier*, 37.

80. "State Agricultural Society," in *Documents of the Assembly of the State of New York, Fifty-Sixth Session* (Albany, NY: E. Croswell, 1833), 28.

81. Zadock Pratt, "The Dairy Farming Region," 415.

82. Wyckhoff, *Developers' Frontier*, 38.

83. Pratt, *Chronological Biography*, 34.

84. Robert S. Rogers, "Forests Dominated by Hemlock (*Tsuga canadensis*): Distribution as Related to Site and Post-Settlement History," *Canadian Journal of Botany* 56, no. 7 (1978): 843–54.

85. Thomas C. Hubka, *Big House, Little House, Back House, Barn: The Connected Farm Buildings of New England* (Lebanon, NH: University Press of New England, 2004), 82.

86. Edward F. Peck, "Description of the Lands of Long Island" (Brooklyn, NY, [1848] 1858), 3.

87. State Geologist William Mather, quoted in Zadock Pratt, "Greene County," *TNYSAS for 1847* 7 (1848): 638–48, 639.

88. Gordon, *Gazetteer of the State of New York*, 58.

89. Sally Gregory Kohlstedt, "Geologists' Model for National Science, 1840–1847," *Proceedings of the American Philosophical Society* 118, no. 2 (1974): 179–95, 179. For Cole's geological paintings, see Rebecca Bedell, *The Anatomy of Nature: Geology and American Landscape Painting, 1825–1875* (Princeton, NJ: Princeton University Press, 2001), 29–33, figs. 13, 16, 17.

90. R. C. Taylor, "Field Notes and Sketches, 1841," "Geology Field Notes," NYSA. For the relationship between internal improvement and the field sciences, see Jeremy Vetter, "Science along the Railroad: Expanding Field Work in the US Central West," *Annals of Science* 61 (2003): 187–211; Michael Freeman, "Tracks to a New World: Railway Excavation and the Extension of Geological Knowledge in Mid-Nineteenth-Century Britain," *BJHS* 34 (2001): 51–65. Mark Adams brought this point to my attention. Rensselaer School, *Rensselaer School Flotilla* (Troy, NY, 1830); Eaton and Romeyn Beck, *Geological Survey of the County of Albany*; Eaton, *Geological and Agricultural Survey of Rensselaer County*; Eaton, *A Geological and Agricultural Survey of the District Adjoining the Erie Canal, in the State of New York*, 2 vols. (Albany, NY: Packard and Van Benthuysen, 1824); Spanagel, *DeWitt Clinton and Amos Eaton*, 23–24.

91. John Smith, *John Smith to James Hall, Lockport, July 18, 1841*, B0561, Letter, Box 1: 1830–41, NYSA, State Museum Director's, State Geologist's and State Paleontologist's correspondence files, 1828–1940.

92. Jacob W. Bailey to Asa Gray quoted in Kohlstedt, "Geologists' Model," 194.

93. Peck, "Description of the Lands of Long Island," 3.

94. Oz Frankel, *States of Inquiry: Social Investigation and Print Culture in Nineteenth-Century Britain and the United States* (Baltimore: Johns Hopkins University Press, 2007), 73–74; Aldrich, *New York State Natural History Survey*.

95. New York (State) Department of State, Distribution lists of state publications [ca. 1813–93], Subseries 7, Name Index to Purchasers and Recipients of "Natural History of the State of New-York" [n.d., 2 vols], NYSA.

96. For a summary of stratigraphy, see Peter J. Bowler, *The Norton History of the Environmental Sciences* (London: W. W. Norton and Co., 1993), 213–18. For stratigraphic controversies, see, e.g., Rudwick, *Great Devonian Controversy*; James A. Secord, *Controversy in Victorian Geology: The Cambrian-Silurian Dispute* (Princeton, NJ: Princeton University Press, 1986).

97. The most recent and comprehensive work on this is Spanagel, *DeWitt Clinton and Amos Eaton*. See also Cecil J. Schneer, "Ebenezer Emmons and the Foundations of American Geology," *Isis* 60, no. 4 (1969): 439–50, 443; Cecil J. Schneer, "The Great Taconic Controversy," *Isis* 69, no. 2 (1978): 173–91.

98. David Roger Oldroyd, *Thinking about the Earth: A History of Ideas in Geology* (Cambridge: Cambridge University Press, 1996), 118; James A. Secord, "The Geological Survey of Great Britain as a Research School, 1839–1855," *History of Science* 24, no. 3 (1986): 223–75. The state survey's final reports were published in four volumes: William M. Mather, *Geology of New-York, Part I, Comprising the Geology of the First Geological District, Natural History of New-York* (Albany, NY: Caroll and Cook, 1843); Ebenezer Emmons, *Geology of New-York, Part II, Comprising the Survey of the Second Geological District* (Albany, NY: W. and A. White, and J. Visscher, 1842); Lardner Vanuxem, *Geology of New-York, Part III, Comprising the Survey of the Third Geological District* (Albany, NY: W. and A. White and J. Visscher, 1842); James Hall, *Geology of New-York, Part IV, Comprising the Survey of the Fourth Geological District* (Albany, NY: Carroll and Cook, 1843); Aldrich, *Natural History Survey*, 35, 51.

99. I have found widely varying accounts of this estimate. This one comes from Ulysses P. Hedrick, *A History of Agriculture in the State of New York* (Albany, NY: NYSAS, 1933), 327.

100. Moses Eames, Moses Eames Political Diary, February 2, 1855, JCHS, Moses Eames Diaries.

101. See Henry S. Randall, *H. S. Randall to James Hall, August 9, 1847*, and *H. S. Randall to James Hall, December 9, 1847*, NIC 851 CCHS, Henry S. Randall Correspondence, 1828–1965.

102. [Orig. *American Agriculturist*], "Making Farm Life Attractive," *Southern Cultivator* 17, no. 3 (1859): 91.

103. Writing under the pen name "Hibernicus," in the *New York Statesman*, DeWitt Clinton saw hope for coal in New York's gas springs and in the arrangement of secondary strata that he called "the habitat of coal." [DeWitt Clinton], *Letters on the Natural History and Internal Resources of the State of New York* (New York: E. Bliss and E. White, 1822), 33.

104. See James A. Secord, "King of Siluria: Roderick Murchison and the Imperial Theme in Nineteenth-Century British Geology," *Victorian Studies* 25 (1982): 413–42.

105. One year after publishing the first stratigraphic map of Britain, canal engineer William Smith published a manual for identifying soil types using characteristic fossils, which, he argued, farmers could use without deep reading. William Smith, *Strata Identified by Organized Fossils* (London: W. Arding, 1816); For Smith's connection to British improvement, see H. C. Darby, "Some Early Ideas on the Agricultural Regions of England," *AHR* (1954): 30–47, 33.

106. Maclure, *Observations on the Geology*, title page.

107. Thomas Dick, *The Christian Philosopher or the Connection of Science and Philosophy with Religion* (Cooperstown: H. and E. Phinney, 1844), 161.

108. Edward Hitchcock, *Elementary Geology* (New York: Mark H. Newman, [1840] 1844), 262.

109. Maclure, *Observations on the Geology*, 34.

110. Eaton, *Geological and Agricultural Survey of Rensselaer County*, 22.

111. Emmons, *Agriculture of New York*, vol 1., 217.

112. Mather, *Geology of New York, Part I*, 5.

113. This was an issue throughout New York. Settlers in Cortland found that their ports into the Susquehanna River system filled up with eroded soil, becoming un-navigable by the early 1820s. H. P. Smith, ed., *History of Cortland County* (Syracuse, NY: D. Mason and Co., Publishers, 1885).

114. Dick, *The Christian Philosopher*, 161.

115. Mather, *Geology of New-York, Part I*, 7.

116. Pratt, "The Dairy Farming Region," 415.

117. Pratt, 415.

118. John Claudius Loudon, "Extracts: Of Grass Lands That Ought Not to Be Broken Up by the Plough," *Cultivator* 5, no. 11 (1839): 192.

119. Mather, *Geology of New York, Part I*, 7.

120. Nahum Capen, *Biography of Zadock Pratt, of Prattsville, N.Y.* (1852?); "Abbott Lawrence and Zadok [sic] Pratt", 8.

121. [Pratt], *Chronological Biography*, 37.

122. William H. Marshall, *The Rural Economy of the West of England: Including Minutes of Practice in That Department*, 2nd ed. (London: G. and W. Nicol, Pall Mall, 1805), 1–2; Darby, "Agricultural Regions of England," 37–38.

123. Jensen, "Butter Making," 820.

124. Pratt did write, "be it remembered to the credit of my farmer and his wife, not a pound of poor butter was made." Johnson and Pratt, "Report on Dairy Farm of Hon. Zadock Pratt," 93.

125. B. P. Johnson and Zadock Pratt, "Report on Dairy Farm of Hon. Zadock Pratt," *TNYSAS for 1857* 17 (1857): 89–94, 91.

126. Pratt, "The Dairy Farming Region."

127. Zadock Pratt, "Greene County," *TNYSAS for 1846* 5 (1846): 441.

128. [From the *Prattsville Advocate*], *Albany Freeholder*, November 29, 1848, AAS.

129. "List of Premiums Awarded," *TNYSAS for 1844* (1845): 19–31, 21.

130. "Ploughing Match," *Albany Freeholder*, October 6, 1847, AAS.

131. Pratt, "The Dairy Farming Region"; "Hon. Z. Pratt's Dairy: For 1863," *New York Observer and Chronicle*, 42, no. 6 (1864): 48.

132. "Report of the Judges Made at the Annual Show of the Society," *TNYSAS for 1848* 8 (1849): 88.

133. Jeffrey L. Pasley, "The Cheese and the Words: Popular Political Culture and Participatory Democracy in the Early American Republic," in *Beyond the Founders: New Approaches to the Political History of the Early American Republic*, ed. Andrew W. Robertson et al. (Chapel Hill: University of North Carolina Press, 2004).

134. Millen, *Bare Trees*, 61.

135. F. B. Hough, *Census of the State of New York for 1865* (Albany, NY: Charles Van Benthuysen, 1867), 320. See also Bruegel, *Farm, Shop, Landing*, 79.

136. Hough, *Census of the State of New York for 1865*, 314, 320, 343.

137. "Orange County Milk and Butter Company," *American Agriculturist* 7, no. 6 (1848): 196.

138. Thomas DeVoe, *The Market Assistant* (New York: n.p., 1867), 400.

139. French, *Gazetteer*, 330.

140. "An Eccentric Farmer and Tanner," *Massachusetts Ploughman and New England Journal of Agriculture* 32, no. 11 (1872): 4; "A Peculiar Man," *Phrenological Journal and Science of Health* 55, no. 6 (1872): 411.

141. "Butter Product of Native Cows," *Ohio Farmer* 60, no. 3 (1881): 37.

142. Stradling, *Making Mountains*, 43.

143. Dunwell, *The Hudson*, 141.

144. Emmons, *Agriculture of New York*, vol. 1, 8.

145. Emmons, 324–26.

146. Emmons, *Geological Survey*, 311.

Chapter Seven

1. "John Jacob Thomas," in *Cyclopedia of American Horticulture*, ed. Liberty Hyde Bailey (London: Macmillan and Co., 1907), 1797.

2. John J. Thomas, "Hints on Describing Fruits," *TNYSAS for 1842* 2 (1843): 269–72, 269.

3. Thomas, "Hints on Describing Fruits," 269.

4. Thomas, 270.

5. See, e.g., Lewis, *A Democracy of Facts*; Drayton, *Nature's Government*.

6. For variety/species distinctions, see Cristiana Oghina-Pavie, "Rose and Pear Breeding in Nineteenth-Century France: The Practice and Science of Diversity," in *New Perspectives on the History of Life Sciences and Agriculture*, ed. Denise Phillips and Sharon Kingsland (New York: Springer, 2016), 53–72.

7. For plant varieties, see Pauly, *Fruits and Plains*; Daniel Kevles, "Fruit Nationalism: Horticulture in the United States—from the Revolution to the First Centennial," in *Aurora Torealis: Studies in the History of Science and Ideas in Honor of Tore Frängsmyr*, ed. Marco Beretta, Karl Grandin, and Svante Lindqvist (Sagamore Beach, MA: Science History Publications/USA, 2008), 129–46; Jack Ralph Kloppenburg Jr., *First the Seed: The Political Economy of Plant Biotechnology* (Madison: University of Wisconsin Press, 2005); Fullilove, *Profit of the Earth*; McCook, *States of Nature*; Emma Spary, "Peaches Which the Patriarchs Lacked: Natural History, Natural Resources, and the Natural Economy in France," *History of Political Economy* 35, Ann. Supp. (2003): 14–41.

8. Here I am influenced by J. A. Secord, "Knowledge in Transit," *Isis* 95 (2004): 654–72.

9. Olmstead and Rhode, *Creating Abundance*. See also Daniel J. Kevles, "Protections, Privileges, and Patents: Intellectual Property in American Horticulture," *Proceedings of the American Philosophical Society* 152 (June 2008): 207–13; Daniel J. Kevles, "New Blood, New Fruits: Protections for Breeders and Originators, 1789–1930," in *Making and Unmaking Intellectual Property*, ed. Mario Biagioli, Peter Jaszi, and Martha Woodmansee (Chicago: University of Chicago Press, 2011); Deborah Fitzgerald, *The Business of Breeding: Hybrid Corn in Illinois, 1890–1920* (Ithaca, NY: Cornell University Press, 1990); and Okie, *The Georgia Peach*, 60–87; see Londa Schiebinger, *Plants and Empire: Colonial Bioprospecting in the Atlantic World* (Cambridge, MA, 2004).

10. For provincial improvisation, see Jaffee, *A New Nation of Goods*; Kelly, *In the New England Fashion*, 175–76.

11. Quoted in William Robert Prince, *The Pomological Manual; or A Treatise on Fruits: Containing Descriptions of a Great Number of the Most Valuable Varieties, Part II* (New York: T. and J. Swords, 1831), vii.

12. Darwin drew on this well-known variability to explain wild variability. James A. Secord, "Darwin and the Breeders: A Social History," in *The Darwinian Heritage*, ed. David Kohn (Princeton, NJ: Princeton University Press, 1985), 519–42.

13. Alan L. Olmstead and Paul W. Rhode, "Biological Innovation in American Wheat Production: Science, Policy, and Environmental Adaptation," in *Industrializing Organisms: Introducing Evolutionary History*, ed. Susan R. Schrepfer and Philip Scranton (New York: Routledge, 2004), 49–51.

14. Prince, *Catalogue of Fruit and Ornamental Trees and Plants* (1823).

15. A. J. Downing, *Fruits and Fruit Trees of America* (New York: Wiley and Putnam, 1845), 2.

16. Prince, *Pomological Manual*, 15.

17. Walter Elder, quoted in Ann Leighton, *American Gardens of the Nineteenth Century: "For Comfort and Affluence"* (Amherst: University of Massachusetts Press, 1987), 65.

18. F. R. Elliott, *The Western Fruit Book* (New York: A. O. Moore and Co., 1859), 17.

19. Thomas Andrew Knight, *A Treatise on the Culture of the Apple and Pear* (Ludlow: H. Procter, 1801), 8, 20; Pauly, *Fruits and Plains*, 66–67; John Lidwell-Durnin, "Inevitable Decay: Debates over Climate, Food Security, and Plant Heredity in Nineteenth-Century Britain," *Journal of the History of Biology* (2018): 1–22 (online December 11, 2018).

20. Pauly demonstrates that American audiences were aware of these theories in the 1860s. Pauly, *Fruits and Plains*, 65–67.

21. John and H. T. Riley Bostock, ed., *The Natural History of Pliny*, vol. 3 (London: Henry G. Bohn, 1855), 484.

22. The standard account of the growth of horticulture and fruit culture in North America is U. P. Hedrick, *A History of Horticulture in America to 1860* (Portland, OR: Timber Press, 1988). See also Daniel Kevles, "Fruit Nationalism," 129–46.

23. S. A. Beach, *The Apples of New York*, State of New York, Department of Agriculture (Albany, NY: J. B. Lyon Company, 1902). See also Alfred W. Crosby, *Ecological Imperialism: The Biological Expansion of Europe, 900–1900* (Westport, CT: Greenwood Press, 1986), 157. Archaeological evidence suggests that Native peoples of Florida adopted peach growing to a large extent as early as 1580. Marvin T. Smith, *Archaeology of Aboriginal Culture Change in the Interior Southeast* (Gainesville: University of Florida Press, 1987), 124–25.

24. Oliver Tillson Papers, Box 8 "Receipts," Kroch Library Manuscript and Special Collections, Cornell, Ithaca, New York.

25. Michael Pollan, *The Botany of Desire: A Plant's-Eye View of the World* (New York: Random House, 2002), 8–9, 31–37; William Kerrigan, *Johnny Appleseed and the American Orchard: A Cultural History* (Baltimore: Johns Hopkins University Press, 2012).

26. Thornton, *Cultivating Gentlemen*; Kevles, "Fruit Nationalism," 132–33; John R. Stilgoe, *Borderland*, 67–93; Pauly, *Fruits and Plains*, 53–54.

27. Henry Ward Beecher, *Plain and Pleasant Talk about Fruits, Flowers and Farming* (New York: Derby and Jackson, 1859), v; Andrew Jackson Downing, "Mr. Downing's Letters from England," *Horticulturist* 5 (1850): 217.

28. A. J. Downing, *Fruits and Fruit Trees*, v.

29. Jaffee, *New Nation of Goods*; Catherine Kelly, *In the New England Fashion*, 10–11.

30. "Fruit Culture," *Massachusetts Ploughman and New England Journal of Agriculture* 9 (1850): 1.

31. William Kerrigan, "Stealing Apples: Markets, Morality, and the Movement to Criminalize Apple-Pilfering in Antebellum Ohio" (paper presented at annual meeting of the Society for Historians of the Early American Republic, St. Louis, July 2013).

32. *Longworth's American Almanac* (New York: Thomas Longworth, 1837); *Doggetts' New York City Directory, 1848–1849* (New York: John Doggett Jr., 1848).

33. Cindy R. Lobel, *Urban Appetites: Food and Culture in Nineteenth-Century New York* (Chicago: University of Chicago Press, 2014), 63; Gergely Baics, *Feeding Gotham: The Political Economy and Geography of Food in New York, 1790–1860* (Princeton, NJ: Princeton University Press, 2016).

34. "Burn Them," "What Shall I Do with My Apples," *Religious Intelligencer* 12, no. 19 (1827): 299; B.M., "Fruit a Substitute for Ardent Spirits," *Christian Spectator* 1, no. 10 (1827): 521.

35. "Apple Trade at Cincinnati," *Ohio Cultivator* 7, no. 23 (1851): 360.

36. Michael Collinson, quoted in William A. Taylor, *The Fruit Industry, and the Substitution of Domestic for Foreign-Grown Fruits* (Washington, DC: Government Printing Office, 1898), 311.

37. Taylor, 344.

38. Paul W. Gates, *The Farmers Age: Agriculture, 1815–1860* (New York, 1960), 257.

39. Gates, 259.

40. Willis P. Duruz, "Notes on the Early History of Horticulture in Oregon: With Special Reference to Fruit-Tree Nurseries," *AH* 15, no. 2 (1941): 84–97, 88.

41. Downing, *Fruits and Fruit Trees*, vi.

42. Lester C. Doubleday, Lester C. Doubleday Diary, October 1, 1851, NYSHA.

43. Thomas, "Hints on Describing Fruits," 270.

44. Andrew Jackson Downing to William Darlington (ca. 1830), Andrew Jackson Downing Papers, NYSL.

45. For gendered and sexualized description of plant bodies, see, e.g., Ann Shteir, "Iconographies of Flora: The Goddess of Flowers in the Cultural History of Botany," in *Figuring It Out: Science, Gender, and Visual Culture*, ed. Ann B. Shteir and Bernard Lightman (Lebanon, NH: Dartmouth College Press, 2006); Londa Schiebinger, *Nature's Body: Gender in the Making of Modern Science* (Rutgers, NJ: Rutgers University Press, 2004), 11–37.

46. For educated visual skill and connoisseurship spreading out from painting to a broader marketplace, see Kelly, *Republic of Taste*, 77–78.

47. John J. Thomas, *The Fruit Culturist, Adapted to the Climate of the Northern States* (New York: Mark H. Newman, 1846), 89.

48. Prince, *Catalogue of Fruit and Ornamental Trees* (1823), v–vii.

49. Leighton, *American Gardens of the Nineteenth Century*, 68. For this role of nurserymen, see Kevles, "Fruit Nationalism," 133–34.

50. Drayton, *Nature's Government*, 139; *Catalogue of Fruits Cultivated in the Garden of the Horticultural Society of London at Chiswick* (London: William Nicol, 1826), vi–vii, 153.

51. Downing, *Fruits and Fruit Trees*, 106.

52. Downing, v.

53. Joan Morgan and Alison Richards, *New Book of Apples* (London: Ebury Press, 1993), 107.

54. Thomas Andrew Knight, *Pomona Herefordiensis* (London: W. Bulmer and Co., 1811), ii.

55. Nurserymen dominated this material: Andrew Jackson Downing wrote for the *Cultivator* and edited *The Horticulturist*; Patrick Barry wrote for the *Genesee Farmer*, as did J. J. Thomas who also edited the *Horticultural Register*. A. J. Downing, "Miscellaneous: Notes on Transplanting," *Cultivator* 2, no. 1 (1836): 157.

56. Michael Floy, George Lindley, and John Lindley, *A Guide to the Orchard and Fruit Garden*, 1st ed. (New York: Collins and Hannay, 1833); William Robert Prince, *A Treatise on Fruits: Containing Descriptions of a Great Number of the Most Valuable Varieties for the Orchard and Garden*, 2nd ed., vol. 1. (New York: T. and J. Swords, 1832).

57. Prince, "Catalogue of Fruits," vi.

58. See, e.g., William Hooker, *Pomona Londoniensis* (London: W. Hooker, 1818); Anne Secord, "Botany on a Plate: Pleasure and the Power of Pictures in Promoting Early Nineteenth-Century Scientific Knowledge," *Isis* 93, no. 1 (2002): 37; on the sometimes gorgeous, always expensive, and often frustrated efforts of American pomologists to produce images of fruit, see Daniel J. Kevles, "Cultivating Art," *Smithsonian* 42, no. 4 (2011): 76–82.

59. Prince, *Pomological Manual*, vol. 2, title page; John Kilbourn, *The Ohio Gazetteer* (Columbus, OH: Kilbourne, 1831), 108–9.

60. For British illustrated volumes, see most notably *The Pomological Magazine* (London: James Ridgway, 1828–39).

61. Prince, *Pomological Manual*, 56–57.

62. Thomas, *The Fruit Culturist*, 127.

63. Floy, Lindley, and Lindley, *Guide to the Orchard and Fruit Garden*, iv. This happened to Mr. Munson, of New Haven, who found that only one of a set of pear trees sold to him by William Prince was *not* a White Doyenne. "Proceedings of the National Congress of Fruit Growers," *Seventh Annual Report of the American Institute of the City of New York* (Albany, NY: Weed, Parson's and Co., 1849), 240.

64. For counterfeiting and confidence, see Halttunen, *Confidence Men and Painted Women*; Mihm, *A Nation of Counterfeiters*.

65. "Northern Spy," *GFGJ* 6, no. 2 (1845): 30.

66. Cheryl Lyon-Jenness, "Planting a Seed: The Nineteenth-Century Horticultural Boom in America," *Business History Review* 78, no. 3 (2008): 381–421.

67. William Prince and Sons, *Annual Catalogue of Fruit and Ornamental Trees and Plants, Cultivated at the Linnaean Botanic Garden and Nurseries* (Jamaica, NY: I. F. Jones and Co., 1835), ii.

68. "The Weather—Transplanting, &c.," *GFGJ* 6, no. 12 (1845): 182.

69. William Prince Receipt, L1 Folder 2, ML, Wadsworth Family Papers.

70. T. J. Jackson Lears, *Fables of Abundance: A Cultural History of Advertising in America* (New York: Basic Books, 1994), 69.

71. Moses Eames, Diary, vol. 1, April 14–15, 1831, JCHS, Moses Eames Diaries.

72. W.W.B., "The Williamson Apple," *GFGJ* 5, no. 16 (1835): 122.

73. William Prince and Sons, *Annual Catalogue 1835*, ii.

74. Mihm, *A Nation of Counterfeiters*, 3.

75. Mihm.

76. Emmons, *Agriculture of New York*, vol. 3, 189.

77. David Henkin, *City Reading: Written Words and Public Spaces in Antebellum New York* (New York: Columbia University Press, 1998), 147.

78. Matthew T. Miller, *Bicknell's Counterfeit Detector and Bank Note List*, 3, no. 14 (1845): 28.

79. William Prince, *Catalogue of Fruit and Ornamental Trees*, ix–x.

80. Thomas, "Hints on Describing Fruits," 269.

81. Thomas, 272.

82. Emmons, *Agriculture of New York*, vol. 3, 9.

83. W.R.Y., "General and Critical Observations on the Cultivation of the Pear," *Gardener's Magazine* 4 (1828): 107–13; W.R.Y., "On Measuring, Delineating, and Describing Fruits, with a Glossary of Terms for the Last Purpose," *Gardener's Magazine* 4 (1828): 230. Charles Dickens's *Household Words* described a naturalist who made an outline from his dessert apple every day, to learn new varieties. See "Strawberries," *Household Words* 18, no. 432 (1858): 60–64.

84. Emmons, *Agriculture of New York*, 9.

85. Halttunen, *Confidence Men and Painted Women*, 40–41; Sandage, *Born Losers*, 113–14.

86. Downing, *Fruits and Fruit Trees*, vii.

87. E.g., "Outline and Description of Two Fine Autumn Apples," *GFGJ* 7, no. 12 (1846): 287.

88. New York State Fair, *Proceedings of the New-York State Fair and of the Pomological Convention Held at Buffalo, September 1848* (Buffalo, NY: Jewett, Thomas and Co., 1848), 1.

89. New York State Fair, 2.

90. Unwilling to be left out, most of the delegates of the first meeting attended the second; the institutions continued in parallel until their union in 1850. The resultant body, the American Pomological Society, remains the professional organization for the judgment of fruit varieties.

91. New York State Fair, *Proceedings*, 2.

92. New York State Fair, 2.

93. New York State Fair, 2.

94. Shawn Johansen, *Family Men: Middle-Class Fatherhood in Early Industrializing America* (New York: Routledge, 2001), 106; Alfred Henderson, "American Horticulture," in *1795–1895:*

One Hundred Years of American Commerce, Consisting of One Hundred Original Articles on Commercial Topics, ed. Chauncey Mitchell Depew (New York: D. O. Haynes, 1895), 248.

95. Sandage, *Born Losers*, 99–105, 129–58.

96. North American Pomological Convention, *Proceedings of the North American Pomological Convention: Held at Syracuse, September 14th 1849* (Syracuse, NY: V. W. Smith and Co., 1849), 18.

97. "Proceedings of the National Congress of Fruit Growers," in *Seventh Annual Report of the American Institute of the City of New York* (Albany, NY: Weed, Parson's and Co., Public Printers, 1849), 225.

98. "Proceedings of the National Congress of Fruit Growers," 225.

99. New York State Fair, *Proceedings*, 2.

100. For the rise of western New York nurseries, see Kevles, "Fruit Nationalism," 140; Cheryl Lyon-Jenness, "Planting a Seed: The Nineteenth-Century Horticultural Boom in America," *BHR* 78, no. 3 (2004): 381–421, 386.

101. New York State Fair, *Proceedings*, 14.

102. New York State Fair, 14.

103. "Proceedings of the National Congress of Fruit Growers" (1849), 216.

104. Devoe, *The Market Assistant*, 370.

105. New York State Fair, *Proceedings*, 11.

106. "Proceedings of the Second Congress of Fruit Growers, Convened under the Auspices of the American Institute, in the City of New York, 1849" (Albany, NY: Charles van Benthuysen, 1850), 212.

107. *Transactions of the Second Session of the American Pomological Society* (Philadelphia: Stavely and McCalla, 1852), 74.

108. North American Pomological Convention, *Proceedings of the North American Pomological Convention* (1850).

109. New York State Fair, *Proceedings*, 12.

110. "Proceedings of the Second Congress of Fruit Growers," 212.

111. Henry David Thoreau, *The Succession of Forest Trees and Wild Apples* (Boston: Houghton, Mifflin and Co., 1887), 69.

112. Thoreau, 69, 72, 78.

113. William Rossi, ed., *Wild Apples and Other Natural History Essays* (Athens: University of Georgia Press, 2002), xviii.

114. D. A. A. Nichols, "Best Fruit for Market Purposes," *GFGJ* 21, no.2 (1860): 59.

115. Lyon-Jenness, "Planting a Seed."

116. Frederick Law Olmsted, *A Journey in the Back Country*, (London: Sampson Low, Son & Co., 1860), 147.

117. As Barbara Hahn has shown, some varieties have no basis in genetic difference; bright tobacco, the kind of tobacco used to make cigarettes, for example, is a product of soils and processing. Barbara Hahn, *Making Tobacco Bright: Creating an American Commodity, 1617–1937* (Baltimore: Johns Hopkins University Press, 2011).

118. For *terroir*, see Amy Trubek, *The Taste of Place: A Cultural Journey into Terroir* (Berkeley: University of California Press, 2008).

Chapter Eight

1. Geo. E. Waring, *The Elements of Agriculture: A Book for Young Farmers, with Questions Prepared for the Use of Schools* (New York: D. Appleton and Co., 1854), 27–28.

2. Waring, 28; Walt Whitman, *Whitman: Selected, with an Introduction and Notes by Leslie A. Fiedler* (New York: Dell Publishing, [1855] 1959), 28. Nineteenth-century American chemists and improvers often used the word "atom" to mean an indivisible particle of an element. Sometimes they followed a colloquial usage, meaning a tiny piece, a jot. Whitman referred to an "atom of blood" much as the geologist and chemist Amos Eaton referred to an "atom of earth." Amos Eaton, "To Gentlemen Residing in the Vicinity of the Erie Canal." For Whitman's poetry in context, see Maria Farland, "Decomposing City: Walt Whitman's New York and the Science of Life and Death," *English Literary History*, 74, no. 4 (2007): 799–827, 810.

3. Emmons, *Agriculture of New York*, vol. 2.

4. The classic accounts of this are Rossiter, *The Emergence of Agricultural Science*; Rosenberg, *No Other Gods: On Science and American Social Thought*. More recently, see Cohen, *Notes from the Ground*; Stoll, *Larding the Lean Earth*.

5. Seward, "Address," 14.

6. For classic accounts of fodder changes, see Ernle, *English Farming, Past and Present*, 176–89; Chambers and Mingay, *Agricultural Revolution*, 66–69. More recently, see Overton, *Agricultural Revolution in England*, 148.

7. J. W. O'Neil, ed., *The American Farmer's Handbook* (Philadelphia: Charles DeSilver, 1859), 110.

8. Sir Humphry Davy, *Elements of Agricultural Chemistry* (New York: Eastburn, Kirk and Co., 1815); J. A. Chaptal, *Chymistry Applied to Agriculture* (Boston: Hilliard, Gray, and Co., 1839); Cohen, *Notes from the Ground*, 82–123; Rossiter, *Emergence of Agricultural Science*, 11; David Knight, "Agriculture and Chemistry in Britain around 1800," *Annals of Science* 33 (1976): 187–96.

9. A. Eaton (pseudonym "The Self Taught Mechanic"), "Old Schoolmaster's Remarks on Troy Teachers," *Daily Troy Press*, August 30, 1833; cutting found in Amos Eaton, *Geological Journal*, Box 4, Folder 1 RPI, Amos Eaton, 1776–1841, Papers 1799–1841; M. Susan Lindee, "The American Career of Jane Marcet's *Conversations on Chemistry*, 1806–1853," *Isis* 82, no. 1 (1991): 8–23; Frank A. J. L. James, "'Agricultural Chymistry is at Present in Its Infancy': The Board of Agriculture, the Royal Institution and Humphry Davy," *Ambix* 62, no. 4 (2015): 363–85, 383–84.

10. Eaton, "Old Schoolmaster's Remarks."

11. Davy, *Elements of Agricultural Chemistry*, 140–56. For humus and mineral theories in continental Europe, see William H. Brock, *Justus von Liebig: The Chemical Gatekeeper* (Cambridge: Cambridge University Press, 2002), 147–50.

12. L.T./R.R.H., "Inexhaustible Sources of Manure," *New York Farmer* 7, no. 2 (1834): 34.

13. For vernacular humoral medicine, see Valencius, *Health of the Country*, 59–63.

14. L.T./R.R.H., "Inexhaustible Sources of Manure," 34.

15. Catherine M. Jackson, "The Wonderful Properties of Glass: Liebig's *Kaliapparat* and the Practice of Chemistry in Glass," *Isis* 106, no. 1 (2015): 43–69, 55.

16. Heermance, *The Farmer's Mine, or Source of Wealth* (New York: Henry Heermance, 1843), 51.

17. Benjamin Hodge, "Reply to Plough Jogger," *GFGJ* 3, no. 35 (1833): 277.

18. Mather, *Geology of New York*, Part I, 11.

19. Plough Jogger, "Use of Plaster," *GFGJ* 3, no. 18 (1833): 140.

20. "B," "Plaster of Paris," *Maine Farmer and Journal of the Useful Arts* 1, no. 36 (1833): 286.

21. Edmund Ruffin, *An Essay on Calcareous Manures* (Petersburg, VA: J. W. Campbell, 1832); "An Onondaga Farmer," "Experience versus Theory," *GFGJ* 6, no. 22 (1834): 173. For Davy on the plaster question, see Rossiter, *Emergence of Agricultural Science*, 17–18.

22. Jennifer Anderson, "'A Laudable Spirit of Enterprise': Renegotiating Land, Natural Resources, and Power on Post-Revolutionary Long Island," *EAS* 13, no. 2 (2015): 413–42.

23. Catherine McNeur, *Taming Manhattan* (Cambridge, MA: Harvard University Press, 2014), 101–7.

24. Richard A. Wines, *Fertilizer in America* (Philadelphia: Temple University Press, 1985), 24.

25. Advertisement, "Poudrette," *Cultivator* 5, no. 3 (1848): 104.

26. Farmer's Visiter, "Farm Accounts," *Farmers Cabinet and American Herd Book* 5, 9 (1841): 290; Wines, *Fertilizer in America*, 26–30.

27. For more on lease agreements, see Drew Swanson, *A Golden Weed: Tobacco and Environment in the Piedmont South* (New Haven, CT: Yale University Press, 2014), 94–96.

28. McNall, *Agricultural History of the Genesee Valley*, 148; Joseph Harris, "Culture of Wheat in Western New York," *GFGJ* 12, no. 7 (1851): 165.

29. Buel, in Carman, *Jesse Buel, Agricultural Reformer*, 102.

30. Buel, in Carman, 102.

31. E.g., Charles Squarey, *A Popular Treatise on Agricultural Chemistry* (Philadelphia: Lea and Blanchard, 1842).

32. Justus Liebig [sic], *Organic Chemistry in its Application to Agriculture and Physiology* (Cambridge, MA: John Owen, [1840] 1841).

33. *Journal: Liebig Manufacturing Company*, Account Book, AAS, Taintor-Davis Family Papers, 1763–1917; E. N. Kent, *Descriptive Catalogue of Chemical Apparatus* (New York: Van Norden and Amerman, 1848).

34. R. Bud and G. K. Roberts, *Science vs. Practice: Chemistry in Victorian Britain* (Manchester: Manchester University Press, 1984).

35. Brock, *Justus Von Liebig*, 128.

36. Ursula Klein, *Experiments, Models, Paper Tools: Cultures of Organic Chemistry in the Nineteenth Century* (Stanford, CA: Stanford University Press, 2003), 66, 218; Ursula Klein, "Shifting Ontologies, Changing Classifications: Plant Materials from 1700 to 1830," *Studies in History and Philosophy of Science* 36, no. 2 (2005): 261–321, 248, 251.

37. A. J. Rocke, "Organic Analysis in Comparative Perspective: Liebig, Dumas, and Berzelius, 1811–1837," in *Instruments and Experimentation in the History of Chemistry*, ed. F. L. Holmes and T. H. Levere (Cambridge, MA: MIT Press, 2000), 273–310; Klein, "Shifting Ontologies," 87.

38. Jackson, "The Wonderful Properties of Glass," 43–69.

39. H. Kamminga and A. Cunningham, "Introduction: The Science and Culture of Nutrition, 1840–1940," in *The Science and Culture of Nutrition, 1840–1940*, ed. H. Kamminga and A. Cunningham (Amsterdam: Rodopi, 1995), 1–14, 3.

40. J. Morrell, "The Chemist Breeders: The Research Schools of Liebig and Thomas Thomson," *Isis* 76 (1972) 182–94.

41. From the Farmer's Visiter, "Farm Accounts," 290.

42. D. Christy, *The Chemistry of Agriculture: Or the Earth and Atmosphere as Related to Vegetable and Animal Life* (Cincinnati: Ward and Taylor, 1852), 113–14.

43. J. Dumas and J. B. Boussingault, *The Chemical and Physiological Balance of Organic Nature: An Essay* (London: H. Balliere, 1844), 1, 114.

44. W. D. Cochran, *Agricultural Book-Keeping: Being a Concise and Scientific System of Keeping Farm Accounts* (Detroit: Daily Advertiser Steam Presses, 1858), 67.

45. Justus von Liebig, *Familiar Letters on Chemistry and Its Relation to Commerce, Physiology, and Agriculture*, ed. John Gardner (New York: D. Appleton and Co., 1843), 171.

46. J. A. Smith, *Productive Farming* (Chambersburg, PA: J. Pritts, 1845).

47. Oz Frankel, *States of Inquiry*, 83.

48. Daniel Lee, *Report of the Commissioner of Patents for the Year 1849, Part II: Agriculture* (Washington, DC: Patent Office, 1850), 8–9.

49. Lepler, *Many Panics*.

50. Stoll, *Larding the Lean Earth*.

51. Mihm, *A Nation of Counterfeiters*; Lepler, *Many Panics*; Temin, *The Jacksonian Economy*.

52. Lee, *Report*, 17.

53. See Erland Marald, "Everything Circulates: Agricultural Chemistry and Recycling Theories in the Second Half of the Nineteenth Century," *Environment and History* 8, no. 1 (2002): 65–84.

54. My thinking on cornucopianism is influenced by Fredrik Albritton Jonsson, "The Origins of Cornucopianism: A Preliminary Genealogy," *Critical Historical Studies* 1, no. 1 (2014): 151–68.

55. Daniel Lee, quoted in "Wheat," *TNYSAS for 1844* 4 (1845): 151–54.

56. "Agriculture," *Albany Freeholder*, October 1, 1845, AAS.

57. Jethro Tull, *Horse-Hoeing Husbandry*, 3rd ed. (London: A. Millar, 1751).

58. Letters republished in "To the Public," *Albany Argus*, December 10, 1839, AAS.

59. *Wadsworth Leases, 1839–1847*, ML, Wadsworth Family Papers.

60. "Agriculture," *Albany Freeholder*.

61. "Farmer's Creed," *Albany Freeholder*, November 25, 1845, AAS.

62. [Originally *Farmer and Mechanic*], "Worn-Out Lands—Average Products," *Michigan Farmer* 4, no. 7 (1846): 116.

63. Davy, *Elements of Agricultural Chemistry*, 140–56.

64. Emmons, *Agriculture of New York*, vol. 1, 227.

65. Though they are not quantitative, noses can subtly distinguish between invisible compounds when visual examination fails. For a recent examination of state of the field in smell, Connie Chiang, "The Nose Knows: The Sense of Smell in American History," *JAH* 95, no. 2 (2008): 405–16; Christine Meisner Rosen and Melanie Kiechle, "Navigating by Nose: Fresh Air, Stench Nuisance, and the Urban Environment, 1840–1880," *Journal of Urban History* (2015): 1–19; see also Valencius, *Health of the Country*; Kiechle, *Smell Detectives*.

66. Kiechle, *Smell Detectives*, 99–103; Martin V. Melosi, *The Sanitary City: Environmental Services in Urban America from Colonial Times to the Present* (Pittsburgh: University of Pittsburgh Press, 2008), 120.

67. Chaptal, *Chymistry*, 47.

68. Farland, "Decomposing City," 799.

69. Farland, 799.

70. Lissa Roberts, "The Death of the Sensuous Chemist: The 'New' Chemistry and the Transformation of Sensuous Technology," *Studies in History and Philosophy of Science Part A* 26, no. 4 (1995): 503–29.

71. Liebig, *Familiar Letters*, 54.

72. John Pitkin Norton, "Letter from Prof. Norton," *Cultivator* 6, no. 1 (1850): 33–34.

73. Liebig, *Organic Chemistry*, 90.

74. James F. W. Johnston, *Lectures on the Application of Chemistry and Geology to Agriculture* (New York: C. M. Saxton, 1850), 27.

75. Waring, *Elements of Agriculture*, 262.

76. Waring, 260.

77. Cultor, "Agriculture, Unscientific and Scientific," *Cultivator* 9, no. 11 (1852): 365.

78. Cultor, 365.

79. Rossiter, *Emergence of Agricultural Science*; Rosenberg, *No Other Gods*; Paul Lucier, "Commercial Interests and Scientific Disinterestedness: Consulting Geologists in Antebellum America," *Isis* 86, no. 2 (1995): 245–67, 249.

80. *Trow's New York City Directory* (New York: John F. Trow, 1857).

81. P. S. and Wm. H. Chappell, "'Chappell's Fertilizer' vs. 'Professor Ducatel' and Others," *American Farmer* 4, no. 12 (1849): 415.

82. Emmons, *Agriculture of New York*, vol. 2, 53.

83. John Pitkin Norton, "On the Analysis of the Oat," *American Journal of Science and Arts* 3, no. 9 (1847): 318.

84. S. W. Johnson, "On the Practical Value of the Analyses of Soils," *Cultivator* 2, no. 8 (1854): 233.

85. James Mapes, "Plaster of Paris," *Working Farmer* 1, no. 1 (1850): 12.

86. "Analyses of Swamp Muck," *American Agriculturist* 5, no. 6 (1846): 178. See also Samuel L. Dana, *Muck Manual for Farmers* (Lowell, MA: Bixby and Whiting, 1843).

87. "Answers to Inquiries," *Country Gentleman* 1, no. 13 (1853): 196.

88. "Answers to Inquiries," 196.

89. Johnson, "On the Practical Value of the Analyses of Soils," 49–50.

90. "Extraordinary Experiment with Wheat," *Maine Farmer* 14, no. 15 (1846): 1. The rise of gluten can be seen, e.g., in Lee, "Wheat," 154.

91. Gregory T. Cushman, *Guano and the Opening of the Pacific World: A Global Ecological History* (Cambridge: Cambridge University Press, 2013), 23–75.

92. Wines, *Fertilizer in America*, 40–46.

93. Wines, 41–50; Edwin Bartlett, *Guano, Its Origin, Properties, and Uses, Showing Its Importance to the Farmers of the United States as a Cheap and Valuable Manure: With Directions for Using It* (New York: Wiley and Putnam, 1845).

94. Bartlett, *Guano*, 28, 33, 46, 50.

95. Wines, *Fertilizer in America*, 50.

96. Cushman, *Guano and the Opening of the Pacific World*, 82.

97. Johnston, *Lectures on the Applications of Chemistry and Geology to Agriculture*, 603; S. W. Johnson, "Commercial Fertilizers: Scale of Price," *Southern Planter* 20, no. 11 (1860): 671.

98. Wines, *Fertilizer in America*, 73. For similar arguments, see Philadelphia Guano Co., *Colombian and Other Phosphatic Guanos, Brought from the Guano Islands in the Caribbean Sea* (Philadelphia: McLaughlin Bros., 1856).

99. Wines, *Fertilizers in America*, 60, 80–90.

100. "Adulteration of Milk in the City of New York," *Farmer's Register* 4, no. 5 (1845): 279.

101. C. H. Peirce, M.D., *Examinations of Drugs, Medicines, Chemicals, &c., as to Their Purity and Adulterations* (Cambridge, MA: John Bartlett, 1852), 226, 105, 15.

102. Kiechle, "Navigating by Nose," 759.

103. Chappell and Chappell, "Chappell's Fertilizer," 415.

104. Margaret W. Rossiter, "Johnson, Samuel William," *American National Biography Online*, February 2000, http://www.anb.org/articles/13/13-01988.html (accessed June 8, 2016).

105. This method for assigning value would come to dominate institutions of chemistry after the Civil War. Alan I. Marcus, *Agricultural Science and the Quest for Legitimacy: Farmers, Agricultural Colleges, and Experiment Stations, 1870–1890* (Ames: Iowa State University Press, 1985), 63.

106. Samuel W. Johnson, *Essays on Peat, Muck, and Commercial Manures* (Hartford, CT: Brown and Gross, 1859), 13.

107. Edmund Monroe Pendleton, *On the Valuation of Commercial Fertilizers* (no city [Georgia]: n.p., ca. 1878): 6.

108. Pendleton, 6.

109. Elizabeth A. Osborne, *From the Letter-Files of S. W. Johnson* (New Haven, CT: Yale University Press, 1908), 109.

110. "Artificial Manures," *GFGJ* 14, no. 7 (1853): 205.

111. "Artificial Manures," 205.

112. The most complex and thoughtful analysis of the Mapes controversy is Ron, "Developing the Country," 97–116.

113. "Improved Superphosphate of Lime," *GFGJ* 13, no. 12 (1852): 366; Wines, *Fertilizer in America*, 102; Marti, *To Improve the Soil and the Mind*, 151.

114. "Artificial Manures," *GFGJ* 14, no. 7 (1853): 205.

115. "Trouble among the Fraternity," *Prairie Farmer* 13, no. 9 (1853): 341.

116. "Prof. Mapes's Farm," *New York Evangelist* 28, no. 13 (1857): 103; Judge French, "High Farming—Prof. Mapes's Farm—Superphosphate," *New England Farmer* 11, no. 6 (1859): 260; Marti, *To Improve the Soil and the Mind*, 151.

117. *Mapes's Nitrogenized Superphosphate of Lime* (New York: n.p., ca. 1859)

118. James A. Secord, *Victorian Sensation: The Extraordinary Publication, Reception, and Secret Authorship of Vestiges of the Natural History of Creation* (Chicago: University of Chicago Press, 2003); Ron, "Developing the Country," 97, 111.

119. Judge [Henry F.] French, "The Progression of Primaries: Analysis of Soil and Fertilizers," *New England Farmer* 10, no. 9 (1859): 410–12, 410.

120. "Agricultural: The 'Prof' Done Over," *New York Observer and Chronicle* 38, no. 1 (1860): 8.

121. French, "The Progression of Primaries," 410.

122. S. F. Halsey's Steam Mills, "Prepared Guano at One Cent per Pound," *Farmer and Mechanic* (1847) unpaginated advertisements.

123. "Trouble among the Fraternity," 341. "Artificial Manures," *GFGJ* 14, no. 7 (1853): 205.

124. Purina Mills (Land O'Lakes), "SteakMaker Feeds," https://www.purinamills.com/cattle/products/steakmaker-feeds; "SteakMaker® 40–20" (accessed August 17, 2017); https://s3.amazonaws.com/media.agricharts.com/sites/658/Feed%20Docs%20&%20Pictures/SteakMaker%20Grower%2038-0_1690210.pdf. Nicole Welk-Joerger's fascinating dissertation in progress, "Feeding Others to Feed Ourselves: Animal Nutrition and the Politics of Health from 1900" (expected date, May 2020), will address the forms of bodily knowledge behind such products.

Epilogue

1. E. Lewis Sturtevant, "Report of the Board of Control of the York [sic] Experiment Station," *Documents of the Assembly of the State of New York*, no. 98 (1883): 9. For the significance and origin of the New York State Experiment Station, see Marcus, *Agricultural Science and the Quest for Legitimacy*. See also Alan I. Marcus, "The Wisdom of the Body Politic: The Changing Nature of Publicly Sponsored American Agricultural Research since the 1830s," *AH* 62, no. 2 (1988): 4–26.

2. Sturtevant, "Report," 8.

3. Sturtevant, 14; Ulysses P. Hedrick, "Edward Lewis Sturtevant," "Sturtevant's Notes on Edible Plants," *State of New York Department of Agriculture Twenty-Seventh Annual Report*, vol. 2, pt. 2 (Albany, NY: J. B. Lyon Company), 4.

4. Sturtevant, "Report," 9–14.

5. Minella, "A Pattern for Improvement."

6. Marcus, *Quest for Legitimacy*, 86.

7. Sturtevant, "Report," 26.

8. Sturtevant, 25.

9. Sturtevant, 13.

10. Hedrick, "Edward Lewis Sturtevant," 10–11.

11. Alan Marcus, "Introduction," in *Science as Service: Establishing and Reformulating American Land-Grant Universities, 1865–1930*, ed. Alan Marcus (Tuscaloosa: University of Alabama Press, 2015).

12. Sturtevant, "Report," 14; Albert B. Costa, "Stephen Moulton Babcock," *Dictionary of American Biography*, 1999.

13. Marcus, *Quest for Legitimacy*, 189; Charles Rosenberg, *No Other Gods*, 153–72.

14. Sturtevant, "Report," 12.

15. Sturtevant, 2.

16. Sturtevant, 11.

17. Sturtevant, 11.

18. Marcus, *Quest for Legitimacy*.

19. Kendra Smith-Howard, *Pure and Modern Milk: An Environmental History since 1900* (Oxford: Oxford University Press, 2017), 86.

20. Marcus, *Quest for Legitimacy*, 93–95.

21. Bruce E. Seely, "Colman, Norman J.," *American National Biography*, 1999.

22. "Grover Cleveland," in *Biographical Record and Portrait Album of Tippecanoe County, Indiana* (Chicago: Lewis Publishing Co., 1888), 117–19.

23. Marcus, *Quest for Legitimacy*.

24. As the list reveals, most of the registering experts have academic posts in land grants. American Pomological Society, "Fruit and Nut Cultivars," http://americanpomological.org/fruit-and-nut-cultivars/ (accessed, May 10, 2018).

25. William A. Broad, *Soil Survey of Greene County, New York*, Sheet Number 9, National Cooperative Soil Survey, 1983; Alex Checkovich, "Mapping the American Way: Geographical Knowledge and the Development of the United States, 1890–1895," PhD diss. (University of Pennsylvania, 2004).

26. Brett Walker, *The Lost Wolves of Japan* (Seattle: University of Washington Press, 2009), 129–57; Donald Roden, "In Search of the Real Horace Capron: An Historiographical Perspective on Japanese-American Relations," *Pacific Historical Review* 55, no. 4 (1986): 549–75; Deborah Fitzgerald, "Blinded by Technology: American Agriculture in the Soviet Union, 1928–1932," *AH* 70, no. 3 (1996): 459–86.

27. Ron, *Summoning the State*.

28. Emmons, *Agriculture of New York*, vol. 1, 349.

29. USDA, "2017 State Agricultural Overview," nass.usda.gov, https://www.nass.usda.gov/Quick_Stats/Ag_Overview/ (accessed May 10, 2018).

30. New York Corn and Soybean Growers Association, "New York Corn & Soybean Yield Contest" (accessed May 17, 2018); https://nycornsoy.org/yield-contests/; New York Corn and

Soybean Growers Association, "New York Corn & Soybean Yield Contest 2017 Results," https://nycornsoy.org/wp-content/uploads/2018/02/2017-Yield-Contest-Results-Double-Sided.pdf (accessed May 17, 2018).

31. Beth Musgrove, "Convention News," "2019 Commodity Classic Trade Show Owens to New Exhibitors May 15," posted May 9, 2018, http://www.commodityclassic.com/for-media/convention-news (accessed May 17, 2018); https://onlinebanking.giffordbank.com/Pages/plotdata.html.

32. Jerome Rodale, *The Organic Front* (Emmaus, PA: Rodale Press, 1948), 23.

33. Rodale, 10.

34. Rodale, 24; Joseph Heckman, "A History of Organic Farming: Transitions from Sir Albert Howard's War in the Soil to USDA National Organic Program," *Renewable Agriculture and Food Systems* 21, no. 3 (2006): 143–50; Christian Feller et al., "Soil Fertility Concepts over the Past Two Centuries: The Importance Attributed to Soil Organic Matter in Developed and Developing Countries," *Archives of Agronomy and Soil Science* 58, no. sup. 1 (2012): S3–S21.

35. Rodale, *Organic Front*, 24; Andrew N. Case, *The Organic Profit: Rodale and the Making of Marketplace Environmentalism* (Seattle: University of Washington Press, 2018); Richard R. Harwood, "A History of Sustainable Agriculture," *Sustainable Agricultural Systems*, ed. Clive A. Edwards et al. (Boca Raton, FL: St. Lucie Press, 1990), 3–19.

36. Case, *Organic Profit*, 44, 50, 49.

37. On reflection, I find that my thinking about belief owes a great deal to Terry Pratchett, *Small Gods* (New York: Harper, 2013).

38. Bill O'Driscoll, "For the Love of the Apples: Movement to Save Heirloom Apples—and Protect Food Diversity," *E: The Environmental Magazine* 22, no. 2 (2011): 12–14; Annie Raver, "He Keeps Ancient Apples Fresh and Crisp," *New York Times*, March 2, 2011. For cultural analysis, see Jennifer A. Jordan, *Edible Memory: The Lure of Heirloom Tomatoes and Other Forgotten Foods* (Chicago: University of Chicago Press, 2015).

39. For "singularization" in the life of a commodity, see Igor Kopytoff, "The Cultural Biography of Things: Commoditization as Process," in *The Social Life of Things: Commodities in Cultural Perspective*, ed. Arjun Appadurai (Cambridge: Cambridge University Press, 1986), 64–91.

40. Mark R. Finlay, "The German Agricultural Experiment Stations and the Beginnings of American Agricultural Research," *AH* 62, no. 2 (1988): 41–50; Rossiter, *Emergence of Agricultural Science*; Mark R. Finlay, "Transnational Exchanges of Agricultural Scientific Thought from the Morrill Act through the Hatch Act," in Marcus, *Science as Service*, 33–60.

41. Fullilove, *Profit of the Earth*, 99–123.

42. Seedshed, "Native American Seed Sanctuary: Preservation, Celebration, and Rematriation," seedshed.org https://seedshed.org/native-american-seed-sanctuary/.

43. Though often they confuse groups—this site posts a Wampanoag-inspired image below a text about the Haudenosaunee. https://www.farmproject.org/blog/2016/3/31/growing-native-american-heritage-the-three-sisters.

44. Ann Vileisis, *Kitchen Literacy: How We Lost Knowledge of Where Food Comes from and Why We Need to Get It Back* (Washington, DC: Island Press, 2008), 62–63; Cronon, *Nature's Metropolis*, 131–34.

45. This is not an innate quality of fruit markets; see Shane Hamilton, "Cold Capitalism: The Political Ecology of Frozen Concentrated Orange Juice," *AH* 77, no. 4 (2003): 557–81.

46. Elyssa Goldberg, "Turning over a New Leaf: Why Southern Farmers Are Cashing in

on Kale," September 9, 2015, bonappetit.com, https://www.bonappetit.com/entertaining-style/trends-news/article/kale-farming-in-the-south (accessed May 17, 2018).

47. Shawn Raymundo, "Mexican Farmers on Alert after Rise in Lime Theft; Prices Have Doubled in the Last Year," March 29, 2014, latinpost.com, http://www.latinpost.com/articles/9738/20140329/mexicos-green-gold-growers-guard-limes-prices-rise.htm; https://cen.acs.org/articles/95/i6/ (accessed May 17, 2018).

48. Recent articles cast Greek yogurt whey as a potential source of animal feed and ethanol, "an untapped goldmine." Britt E. Erickson, "Acid Whey: Is the Waste Product an Untapped Goldmine?," *C&EN* 95, no. 6 (February 6, 2017): 22–30, Acid-whey-waste-product-untapped.html; Geoff Herbert, "NY Dairy Farmers Struggling as Milk Prices Hit 10-Year Low," March 8, 2018, syracuse.com (accessed May 20, 2018).

49. Sasha Khokha, "Cattle Ranchers Lock Horns with Almond Investors," May 8, 2015, KQED.org, https://www.kqed.org/news/10518234/cattle-ranchers-lock-horns-with-almond-investors (accessed, May 20, 2018). For the new Australian version of this boom, see Lucy Craymer, "Harvard's Nutty Idea: Cracking into the Almond Market," March 18, 2018, wsj.com, https://www.wsj.com/articles/health-nuts-pensions-and-endowments-scoop-up-almonds-1521374401 (accessed May 20, 2018); Heather Davis, "Almonds: Harvesting Value beyond the Farm," August 7, 2013, tiaa.org, https://www.tiaa.org/public/assetmanagement/insights/commentary-perspectives/perspectives/almonds-harvesting-value (accessed May 20, 2018).

50. Forgive this diversion into the present. Other kinds of historians of the future often have the advantage that their readers have a sense of which futures have come to pass and which have not, without being told. Videophones exist; hovering skateboards, not as yet. But the depopulation of the American countryside renders modern agricultural development and the traces of past agriculture essentially invisible to most people.

51. SweeTango Apples, "The SweeTango Story," https://sweetango.com/about/our-story/ (accessed, May 15, 2019); S. K. Brown and K. E. Maloney, "Making Sense of the New Apple Varieties, Trademarks and Clubs: Current Status," *New York Fruit Quarterly* 17, no. 3 (2009): 9–12.

52. USDA, "America's Diverse Family Farms, 2017 Edition," Economic Research Service, Economic Information Bulletin no. 185, December 2017, 2–5. For corporate family farming, see Seth Holmes, *Fresh Fruit, Broken Bodies: Migrant Farmworkers in the United States* (Berkeley: University of California Press, 2013), 48–54; for tax law and hobby farming, see Carol Ekarius, *The Essential Guide to Hobby Farming: The How-To Manual for Creating a Hobby Farm* (Irvine, CA: i5 Publishing, 2015), 19.

53. For critiques of the family farm, see Rosenberg, *The 4-H Harvest*; Margaret Gray, *Labor and the Locavore: The Making of a Comprehensive Food Ethic* (Berkeley: University of California Press, 2013); Rebecca L. Schewe, and Bernadette White, "Who Works Here? Contingent Labor, Nonfamily Labor, and Immigrant Labor on US Dairy Farms," *Social Currents* 4, no. 5 (2017): 429–47; Micah P. Johnson, "Cultivating Injustice in the Richest Fields: How US Labor Law Leaves Migrant Child Workers Unprotected on the American Farm," *Drake Journal of Agricultural Law* 22 (2017): 359; agricultural scientist Sarah Taber consistently produces some of the most thought-provoking critiques of these myths on her Twitter feed. Like many, I eagerly await her book.

Index

Adams, John Quincy, 103, 113
adaptation, 131, 134; and animals, 152; and butter, 142, 144; and chemistry, 201; and climate, 138; and divine intention, 135–37, 190; justifying slavery, 137; justifying white settlement, 137; and landscape, 135–40; maps of, 149
Adriance, John, 50
adulteration, 17, 216; city milk, 212
African Americans: Black abolitionists, 9; and appropriated knowledge, 58, 73; free Black farmers, 57; laborers, exclusion from credit, 57, 59. *See also* Brown, James F.; Humphrey, Moses; Northup, Anne; Northup, Solomon; slavery
agrarian reform, 79
agribusiness, changing tastes, 227
agricultural accounting, 18, 25, 78–79, 200; with animals and plants, 199; chemical tables, 199; eating, as measurable transaction, 198; as experimental writing, 75–77; witnessing, as form of, 77
Agricultural and Horticultural Society of India, 137
Agricultural Book-Keeping (Cochran), 199
agricultural capitalism, 37, 227, 242n27; features of, 25
agricultural chemistry, 192, 203, 206; accounting practices, 207; ash analyses, 208; consulting chemists, 207; and nitrogen, 207–8; and smell, 204–5
agricultural experimentation, 63–64, 79; borrowings of, 65; and capitalism, 66; experiment stations, 220; field trials, legacy of, 79–80; product testing, 261n4; social novelty, evoking of, 65–66; trial culture, roots in, 65, 80; as visible luxury, 77
agricultural experiment stations, 221

agricultural fairs, 13, 37–38, 43, 59; as public entertainment, 60
agricultural giants, 2–5, 10, 11, 19, 39, 116, 119, 153, 165, 222
agricultural improvement, 5, 11–12, 18, 33, 35–36, 60, 89, 134, 204, 229; and chemistry, 196; commercial goods, turn to, 82; diversity of, growth in, 38; expansion of, 13–14, 38; future, at root of, 129, 219; industrial agriculture, 223; landscape of, 9–10; machinery, 82; middling farmers, 49–50; popularity of, 39; products, centering on, 97–98; scientific knowledge, 6. *See also* machinery, principles of; machinery makers
Agricultural Institute, 215
agricultural journals, 5–6, 13, 25, 26, 32–33, 37–39, 43, 47, 50, 73, 82, 98, 108, 140, 221, 224; attacks on, 220; book farmers, 7; experimental farms, advocating for, 69; and experimentation, 71; middling farmers, targets of, 48–49; mulberry trees, advertising of, 113; physical labor, 56; refinement, instruction in, 53–54; refinement, mixed feelings over, 54–55
agricultural knowledge, 45, 227; and genealogy, 225–26
agricultural landscape: social structures of, 19; as unstable, 19
agricultural machinery, 46, 135; progress, racialized notion of, 92–93; public testing of, 82–83; reaper test, 83; trials of, 81, 83; trials of, as public entertainment, 65
agricultural press, 190
agricultural prodigies, 222
agricultural revolution, 11, 25; and capitalism, 10
agricultural schools, 56
agricultural science, 18, 37, 206, 222; experiment stations, 220; organic chemistry, 190

agricultural societies, 5–7, 10, 12–13, 37–38, 54, 69, 71, 128–29, 221, 241–42n21; British counterparts, modeling on, 11; experimental farms, advocating for, 69; landlords, centrality to, 35; museums of, 91–93; state, functions of, 222; urbanites, visible force of, 43
agricultural writing, 68–69
agriculture, agriculturist: biological innovations, 162; manias of, 104; as term, 45; varieties, as different entities, 162
Agriculture of New York (Emmons), 155–56, 190
Alabama, 113
Albany, NY, 13, 37, 43, 98, 113, 195–96, 206
Albany Freeholder (journal), 33–36, 43–44, 79, 140, 145, 201
alfalfa, 191
Algonquian people, 8, 12, 74, 137, 226–27
Allaire, Jasper, 66
Allen, A. B., 89
Allen, Jeremiah, 35
Allen, Lewis Falley, 179, 222
almonds, as investment crop, 228
Amenia, NY, 51–52
American Agricultural Association of New York, 209
American Agriculturist (journal), 49, 56, 89, 210
American Farmer (journal), 109
American Guano Company, 211
American Institute, 2, 46, 88, 109–10; Farmers' Club, 215, 221
American manufacturing, agricultural implements, 46
American Museum, 1–2
American Pomological Society, 222, 279n90
American Revolution, 23, 26, 107, 168, 170
American Soybean Association, 223
"Analysis of the Oat" (Norton), 207
analytic tables, 190, 207, 220, 224; as expressions of indebtedness, 199; limitations of, 207
antebellum period, 5–7, 16, 26, 44–45, 51, 65, 68, 95–96, 126, 137, 216–19, 222, 224–25, 227, 229, 245n55; accountable landscape, as transparent, 200; adulterants, worry over, 212; agricultural monsters, 2, 4; agricultural science, 19, 71, 79–80; Northern vs. Southern farmers, 40; postmillennial evangelicalism, 9, 135; and silk, 105; speculative booms and busts of, 104; value, nature of, 17–18
Anti-Rent Committees, 33–34
Anti-Renter (journal), 34
Anti-Rent Party, 34, 201
Anti-Rent movement, 13, 23, 33–34, 36–38, 79, 140, 203; legitimacy, seeking of, 35
Anti-Rent wars, 41, 202
apples, 63, 97, 169; apple detectives, 225; apple orchards, 19, 168; apple seeds, 164; new patent varieties, 228; Northern Spy, 173–74, 182–83; Pink Lady, 225; Red Astrachan, 185–86; Red Delicious, 225
Armenia, 107–8
Armstead, Myra B. Young, 58
Assam, 10–11, 137
Association of Equipment Manufacturers, 223
Atkins' automaton, 81–86, 94–95, 97
Auburn, NY, 92
Auburn State Prison, 66
Aurora, NY, 138
Australia, 5, 10, 28–29
Austria, 148
authenticity, 47–48, 54, 60, 173; labor, public performances of, 55–56
automata. *See* robots

Babcock, Stephen Moulton, 220; butterfat tester, 221
Backus, Culver, 49
Bacon's Rebellion, 107
Badger, Harvey, 76–77
Bailey, Jacob, 147
Bakewell, Robert, 25
Ball, J. W., 142
Baltimore, MD, 210
Bancroft, George, 39–41, 48–49, 59
Bank of England, 113, 114
Banks, Joseph, 27–28
banks and banking, 42, 44, 49; and bank notes, 133–34, 163, 175; as metaphor for soil, 196, 200, 202, 207; and volatility, 17, 45, 66, 104, 118, 125
Bank War of 1834, 200. *See also* Bank Wars
Bank Wars, 66, 103, 113–14, 128
Barclay, Sonya Marie, 254n65
Barnum, P. T., 1, 215
Barry, Patrick, 182, 184–85, 187, 221
Bartlett, Edwin, 210
Beauchamp, Christopher, 96
Beckert, Jens, 15, 245n58
Beecher, Henry Ward, 167
Belgium, 170
Benson, Alfred G., 211
Berkeley, William, 107
Berlin, NY, 11
Berzelius, Jacob, 198
Biddle, Nicholas, 103
Bloomfield, Joseph E., 142–43
Board of Agriculture, 29
Bolivia, 227
book farming, 7, 41, 47, 53, 70, 97
Book of Fruits (Manning), 171
Boston, MA, 103, 183
Bostwick, Salmon, 48, 76
botany, botanical networks, 27, 106–7, 111–12, 129, 170–71, 187, 224; compared with horticulture,

INDEX   291

165; and global trade, 11, 12; and refinement, 163, 172
Boyle, Robert, 73
Brazil, 148
Brewster, Jarvis W., 67–70, 72–74, 78, 81, 94, 98, 117, 151
British Board of Agriculture, 27
British Empire: mass displacements in, 26; role of agricultural improvement in, 10–11, 137
British Geological Survey, 148
British West Indies, 224
Broome County, 143
Brown, James F., 58
Bruce, Charles, 137
Bruegel, Martin, 51, 140
bubbles: differences between, 126; epistemic crisis, 264n4
Bucktails, 28
Buel, Jesse, 43, 46–48, 78, 123, 139, 196, 202
Buffalo, NY, 182, 222
butter, 141, 150, 154, 227; and adaptation, 142, 144; butter districts, 16; butter landscape, as physical place, 142; butter money, 133–34; butter production, and women, 140, 142–43, 151, 155–56; and experimentation, 152–53; and grazing, 146–47

Cahours, Auguste André Thomas, 198–99
California, 130, 228
Canada, 175, 180
Cannadine, David, 25
Cape Colony, 5
capitalism, 19, 25, 95–96; active storytelling, as central to, 15; agricultural experiments, 66; agricultural revolution, 10; failure, as standard feature of, 16; paper money, 17, 125; rural, 239–40n15; science, relationship to, 14–15, 17; urban, 9, 18
capital of nature, 210, 213
Caribbean, 25
Carolinas, 51, 113
Carr, Isaac, 139
Cato, 195
Catskills, 131–33, 137, 141–42, 146, 149, 153, 155–57
cattle, 2, 4, 10, 12, 14, 25, 45, 49, 52, 59, 68, 75, 77–78, 85, 87, 91, 105, 128, 130, 133, 137, 139–40, 144, 146, 152, 154, 157, 164, 166, 176, 186, 191–92, 198, 208–9, 211–12, 215–17, 222, 226; cattle shows, 27, 39; lactometer, 141
Cayuga people, 166
Central New York Farmers' Club, 221
Chamberlin, Calvin, 51
Chambers, Robert, 215
Chandler, Lyman, 145
Chapin, Heman, 50
Chapin, Oliver, 182–83

Chapman, John, 166
Chappell, P. S., 213
Chappell, William H., 213
Chaptal, Jean-Antoine, 194, 204
Chartist movement, 33–34, 79
chemistry, 201, 212, 281n2; chemical tables, and accounting practices, 199; eating, as measurable transaction, 198; nourishment, foods and soils, value to, 199; value, assigning of, 284n105
Chemung County, 58, 136, 141, 143–44, 152–54
Chicago, IL, 227
Chile, 5, 124
Chilton, James, 208, 212
China, 10, 103, 105, 112, 118, 137, 141, 148
Chiswick Garden, 170–71
cholera, 44
chorography, 72
Christian Philosopher (Dick), 149–50
Christy, David, 198
Cincinnati, OH, 113, 168, 172
Cincinnatus, 44
Civil War, 5–7, 58, 69, 132, 154, 215, 217, 220, 284n105
Clarke, George, 32
Clarke, Jonathan, 118–19
Clermont (steamboat), 29
Cleveland, Grover, 222
climate change, 138, 229
Clinton, DeWitt, 28, 112, 274n103
Cobb, Jonathan, 118
Cobbett, William, 34, 45
Cochran, William D., 199
Cohen, Joanna, 268n94
Coke of Norfolk, 25
Cole, Thomas, 9, 147
Colman, Norman Jay, 221–22
Columbia County, 36, 56, 194
Columella, 195
commodification, 18
Commodity Classic, 223
confidence men, 17, 41
Connecticut, 108, 113–14, 123
Connecticut Agricultural Experiment Station, 224
Connecticut State Agricultural Society, 213
conspicuous consumption, 43; as experiment itself, 72; as foil for experiment, 73
conspicuous production: definition, 55; in machine trials, 95; rhetorical usefulness of death in, 55; role in experimental writing, 72
consumer culture, 163
Coons, Herman, 7, 56–57, 144
Cooper, James Fenimore, 9
Cooper, William, 26–27
Cooter, Roger, 251n73
Copeland, J. S., 56
Cornell Experiment Station, 222–23

Corning, Erastus, 48
Corn Laws, 8, 25, 84
Cortland, NY, 92, 274n113
Cortland County, 154
Cortlandts, 11
cotton, 109, 128, 162
counterfeiting, 150, 182; in butter, 150; counterfeit detectors and skilled sight, 175; in fruit, 174; role in changing notions of value, 128
Country Gentleman (journal), 221, 224
country living: gentlemanly identity of, 44; as republican virtue, 44. *See also* domesticity
countryside: country and town, symbolic lines between, 48; depopulation of, 226, 288n50
county societies, 32, 39
Course of Empire, The (Cole), 9
Course of Experimental Agriculture (Young), 68, 74
Crimean War, 8
Crystal Palace, 86
Cullen, William, 192
Cultivator (journal), 1, 5–7, 11, 32, 35, 43, 49, 56, 67, 70, 75, 89, 122–23, 196, 205, 213, 224; "Design for a Genteel Farm House," 53–54

dairying, 140; dairy farms, 19, 227–28; expansion of, 146
Dana, Samuel L., 190, 208
Darlington, William, 145, 169
Darwin, Charles, 64, 276n12
Davies, William Augustus, 49
Davy, Humphry, 190, 192–94, 197, 210; analytic methods of, 203
De Candolle, Augustin, 71, 194
Delafield, John, 44
Delaware, 113
Delaware County, 150, 156
Democratic Party, 253n35
Dennis, Jonathan, 118, 120
Deringer, William, 264n4
Derrickson, George, 76
"Deserted Village, The" (Goldsmith), 34
Devoe, Thomas, 154
Devyr, Thomas Ainge, 34, 79
De Witt, Simeon, 27
Dibble, Abraham, 49
Dick, Thomas, 149
Dickens, Charles, 279n83
differentiation, 136
Dix, John A., 55
Dobbs, J., 87
domesticity: as feminizing, 44; as performance, 54; sentimental literature of, 44. *See also* country living
dot-com booms, 126
Doubleday, Lester, 168

Downing, Andrew Jackson, 42, 58, 164, 167–71, 177, 179, 182, 204
Dumas, Jean-Baptiste, 197–99
Durham Ox, 11
Dutchess County, 36, 49, 56, 194; inequality in, 51–52; slaveholding families in, 51; wealth of, 51–52
Dutchess County Agricultural Society, 49
Dutch Republic, 10
Dyer, Oliver, 179

Eames, Moses, 54, 174
Eastern Woodlands, 8
Eaton, Amos, 29–30, 138, 149, 192, 281n2
economic practices: and accounting, 18; consumer desire, creation of, 18; management, 18
"Economy" (Thoreau), 67–68, 72–73
Edinburgh (Scotland), 92
Ejectment Acts, 25
Elder, Walter, 164–65
Ellwanger, George, 187, 221
Ellwanger and Barry, 168
Elmira Farmers' Club, 221
Emery, Horace, 89, 91
Emery and Sons, 208–9
Emmons, Ebenezer, 137–39, 155–56, 172, 177, 190, 192, 206–7, 209, 217; trope of knowing clerk, 175
enclosure movement, 24–25
Endicott and Co., 133
England, 34, 106–7, 118, 256n97. *See also* Great Britain
Erie Canal, 5, 7, 25, 27, 29, 130, 135, 147–48, 194, 201; rural landscapes, shifts in, 8
Essay on the Management of Small Farms (O'Connor), 79
Europe, 65–66, 107, 111–12, 125, 144, 163, 190, 194, 197, 261n4
Evans, George, 34
experimentation, 64, 98; accounting system, 75; decentralized knowledge, 71; experience, link with, 65; experimental agriculture, 68; experimental culture, and knowledge, 73–74; experimental farms, 71; experiment stations, 220, 222, 226; experimental trials, and witnessing, 97; experimental writing, 67–68, 77; invisible technicians, erasure of, 73; and labor, 72; patterns of, 69; profit, centrality of, 74–75; as prolix, 69–70; punctuality, as virtue, 72; urbanites and landlords, as task of, 78–79; witnesses of, 70. *See also* science
Exposition Universelle, 96
Eyck, Jacob Ten, 48

Familiar Letters on Chemistry (Liebig), 199
family farm: romantic idea of, 228–29; USDA definition of, 228

Faraday, Michael, 64
Farmer and Mechanic (journal), 215
farmers, 59, 138; American distinctiveness, symbol of, 40; authenticity, questions of, 47–48, 54–55, 60; book farmers, 47; butter production, 141; definition of, as in flux, 6; divinity of, 40; "Farmers' Creed," 202; gentility, evoking of, 60; gentlemanly identity of, 46; idea of, as political performance, 42; images of, 16, 41; imagined past of, 228; multiple farming identities of, 41–42, 60; nostalgia, for countryside, 41; "real farmers," 41; republican virtues of, 60; rural virtues, 40–41; terms for, 41; virtues of, 39–41; working class, 48. *See also* middling farmers
Farmers' Club of the American Institute, 211
farmers' clubs, 6–7
farmer societies, 27–28
farming, 191, 207; and chemistry, 192; eating, as measurable transaction, 198; farm labor, and authenticity, 195; farm nutrition, 195; fertility, as money, 198; green manures, 191; mercantile failure, high rates of, 45; soil exhaustion, 195
farms: accounting practices, and organic analysis, 191; as complex assemblages, 6; idealization of, 19; myths associated with, 19; nostalgic mist surrounding, 19; urbanites, presence of, 42–43
fertilizer, 14, 32, 50–51, 91, 97–98, 139, 191, 195–97, 208–10, 212, 215–16; and cultivators, 93; fraud in, 213–14; NPK formula, 224
flockmasters, as term, 45
Florida, 113–14, 277n23
Floy, Michael, 173
Flushing, NY, 111–12, 116
food riots, 45
Foote, Eunice, 31
Foster, Herman Ten Eyck, 52–53, 55, 58, 72
Fourierists, 9
Fowler, John, 42
Fowler, Orson, 9
France, 28–29, 32, 103, 106–8, 111–12, 193
Franklin, Benjamin, 193
fraud, 142, 174, 187, 191, 215; chemical, 212; in fertilizer industry, 213–14
Free Soil Party, 13, 37, 201
free trade, justifications for, 136
French, Henry F., 215
French Guiana, 111
Frost, E. C., 143–44
fruit: consumer taste, 184; and flavor, 177; fruit culture, 161–62, 171; fruit landscape, 163; fruit profiles, and confidence men, 177; fruit seeds, 164; good fruit gardens, 170; identities, as fixed, 181; metropolitan market for, 183–84; naming practices, 165–66, 169–70, 173, 180, 187; reputations, creation of, 181–83; sexual imagery, 169. *See also individual fruits, varieties*
Fruit Culturist, The (Thomas), 169
fruit grafting, 166–69; public good, 184, 186
fruit profiles, 177
fruit trees, 162, 165–68, 171; as consumer good, 176
fruit varieties, 162; valuation of, 188, 190
Fullilove, Courtney, 226
Fulton, Robert, 29
Fussell, G. E., 257n9
future predictions: Anti-Renter predictions, 35–37, 79; bubble forecasting, 123; centrality to improvement, 4–5; criticism of profit predictions, 50; declensionism and progressivism, 9; investor predictions, 45–47; landlords' predictions and speculation, 24, 27–29, 78; multiple parallel predictions, 9, 13, 18; predictions of growth, 109, 119, 123; predictive sciences, 15; rapid succession of, 5; reproductive calculations, 119–20; retrospective characterization, 16; role in definition of value, 17; role of foreign models, 10, 29, 105; value of failure in studying, 139. *See also* natural theology
futurism, 229

Gage, Matilda, 58
Garbutt, William, 194
Genesee Farmer (journal), 32, 54–55, 78, 187, 214, 216
Genesee Valley, 131, 138–39, 209
Geneva, NY, 81, 83
Geneva trials, 84–87, 93–95
geology, 146, 192, 203; agricultural improvement, links to, 148; Champlain Group, 148; Erie Group, 148; as fashionable, 147; Helderberg Series, 148; land values, 147; mapping of, 190; place names, 148; and railroad cuttings, 147; shifting soils, 149–50
George, Samuel K., 210
George III, 28
Georgia, 107, 113–14
Georgics (Virgil), 145
Germanic states, 105–6
Germantown, PA, 114–15
Germany, 10, 28, 68–69
Glasser, Joseph, 65
global trade, 11, 130, 210
gluten, 209, 212
Godey's (magazine), 14
Gold, Thomas, 47
Goldsmith, Oliver, 34
Gould, Jay, 48
grafting, 165, 166; grafting season, 174
Grahamites, 9
Gray, Asa, 147

grazing, 146–47
Great Britain, 5, 8, 12, 27–30, 44, 45, 60, 64, 68–69, 80–81, 94, 108, 131, 170–71, 177, 195, 210–11; agrarian revolt in, 33; agricultural improvement, as international center of, 10; agricultural revolution of, 11; foddering, 191; internal colonialism, 10; Irish migrant workers, reliance on, 83–84; laboring classes, 58, 79; landed elites, 25–26; landed power, rise of, 24–26; machinery trials in, 86–87; ploughing competitions, 261n17; reapers, importance of to, 83–84, 93; rural wars, 34; and silk, 105–6; stratigraphic map of, 274n105; tenants, eviction of, 25. *See also* England; Scotland; Wales
Great Exhibition (1851), 84–85; reaper test at, 83
Greek yogurt, acidic whey, as pollution, 227
Greeley, Horace, 215
Greene County, 144–46, 150–51, 154–56
Greene County Agricultural Society, 134, 136–37
Greene County Fair, 139, 142, 152
guano, 69, 209–13, 216; artificial guanos, 212, 214, 224
Guano Islands Act, 211, 215
Gustav III, 106
gypsum, 193–94

Hahn, Barbara, 280n117
Hall, Francis, 144
Hall, James, 147
Halsey, Jehiel, 47, 71
Halsey, S. F., 216
Handy, Abel, 87
Hanson, Joseph, 76
Harris, Ira, 36
Harris, Joseph, 212
Harrison, William G., 120
Harrison, William Henry, 47
Harwood, Jonathan, 261n4
Hatch Act, 221–22
Haudenosaunee people, 4, 8, 11–12, 23, 26, 131, 137–38, 166, 226–27; as sources of appropriated knowledge, 73, 137–38
Hawley, Gideon, 133, 142–43, 151
Hawthorne, Nathaniel, 7–8
Hazely, William, 58–59
hemlock, 132–33, 139, 144–46, 151, 153
Henke, Christopher, 79–80
Henkin, David, 175
Henry, Joseph, 211
Henry, Mrs. G. W., 59
Henry IV, 105
Herkimer County, 8
"Hints on Describing Fruit" (Thomas), 161, 176
Hitchcock, Edward, 149
Holbert, John, 58, 143, 152–53
Holbrook, Frederick, 88

Holland, 126
Holland Land Company, 136
Holland Purchase, 145
Homestead Act, 37
hops, 8, 19, 32, 128, 129, 131
Horace, 43–44
Horticulturist (journal), 58
horticulture, naming practices, 170
Household Words (Dickens), 279n83
Hovey, Charles, 184
Howard, Albert, 224, 226
Hudson Valley, 11, 26, 29, 36, 42, 51, 57, 130–31, 144, 166, 168, 183, 195, 242n27
Hull, Hezekiah, 11
Humboldt, Alexander von, 210
Hume, David, 136
Humphrey, Moses, 57, 59
Hurricane Irene, 269n5
Hussey, Obed, 84
Huston, Reeve, 37

Illinois, 46, 85, 103, 130–31
improvement, 6, 42, 58, 188, 192–93, 195–96, 217, 222, 225–26, 251n73, 281n2; accounting systems, 76–77, 199; adapted landscape, 136–37; agricultural capitalism, 227; agricultural regions, boundaries of, 71; agricultural transmutations, 191; book learning, displaying of, 72; and butter, 142; chemistry, rise of, 190–91; and experimentation, 70, 78; fruit culture, 162; as futuristic, 15, 229; as global phenomenon, 5; as hobbyists, 48; illusion of, 228; institutions, rural differences, 53; and landscapes, 163; and *multicaulis*, 116; new inventions, 25; organic analysis, 191; place, effect of, 70–71; politics, future in, 46; practices of, as expensive, 50–51; profit, aim of, 18; value determination, concern with, 16–17; vitalist traditions, 223–24; warehouses, and knowledge exchange, 92–93; white settlement, 137–38
India, 25, 141, 224
Indiana, 113, 242n27
Industrial Revolution, 25. *See also* agricultural machinery; machinery, principles of; machinery makers
Ingalls, William, 222–23
intellectual property rights, 228
intercropping, 73–74
Ireland, 12, 25–26, 84, 106, 109; famine in, 34
Iroquois. *See* Haudenosaunee people
Italian states, 107–8
Italy, 10, 12, 103

Jackson, Andrew, 66, 153
Jaffee, David, 167
James I, 106
Jamestown, VA, 106–7

INDEX

Japan, Meiji government, 222
Jasanoff, Sheila, 245n55
Jay, John, 59
Jefferson, Thomas, 112
Jefferson County, 27
Jensen, Joan, 140
Johnson, B. P., 83, 141–43, 146, 151, 154
Johnson, Daniel, 76
Johnson, Edward, 67
Johnson, Samuel W., 207, 213–16
Johnston, James F. W., 136, 190, 205
Journal of Commerce (journal), 114–15, 119, 123–25
Judson, Andrew, 113
Justinian (emperor), 105

kale, as golden leaf, 227
Kelly, Catherine, 54, 167
Kennicott, J. A., 180
Kenrick, William, 115, 125
Kentucky, 107, 113
Ketchum, William, 95
Khan, Zorina, 96
Kiechle, Melanie, 203–4
Kim, Sang-Hyun, 245n55
King, Irving P., 221
Knight, Thomas Andrew, 165, 170–71, 186
knowledge-making practices, 223; commercial network of, 82; knowledge makers, 98; state institutions, 18; and warehouses, 92. *See also* warehouses
knowledge movement, 81

labor: dignity of, 59; and experiments, 72; power of, 201; value, creating of, 201–2
laborers, 10, 11, 25, 33, 34, 41, 43, 51, 53, 55, 56, 57–58, 79, 83–84, 147, 209; concealed, 72–73, 226; farm laborers, 256n97
labor theory of value, 35–36
lactose, 227–28
land bubble, 45
land-grant colleges, 38, 220
land grants, 221
landlordism, 23–24, 26, 28, 32–33, 36–37; decline of, 38, 42; tenants, clashing with, 41
landlords, 26, 28, 32–33, 59–60, 104, 108, 166–67, 200, 251n73; agricultural societies, centrality to, 35; and experimentation, 78–79; manors, building of, 23–24, 42; profit-orientation of, 24–26; refinement, concern over, 53; status and space, 42; and tenants, 41, 80, 203
landscape, 5, 16, 18, 29, 35, 46, 51–52, 65, 77, 79, 87, 107–8, 125, 137, 140, 142, 146, 149, 152, 154, 156, 161–62, 171, 175, 188, 200–201, 204–5, 208, 217–18, 227; accounting practices, 198; agricultural, 7–8, 17, 19, 139, 190; apples, 186–87; British-style, 12–13, 167; capacities of, 131, 155;

capitalist, 25; and commerce, 136; commercial orchards, 168; cultivated, 135; economic, 129; as elusive, 138; of exploration, 14–15; and fruit, 166; functions of, 138; and herbicide, 226; as human-built, 4; idealized, 36; imperial, 12; and improvement, 163; of manufacturers, 88; mapping of, 15; of market development, 132; patented principles, 95; and progress, 97; providential intention of, 135; rural retirement, 43; as shifting, 130, 187; traditional, restorations of, 229; uncertainty, conditions of, 245n58; vanished, 132, 186; wild, 135
Last of the Mohicans, The (Cooper), 9
Leaves of Grass (Whitman), 189
Lee, Daniel, 199–201
Lee, Hannah Farnham Sawyer, 66
Lentilhon, Pauline, 52
Lewis, Andrew, 245–46n61
Liebig, Justus von, 190, 199–200, 204–6, 209, 211, 216–17, 220, 223–24; analytic methods of, 198; as anti-vitalist, 198; enthusiasm toward, 196–97; *Kaliapparat* apparatus, 197–98; and nitrogen, 207–8
limes, as "green gold," 227
Lincoln, Abraham, 96, 220
Lindley, John, 164, 173
Linnaean Garden, 111–12, 124–25, 172
Linnaean Society of Paris, 111–12
Linnaeus, Carl, 106, 111–12, 170
Little Ice Age, 138
Livingston, Robert, 27–29, 32, 59, 110
Livingstons, 11, 23, 26
Llewelling Nursery, 168
local food movement, 227
Lock Navigation Company, 26
Lodi Manufacturing, 195
London (England), 10
London Company, 106
London Crystal Palace, 46
London Horticultural Society (LHS), 170, 176
Long Island, 195, 201, 210
Lorain, John, 74
Loudon, John Claudius, 150
Louis XIV, 105
Louis XVI, 28
Lovisa Ulrika (queen), 106
loyalists, confiscation of lands, 23
lucerne. *See* alfalfa
Lucier, Paul, 245–46n61
Lyon-Jenness, Cheryl, 187

machinery, principles of, 95–97
machinery makers: commercial networks, 89, 91–93; implement makers, 88–89, 92; as improvers, 80; patents, 82, 96; plow construction, 87; reaper trials, 83–86; reaping machines, 81

machine trials, 80–81, 96–98; conspicuous production, opportunity for, 94; experimenting, phases of, 86–87; and international patent-licensing markets, 84, 95–96; performance, measurement of, 93–95; as persuasive events, 88–89; private machine trials, as form of advertising, 99–100; public competitions, 87; public demonstrations, 86–87; public trials, popularity of, 99–100; of reapers, 83–86
Maclure, William, 138, 148–49
MacNeven, William, 44
Maine, 113
Maine Cultivator (journal), 89, 92
Manchester, CT, 119
Manila (Philippines), 112
Manning, Ann B. Harned, 58, 98–99
Manning, Robert, 171
Manny, J. H., 86, 95
Manny and Co., 95–96, 98–99
Mapes, James J., 192, 208, 214–16, 224–25
Marcet, Jane, 192
Marcus, Alan, 222
Marshall, William, 151, 155
Maryland, 113
Massachusetts, 27, 113–14, 123
Mather, William, 146–47, 149–50
Matoaka, 106
Mayher and Co., 95
McCormick, Cyrus, 84, 87, 94–95; patent rights, 96
McCormick v. Talcott, 97
McGlincy, K. P., 134
McLallen, James, 58–59, 77
McMurry, Sally, 49, 140
Meacham, Thomas, 153
Mechi, John Joseph, 84
Menard, Russell, 12
Mennonites, 226
Mercantile Agency, 181
merino sheep, 28–29, 49, 131; mania over, 104, 110, 137, 144
Merino Society, 28
Merrill, L. L., 94
Mexican War, 8
Mexico, 118, 124, 201
miasmatic theory, 44, 70, 139, 203–4
Michaux, François André, 113
Michigan, 130–31, 168
Middlesex Company, 59
middling farmers, 50, 60, 80, 200; artificiality, deprecation of, 55; and education, 56; multiple farming identities, 42; personal exchanges, 76; physical labor, dignity of, 53, 55–57; rural refinement, expansion of, 53–54, 56–57; targets of, 48–49
Miller, Elnathan, 50

Miller, William, 9
Miner, T. C., 55
Mississippi River, 29
Mitchell, Samuel, 266n45
Mobile, AL, 103, 112
modernity, 5, 23–24, 33, 38, 110, 225
Mohawk Valley, 131, 138–39, 143, 195
Mohican people, 9
Monroe County, 50, 56, 79
Monsanto, 223
monsters, 1–5. *See also* agricultural giants
Montreal (Quebec), 112
Morrill Act, 220
Morus multicaulis, 103–4, 106, 111, 113–14, 124, 129, 200, 226; advertisements for, 120; arrival of, 108; bubble, end of, 125–26, 128; characteristics of, 112; failures of, 122; leaves of, 116, 118; multiplication of, 119–20; and silkworms, 116, 119; and speculators, 119–20, 122–23, 125–26, 128; speedy growth of, 112; US, arrival of in, 112. *See also* mulberry trees
mulberry farms, 45
mulberry trees, 15, 50, 106, 108–9, 110, 116, 119, 124–25, 128; advertising for, 113; auction sales, 122–23; characteristics of, 111; mania for, 16, 113–15; mulberry bubble, 103–4, 105; rise of, 103–5; speculation on, 104. *See also* silk culture/industry
museums, 1–2, 6, 86, 147, 219; agricultural, 13, 32, 148; displays of progress in, 93; warehouses as, 14, 82–83, 89–92

Napoleon, 29
Napoleonic Wars, 10, 25, 27–28
National Congress of American Fruit Growers, 179
National Corn Growers Association, 223
Native Americans, 195. *See also* Algonquian people; Haudenosaunee people; Mohican people; Oneida people; Seneca people; St. Regis Mohawk/Akwesasne Tribe
Native American Seed Sanctuary, 227
natural history, 163
natural knowledge: and commerce, 162; and variety, 162
natural law, 1, 65
natural theology, 97, 135, 149–50; benevolent design, 139; divine intention for profit, 15, 18, 131; equivalence of human and natural design, 97; natural law as God's law, 216. *See also* Paley, William
Nelson, David, 59, 60
Newark, NJ, 214
Newcastle (England), 34
New England, 107, 164, 166, 195, 209, 242n27, 259–60n62

INDEX

New Hampshire, 113
New Horse-Hoeing Husbandry (Tull), 201–2
New Jersey, 114, 143, 216
New World, 24
New York and Erie Railroad, 143–44, 154
New York Athenaeum, 44
New York Banking Company, 44
New York Board of Agriculture, 27
New York City, 8, 13, 113, 133–34, 153, 179, 181, 210; banking power, move to, 200; chemists in, 206; cholera epidemics in, 44; city milk, 212; Manhattan, 103, 111–12
New York Corn and Soybean Growers Association, 222–23
New York Crystal Palace, 46
New York Experiment Station, 219–20; as national model, 221
New York Farmer (journal), 32, 108
New York Horticultural Society, 113
New York State, 5, 16–17, 25, 29, 38, 45, 48, 55, 57, 81, 93, 110, 128, 130–32, 135, 137–38, 140–41, 143, 146, 150, 161–62, 170, 172, 180, 182, 192, 194, 206–7, 210–11, 229, 241–42n21, 242n27, 246n4, 274n103; accounting practices, 76–78, 200; agricultural capitalism, 37; agricultural improvement in, 10, 12, 14, 18, 23, 28, 32–33, 42, 46, 56, 59, 82, 73–74, 139, 215, 219–21, 226; anachronism, accusations of, 24; Anti-Renter voters, courting of, 34; Anti-Rent movement, 36–37; aristocratic landlords in, 23–24; British aristocrats, imitating of, 24; British-style improvement in, 11; as center of improvement, 7; chemistry, presence in, 192; and coal, 274n103; dairy farms, 227, 228; experimental farms, 71; farming, in state of flux, 9; Federalist elite in, 23; forests of, 144–45; geological survey of, 147–48, 155–57; growth of, 7–8; implement makers, 88; Indian land rights, 137; indigenous crops, growth of, 12; land, meaning of, 60; landlords in, 23–24, 26, 28, 32–33, 104, 108, 166–67, 200; land speculation, 26–27; manors in, 23–24, 42; market revolution in, 13; merino sheep, importing of, 28; mestizo agriculture, as site of, 12; midge, life cycle of, 164; New Englanders, wave of, 23; orchards in, 168; refinement, concern over, 53–54; silk production, 108; social structures of, 12, 41; soil exhaustion, 195–96; Southern Tier, 201; Southern Tier, as butter land, 154; speculation in, 42; tenants, power tactics of, 33; urbanites in, 42–43; wealth of, 9; white settlers, domination by, 9. *See also individual counties*
New York State Agricultural Society, 7, 13, 32, 39, 83–84, 104, 176, 179, 180, 205, 221–22
New York State Fair, 190
New York State Society, 221

New Zealand, 10
Niagara County, 58
Niblo's Garden, 13, 116
Nichols, D. A. A., 187
nitrogen, 189, 199, 217; as commodity, 213–14; controversies about, 207, 211, 216; as flesh-forming, 207, 209
North America, 112, 163, 166, 195; ploughing competitions, 261n17
North American Sylva (Michaux), 113
Northampton Gazette (newspaper), 123
Northern Farmer (journal), 55
Northup, Anne, 48
Northup, Solomon, 48
Norton, John Pitkin, 205–7
Nova Scotia, 194
nurserymen, 14, 110–12, 114–15, 122, 163, 178, 180, 183–86, 188, 225; as experts, 92, 112, 163, 169, 170–72, 176–78, 180, 225; fraud, accusations of, 174–75, 187
nutrition, 190, 208; and fertilizer, 212–13; and guano, 211; and money, 218; nutritional value, food and money, connection between, 189; organic matters, 192; and phosphorus, 211; soil, nutritional value of, 209, 212, 214, 217–18; soil, separate from, 192–93; value in, 201. *See also* agricultural chemistry

O'Connor, Feargus, 79
Ohio, 107, 130–31, 209, 242n27
Olmsted, Frederick Law, 187, 204
Oneida Community, 9, 66
Oneida County, 8, 67, 70
Oneida people, 9, 137
Orange County, 56, 147, 154–55, 272n58; butter, manufacturing technique, 142; butter debate in, 141–44; butter in, 133; butter money, 133–34
"ordinary farmers," myth of, 40
Oregon, 168
organic agriculture, 226
organic chemistry, 191
Organic Chemistry in Its Application to Agriculture and Physiology (Liebig), 196
Organic Front, The (Rodale), 224
Organic Gardening and Farming (journal), 223
organic life, 189–90
Osterud, Nancy, 140
Oswego County, 142–43, 153
Otsego County, 27, 32, 131, 142
Owen, Aaron, 52

Paine, Thomas, 35
Palermo (Italy), 105
Paley, William, 135
Panic of 1819, 8, 27, 33
Panic of 1837, 8, 33, 44–45, 104, 113–14, 118, 128, 181

Panic of 1839, 8, 104, 128
Panic of 1857, 8
Paris (France), 111
Parmentier, Sylvia Marie, 112–13, 116
Pascalis, Felix, 110, 112–13, 116
patents, 58, 81, 84, 87–88, 228
Peck, Edward, 146–47
Pell, Robert, 59, 209–10, 212
Pendleton, Edmund, 213
Pennsylvania, 107, 113, 143, 148
Perdue, Sonny, 223
Perrottet, Georges, 11–12, 115
Perth (Scotland), 92
Peru, 124, 201, 210–11, 227
Peters, Richard, 92
Philadelphia, PA, 27, 103, 107, 124, 200
Phoenix Bank, 44
phosphorus, 199, 201, 208, 211, 215–16
Pierce, Franklin, 211
place: and chorography, 72; and experimentation, 70–71
plants, commercialization of, 188
plaster of Paris, 32, 49, 50, 52, 63, 67, 193–94, 202, 210, 211–12
Plato, 136
Pliny the Elder, 84, 166, 195
Plutarch, 136
poaching, punishment for, 25
Pocahontas. *See* Matoaka
poetry, 85, 111, 118, 153, 169, 188–89, 204
Pomological Manual (Prince), 171–72
pomology, 98, 169–72, 176–77, 179–80, 182, 184–85, 188, 203; naming practices, 186; and pandemonium, 182; and pomologists, 163, 169, 170–71; and print culture, 161, 163, 171–73
Poughkeepsie, NY, 57, 63, 92
Poughkeepsie Fair, 39, 41, 59, 113, 152
Prairie Farmer (journal), 216
Pratt, Zadock, 131, 133–34, 136–37, 139–42, 144–46, 149–50, 153–57, 269n5, 275n124; adaptation, and animals, 152; background of, 132; experimentation, in butter making, 152; farm, publicity of, 151–52
Pratt's Rock, 132–33, 144, 150
Prattsville, NY, 131–34, 150–51, 153, 222
Prattsville News (newspaper), 153–54
Prince, L. Bradford, 115
Prince, William, II, 112, 114, 119, 120, 124, 125, 173–74
Prince, William R., 112, 115, 120, 124, 125, 164, 171–72, 174–76, 180, 182, 225
Protestantism, 135
Prussia, 5, 106
public good, 186
public space, 144

Pumfrey, Stephen, 251n73
Purina, "SteakMaker" line, 216–18

Queens County, 8
Queen's County Agricultural Society, 70
quinoa, 227

radical agrarianism, 34; among Anti-Renters, 33, 79; and arguments about soil fertility and the value of labor, 202–3; transatlantic connections of, 11, 33–34, 79
Randall, Henry S., 148, 253n35
rationalization, 222
ready-made goods, 13–14
Reaper, The (pamphlet), 96
reapers, 93, 95; reaper trials, 83–86, 98
recreational farming, 44–45. *See also* book farming; conspicuous production; rural retirement
regionalization, 130–32, 138
Remington Agricultural Company, 221
Rensselaer, Stephen Van, III, 23–24, 29–33, 41, 59, 108–9
Rensselaer County, 27, 88; surveying of, 149
Rensselaer County Fair, 35
Rensselaer Institute, 30, 37, 192, 207
Rensselaers, 11, 23, 26, 35, 37, 59
Rensselaer School Flotilla, 30, 147
Republican Party, 253n35
Rhode Island, 113
Richmond, VA, 103
Roberts, Della A., 58
robots, 88, 262n29; actual automata, 82, 97; metaphorical automata, 56, 97
Rochester, NY, 9, 13
Rochester Fair, 7
Rochester Seed Store, 89, 92
Rodale, Jerome I., 223–24
Roger (king of Sicily), 105
Rolfe, John, 106
Rolfe, Rebecca. *See* Matoaka
Roman Empire, 10; literature influence on improvers, 43–44, 84, 145, 166, 195
Rome, NY, 9, 92
Ron, Ariel, 222
Rosenberg, Charles, 80
Rosenfeld, Sophia, 69
Rothamsted Agricultural Station, 69, 224
rotten boroughs, 25
Royal Agricultural Society, 86
Royal Society of London, 26–27, 69, 74, 192
Rural Economy of the West of England (Marshall), 151
rural life: as litigious, 65; rural investment, 46; rural refinement, expansion of, 53
rural retirement, 43, 45–46; commercial failure, as

INDEX

honorable mask for, 44; mental cultivation, as marker of, 44
Rush, Benjamin, 48
Russia, 5, 105–6, 226. *See also* Soviet Union

Saxony, 109
Schuylers, 23, 26
science, 30–31, 65, 162, 213; capitalism, relationship to, 14–15, 17; commercial funding for, 14, 32, 82, 89–93, 223; decentralized knowledge and, 71; definition of, 241n19; democratic access and, 140; field experimentation, 79–80; improvement as a particular style of, 5–6, 18, 19, 80; monopolization by chemistry, 190, 205–6; private funding for, 29–31; profit as epistemological claim, 74; scientific knowledge, Strong Programme, 245n59; scientist, as term, 64, 220–21; state funding for; 13, 147–48, 220; theory, suspicion of, 69; women's participation in, 31. *See also* botany, botanical networks; chemistry; experimentation; geology; natural history; natural law; natural theology; pomology
Science (journal), 220
Scioto Valley, 209
Scotland, 10, 12, 92, 192, 207; crofters, eviction of, 33; famines in, 33. *See also* Great Britain
Searle, Llerena, 269n3
Second Bank of the United States, 103
Second Coming, 9
Sellers, Charles, 251n6
Seneca County, 52, 145
Seneca people, 9, 137, 166
Senegal, 111–12
settler colonialism, 10–11, 106, 166
settler societies, 137
Seward, William, 37, 96–97, 135, 190, 211
Shapin, Steven, 69
Shapiro, Barbara, 65
Sherwood, Mary, 59
Silk Company, 115
silk culture/industry, 19, 50, 103, 105–6, 111–13, 116–17, 119–20, 123, 128–29, 264n4; failure of, 107–8; homegrown, advantages of, 109–10; journals of, 108; mania for, 113, 118; qualities of, 118; silkworms, 118, 124–26; white women, labor of, 110
Silliman, Benjamin, 266n45
Sinclair, John, 26–27
Sing-Sing Prison, 110
skateboards, flying, 288n50
slavery, 25, 51, 58, 76, 109, 172, 193; in Anti-Renter rhetoric, 37; and capitalism, 5, 76; knowledge of enslaved people appropriated, 52, 73, 226; as model for 21st-century agriculture, 227; opposition to, 9, 37; prevalence in rural New York, 51; in silk culture, 110; slaveholders as participants in improvement, 172, 193
smell, 283n65; agricultural chemistry, 204–5
Smith, Adam, 136
Smith, Gideon B., 119
Smith, John, 147
Smith, Mahlon F., 221
Smith, William, 274n105
Smithsonian Institution, 211
social stratification, rural, 24–25, 38, 60; among "ordinary" farmers, 51
Society for the Encouraging of Arts, Manufacturing, and Commerce, 26
Society for the Promotion of Agricultural Science, 221
Society for the Promotion of Agriculture, Arts, and Manufactures (SPAAM), 26–27
Society for the Promotion of Arts, Agriculture, and Oeconomy, 26
Society of Friends, 66
sociotechnical imaginary, 245n55
soil, 6, 8, 13, 67–70, 91, 130, 134–40, 146–48, 150, 155, 161, 176, 188, 190–94, 198–200, 205–8, 212, 214, 216, 219, 222, 224, 226; damage of, 201; and degradation, 1; diversity of, 71; exhaustion of, 50, 109, 195–96, 202; nutritional value of, 209; shifting of, 149; smell detectives, 203–4; and stench, 203; thinness of, 63, 145; wasting, 203
Soil Survey of Greene County, New York, 222
Sotham, William H., 91
South Africa, 10
South America, 132
South Carolina, 107, 114
South Sea bubble, 126
Soviet Union, 222, 261n6. *See also* Russia
Spain, 28
species: domesticated, and variability, 164; varietal description, 179. *See also* varieties
speculation, 5, 42, 65, 103, 119, 120, 123, 125, 128; land, 26, 33; on living organisms, 126
Speed, J. R., 56
Spence, Thomas, 35
Sprague, George T., 77
Stanton, Edwin, 96
State Agricultural Fair, 37
State Grange, 221
Steiner school of biodynamics, 226
St. Helena, 109
Stirling (Scotland), 92
St. Regis Mohawk/Akwesasne Tribe, 227
Sturtevant, E. Lewis, 219–21
Sturtevant, Hattie, 220
Suffolk County, 8
Sullivan County, 150

Sullivan expedition, 11–12, 116
supermarkets, 227
Sweden, 28, 106, 109
Swing Riots, 33–34
symmetry, 245n59
Syracuse, NY, 9, 92

Tallmadge, Nathaniel, 65, 66
Tappan, Lewis, 181
tenants, 37, 60; Anti-Rent committees of, 33; ladder hypothesis, 250n68; landlord claims of benevolence, attacking of, 36; landlords, clashing with, 41, 80, 203; landsite doctrine, 203; power tactics of, 33; and radicalism, 34; resistance of, 36; soil exhaustion, dangers of, 202, 203
terroir, 188, 227
Texas, 124, 130
"This Compost" (Whitman), 204
Thomas, John J., 161–62, 165–66, 168–69, 176–77, 179–80, 182, 186, 221
Thoreau, Henry David, 67–69, 71–74, 117, 186–87
Thornton, Tamara Plakins, 44
Thorpe, John and Jeanette, 51
Three Experiments of Living (Lee), 66
Tillson, Oliver, 48
tobacco, 106–7, 212, 280n117
Tompkins County Fair, 58–59
Tone, Peter, 78–79
Torrey, John, 266n45
Townsend, William, 143
Townshend, "Turnip," 25
trials, judicial, connections to experiment, 65–66, 82, 96–97, 98
Troy, NY, 9, 113, 195
Troy Female Seminary, 31
Trump, Donald, 223
Trustees of Georgia Colony, 107
Tucker, Luther, 89
tulip mania, comparison with *multicaulis* mania, 126–27
Tull, Jethro, 201–2
Turkey, 103

Ulster County, 48, 150
United States, 1, 6–7, 12, 23, 29, 34, 64, 68, 81, 83, 86–87, 108, 111, 114, 136, 138, 162, 176, 196–97, 206, 211, 222, 227; dairy region of, 143; first geological map of, 148–49; freeholders, as community of, 37
United States Guano Company, 211–12
urbanites, 45; and experimentation, 78; rural retirement, 46
urbanity, 42–43
US Department of Agriculture (USDA), 37, 220, 222; family farm, definition of, 228

Vail, George, 59
Van Buren, Martin, 28
Van Mons, Jean Baptiste, 165, 170–71, 186
Van Schoonhoven, William, 35
varietal lists, 163, 180–81, 183, 187
varieties, 4, 28, 43, 106; books of description, 171–73; as changing when moved, 16, 175–76; as collectible, 167; competing varieties of mulberry, 114, 116; counterfeiting and fraud, 173–74, 175, 187; as divinely adapted to particular places, 139; informal circulation, 174; as more significant than species, 162; problems in description and naming, 161–62, 169–70, 173; as productions of culture, 165; rating systems for, 177–78, 181–86; as refined commodity, 163, 168; and the seedling landscape, 163, 164, 166–67, 171, 181–82; testing gardens for, 170–71; truth in, 165. *See also* fruit; fruit profiles; pomology; varietal lists
Varro, Marcus Terentius, 195
Vermont, 113
Verplanck, Mary Anna, 58
Vestiges of the Natural History of Creation, The (Chambers), 215
Victoria (queen), 148
Virgil, 145
Virginia, 122
vitalism, 193, 196, 197, 198, 208; persistence of, 224

Wadsworths, 26, 32, 59, 78–79, 87, 202
Wales, 256n97. *See also* Great Britain
Walsh, Alexander, 91–92
Ward, Alson, 63, 74, 98, 145
warehouses, 6, 97, 104, 111, 202, 208; as drivers of improvement, 14, 32, 82; and machinery, 95; modern descendants of, 222–23; partnership with journals, 89–91; as sources of knowledge, 92; as sources of refined goods, 54
Waring, George, 192, 197, 204–8, 211, 213–14, 217; nitrogen atom, 189–90
War of 1812, 28
Washington, George, 26, 44
Watson, Elkanah, 27, 47, 137; Modern Berkshire System, promoting of, 43
Watts, James H., 183–84
Webster, Daniel, 153
Weed, Thurlow, 46
Weeks, John B., 2, 52–54, 65, 76–77, 98–100
Weeks, Levi, 52–54, 65, 76–77, 98–100
Weld, Eliakim, 46
Wells, David A., 209
Westchester, NY, 57
Western Inland Lock Navigation Company, 26
Western New York Horticultural Society, 221
West Indies, 27, 124, 141

INDEX

West Virginia, 145
wheat, 8, 24, 45, 50, 52, 63, 67, 78, 87, 104, 109, 136, 145–46, 155, 165, 194–96, 199, 202, 209, 212, 226–27; prices, collapse of, 27; ups and downs of, 93; wheat midge, 9, 94, 164
Whig Party, 46–47, 253n35
whiteness: and authorship claims, 58–59; and claims about breeding, 4; and financial credit, 51; and Haudenosaunee erasure, 4; and settler colonialism, 11, 13; and wealth, 57; white settlers, and adaptation, 137–38; white women, nostalgia, and delicate labor, 110
Whitman, Walt, 189, 197, 204, 281n2
"Wild Apples" (Thoreau), 186
Willard, X. A., 143
William Prince and Sons, 112, 120
Williams, Richard M., 73
Wines, Richard, 211
Wisconsin, 46, 130, 139
Wöhler, Friedrich, 197

women, 9, 14, 31–32, 75; butter production, 140, 142–43, 151, 154–55, 226; concealed knowledge of, 58, 132, 142–43, 151, 155, 226; labor of, 54–55, 58–59, 73, 75, 110; and refinement, 14, 54–55, 59; as scientists and inventors, 31–32; silk production, 13, 110
Working Farmer (journal), 214–15
Workingmen's Party, 201
World War I, 226
World War II, 74, 226
Wright, J. S., 81, 84, 86, 94
Wright, Silas, 34, 55
Wynkoop, A. J., 143–44

Yale Analytical Laboratory, 214
Yoder, Frank, 250n68
Young, Arthur, 26–27, 69, 78, 99; British improvement, publicizing of, 68; correspondence with Washington, 26; as definer of experimental norms, 68–70, 74–75, 78

Made in the USA
Columbia, SC
14 June 2023

18067291R00170